Digital Nonlinear Editing

New Approaches to Editing Film and Video

Thomas A. Ohanian

Focal Press

Boston   London

For Aram and Susan

Focal Press is an imprint of Butterworth–Heinemann.

∞ Recognizing the importance of preserving what has been written, it is the policy of Butterworth–Heinemann to have the books it publishes printed on acid-free paper, and we exert our best efforts to that end.

Cover photo by Michael E. Phillips.

Library of Congress Cataloging-in-Publication Data
Ohanian, Thomas A.
    Digital nonlinear editing : new approaches to editing film and video.
        p.   cm.
    Includes bibliographical references and index.
    ISBN 0-240-80175-X
    1. Motion pictures—Editing. 2. Video tapes—Editing. I. Title.
    TR899.042   1992
    778.5'235'0285—dc20   92-32378

British Library Cataloguing-in-Publication Data
A catalogue record for this book is available from the British Library.

Butterworth–Heinemann
80 Montvale Avenue
Stoneham, MA 02180

10 9 8 7 6 5 4 3 2 1

Printed in the United States of America

Contents

Preface ix

Acknowledgments xi

1 **Word Processing, Linear and Nonlinear Editing, and the Promise of Digital Nonlinear Editing** 1
WORD PROCESSING: NONLINEAR EDITING WITHOUT PICTURES AND SOUNDS 2
THE EVOLUTION OF DIGITAL MEDIA PROCESSORS 3
The Turing Machine 4
THE INCREASING COMPLEXITY OF THE EDITING PROCESS 4
THE COMING TOGETHER OF FILM AND VIDEO EDITING 6

2 **The Editing Process: Film and Videotape Post-Production Procedures** 7
FORMATS AND STANDARDS 7
FILM EDITING PROCEDURES 9
Shooting and Preparing for Editing 9
Synchronizing Dailies 9
The Nonlinear Mind Set 10
Editing the Film 11
THE DEVELOPMENT OF VIDEOTAPE 15
VIDEOTAPE EDITING PROCEDURES 17
Trimming Shots 18
Completing Audio for Video as a Separate Stage 18
Integrating and Orchestrating Equipment via Timecode 19
LINEAR VERSUS NONLINEAR EDITING 21
The Linear Process of Videotape Editing 22
Understand the Needs of the Program before Choosing the Editing System 24

3 **Defining the Electronic Nonlinear Random-Access Editing System** 25
THE RISE OF DIGITAL SUPPORT, DIGITAL MANIPULATION, AND DIGITAL STORAGE OF VIDEO 26
Timebase Correctors Manipulate a Signal's Waveform Characteristics 27
Digital Framestores Process Video Information 28
Digital Video Effects Devices Manipulate Moving Video in Real Time 28
Digital Still Store Devices Store and Recall Frames 29
Images Are Stored to Computer Disk 30
Digital Disk Recorders Provide Simultaneous Digital Playback and Recording Capabilities 30
TALKING WITH EDITORS 32
What Do You Try to Do for Directors or Clients When You Review Their Footage? 32

4 **Offline and Online Videotape Editing** 35
THE OFFLINE–ONLINE LINK 36
THE REASONS FOR OFFLINE EDITING 37
THE PROCESS OF OFFLINE EDITING 38
Step 1: Window Dubs 39
Step 2: Logging Footage 39
Step 3: Paper Edit List 39
Step 4: Offline Editing 41
Step 5: List Cleaning 42
THE PROCESS OF ONLINE EDITING 43
The Steps of Online Editing 44
Audio Editing in the Online Room 47
THE REASONS FOR ONLINE EDITING 48
THE UNFULFILLED PROMISE OF LINEAR OFFLINE EDITING 49
THE EVOLUTION OF OFFLINE VIDEOTAPE EDITING 50

EDITING MODES: DIFFERENT FORMS OF
THE EDIT LIST — 51

5 The Operational Aspects of Film and Videotape Editing — 55

THE CREATIVE PROCESS OF FILM VERSUS
VIDEOTAPE EDITING — 56

Can Film and Videotape Editing Be Improved? — 57

ELECTRONIC NONLINEAR EDITING SYSTEMS — 58

Common Stages of the Electronic Nonlinear
Editing Process — 58
Work Products of Nonlinear Editing — 60

TALKING WITH EDITORS — 62

How Are the Film and Video Editing Cultures
Alike and Different? — 62
Is There a Bias Against Videotape Editors? — 63

6 Videotape-Based Systems — 65

THE CMX 600 — 65

THE FIRST WAVE — 67

THE PLAYLIST — 68

TRAFFICKING — 72

VIRTUAL RECORDING — 72

MULTIPLE VERSIONS — 73

USER INTERFACE — 74

THREE SYSTEMS OF THE VIDEOTAPE-BASED
WAVE — 75

Montage — 75
Ediflex — 76
BHP TouchVision — 77

APPRAISING THE FIRST WAVE — 79

ECONOMIC BENEFITS OF THE FIRST WAVE — 80

EVOLUTION OF THE FIRST WAVE — 81

TALKING WITH EDITORS — 81

How Did Film and Videotape Editors React to
the First Appearance of Nonlinear Systems? — 81

7 Laserdisc-Based Systems — 83

TYPES OF LASER VIDEODISCS — 83

TRANSFERRING VIDEO AND AUDIO TO
LASERDISC — 84

Discovision Associates (DVA)Code — 85
Disc Copies — 86

OTHER CHARACTERISTICS OF LASERDISCS — 86

THEORY OF OPERATION OF LASERDISC-
BASED NONLINEAR SYSTEMS — 87

Transferring Material — 88
Storage, Discs, and Disc Players — 88

TYPICAL SECOND-WAVE EDITING SYSTEM
DESIGN — 89

TRAFFICKING — 91

PRE-VISUALIZATION TOOLS — 91

GRAPHICAL USER INTERFACE — 93

GRAPHICAL USER INTERFACE AND
AUDIO EDITING — 95

Traditional Film Sound Editing — 95
Second-Wave Audio Editing — 96

WORK PRODUCTS — 99

APPRAISING THE SECOND WAVE — 100

EVOLUTION OF THE SECOND WAVE — 100

8 Digital-Based Systems — 103

HOW DIGITAL NONLINEAR SYSTEMS WORK — 104

General System Objectives — 104
System Work Flow — 105

PARADIGMS OF THE DIGITAL NONLINEAR
EDITING SYSTEM — 108

The Clip — 108
The Transition — 109
The Sequence — 111
The Timeline — 111

DIGITIZING AND STORING MATERIAL — 112

Digitizing the Footage — 113
Playback Speeds — 114
Storing to Disk — 115
Digitizing Parameters — 116
Storage — 119
Storage in Early Digital Nonlinear Systems — 121
How Image Complexity Affects Storage
Requirements — 122

THE USER INTERFACE — 124

GENERAL DESIGN OF DIGITAL NONLINEAR
EDITING SYSTEMS — 126

Representing the Footage — 127
The Editing Interface — 129

SAMPLE PROCEDURE FOR DIGITAL
NONLINEAR EDITING — 131

Step 1: Transferring Material from Film to
Videotape — 132
Step 2: Digitizing the Material — 133
Step 3: Editing — 135
Step 4: Output — 138
Step 5: Transferring to Film — 139
Step 6: Cutting the Negative — 139
Step 7: Producing the Finished Products — 140
Conclusion — 140

TALKING WITH EDITORS — 141

How Did Editors React to the First Appearance of
Digital Nonlinear Editing Systems, and What
Predictions Do They Make for the Future of
Nonlinear Editing? — 141

9 Editing on a Digital Nonlinear System 143

THE INPUT STAGE 143

THE EDITING STAGE 144

Splicing Shots Together 145
Trimming a Shot 145
Customizing the Timeline 146
Rearranging the Order of Clips 146
Splicing New Material into an Existing Sequence 148
Adding and Deleting Material without Affecting
the Length of the Sequence 149
Combining Pictorial and Graphical Views of the
Sequence 150
Audio Editing 150
Basic Digital Nonlinear Editing Operations 151

THE OUTPUT STAGE 151

WILL THE PROJECT BE FINISHED IN
LESS TIME? 151

SYSTEMIC ISSUES 152

The Operating System versus the Software
Program 152
File Organization 152
Documentation 152

THINKING NONLINEARLY 153

**10 Digitization, Coding, and Compression
Fundamentals** 155

BASIC TERMS 155

AN EARLY SAMPLING EXPERIMENT 155

Pixels and Sample Points 157

THE FLASH CONVERTOR 157

How Does a Flash Convertor Work? 158
Frequency Definition and Amplitude Definition 159

COMPUTERS AND VIDEO 161

Computer Limitations 162
Evolution of the Flash Convertor 163
Approaching Real-time Digitization 164

COMPRESSING AND CODING 165

The Bit 165
Compressing a File 166
Coding Techniques 167

SAMPLING 169

Sampling, Subsampling, and the Sampling
Theorem 170
Nyquist Limit 171
Affect of Sampling and Selective Removal
of Samples on the Message 172
Decimation 173
Error Masking Techniques 174

DIGITIZING THE AUDIO SIGNAL 176

BANDWIDTH AND STORAGE 177

Basic Storage Terms 177

INTRODUCTION TO COMPRESSION
TECHNIQUES 178

Lossless Compression 178
Lossy Compression 179

PRODUCTS AND CAPABILITIES BASED ON
DIGITAL MANIPULATION 180

11 Digital Video Compression 183

EDITING FULL-RESOLUTION FULL-
BANDWIDTH DIGITAL VIDEO 184

ANALOG COMPRESSION 185

DIGITAL VIDEO COMPRESSION 186

HARDWARE AND SOFTWARE COMPRESSION
METHODS 188

SOFTWARE-ONLY COMPRESSION 188

Software Compression Methods, 1987–1989 189
External Disk Storage 191
Software Compression 191
Results of Subsampling 192

HARDWARE-AIDED COMPRESSION 193

JPEG COMPRESSION 193

QUANTIZATION 197

Quantization Table Elements, Q Factor, and
Quantization Frequency Array 198
Zero Packer 199
Huffman Coder 199

SYMMETRICAL COMPRESSION VERSUS
ASYMMETRICAL COMPRESSION 200

FIXED FRAME SIZE VERSUS VARIABLE
FRAME SIZE 200

Fixed-Frame-Size Technique 201
Variable-Frame-Size Technique 201
Intraframe Coding 202

MPEG COMPRESSION 203

Interframe Coding 203
Implications For the Digital Nonlinear Editing
Process 204

DIGITAL VIDEO INTERACTIVE COMPRESSION 206

Presentation-Level Video 206
Real Time Video 206
PX64 207

DEVELOPING AND EMERGING COMPRESSION
TECHNOLOGIES 207

Fractals 207
Wavelets 209

EVOLUTION OF DIGITAL VIDEO
COMPRESSION TECHNIQUES 210

COMPRESSION TECHNIQUES FOR HIGH-
DEFINITION TELEVISION 210

CHARACTERIZING THE RESULTS OF DIGITAL
VIDEO COMPRESSION 210

STATE-OF-THE-ART DIGITAL VIDEO
    COMPRESSION 1989–1992   213
Software-Only Compression, 1989–1991   213
Hardware-Assisted JPEG Compression, 1991–1992   215
Second-Generation Hardware-Assisted JPEG
    Compression, 1992   216
TIMEBASE CORRECTORS AND THE IMPORTANCE
    OF A PROPER INPUT SIGNAL   217

**12 Storage Devices for Digital
Editing Systems   219**

COMPUTER STORAGE DEVICES   219
Disk Characteristics   219
Data Rate   220
kB/Frame   220
Formula for Determining Data Rate and Storage
    Capacity   221
DISK TYPES   222
Floppy Disks   222
Magnetic Disks   223
Disk Arrays   225
Evolution of Capacity and Cost of Hard Disks   226
Optical Discs   227

13 Transmitting Video Data   233

METHODS OF TRANSMITTING DATA   234
Modem   234
Integrated Services Digital Network   234
Computer Networks   235
Ethernet   235
T1 and T3   236
Fiber Digital Data Interconnect   236
TRANSMITTING PICTURES AND SOUNDS
    AMONG EDITING SYSTEMS   237
TRANSMITTING VIDEO DATA FOR THE
    BROADCAST INDUSTRY   238
IMPROVEMENTS AND STANDARDIZATION
    REQUIREMENTS OF DIGITAL VIDEO
    COMPRESSION METHODS   240

**14 Logging for Digital Nonlinear
Editing Systems   243**

THE VIDEO SCRIPT   244
The Log Sheet   244
THE FILM SCRIPT   246
Marking up the Script   248
The Continuity Sheet   248
The Lab (Telecine) Report   249
The Film Transfer Log Sheet   250
Camera and Sound Reports   251
Additional Notes   251

THE COMPUTER DATABASE   252
Searching with the Computer   254
Script Integration   255
Logging During or After Shooting   255
Preparing for Editing   256
Avoiding Duplicative Work   256
EVOLUTION OF THE LOGGING PROCESS   257
TALKING WITH EDITORS   257
How Has Logging Software Affected the
    Editorial Process?   257

15 The Digital Media Manager   259

INTRODUCTION TO THE DIGITAL MEDIA
    MANAGER   259
File Incompatibility   260
Dedicated Systems versus Software Modules   261
DEVELOPMENT OF THE TAPE- AND
    LASERDISC-BASED WAVES   262
DEVELOPMENT OF THE DIGITAL-BASED
    WAVE   263
Keying   264
Advanced Audio Editing   264
Digital Video Effects   264
Compositing   265
Journaling   266
Musical Instrument Digital Interface   267
HORIZONTAL AND VERTICAL EDITING
    CONCEPTS   268
FILE COMPATIBILITY   269
ARE DIGITAL NONLINEAR SYSTEMS
    OFFLINE OR ONLINE?   270

16 The Film Transfer Process   273

CREATING VIDEOTAPE MASTERS FROM AN
    ASSEMBLED FILM NEGATIVE   274
HOW FILM, VIDEOTAPE, AND COMPUTERS
    COEXIST   275
FILM TO TAPE   275
FILM AND VIDEOTAPE SPEEDS   275
Fast Motion (Undercranking)   276
Slow Motion (Overcranking)   276
CORRELATION OF FILM AND VIDEOTAPE   276
Direct Correlation of Film Frames to
    Video Frames   277
FILM TO TAPE TO FILM   277
Telecine   278
EDGE NUMBERS   278
PULLDOWN   279
2-3 and 3-2 Pulldown   280
Pulldown Mode   280

Pulldown Mode Identification 281
SYNC POINT RELATIONSHIPS 281
Punching the Printed Film 282
Automatic Key Number Readers 283
EDITING AND DELIVERY ON VIDEOTAPE
    AND FILM 286
Film Cut Lists 286
Conforming the Negative 287
Sound 288
FILM TO TAPE TO FILM TO TAPE 288
EDITING AT 24 FPS 289

**17 Evaluating Electronic Nonlinear
     Editing Systems 293**

TECHNICAL CONCERNS 293
Anticipated Use 293
Storage 294
Audio Quality and Number of Channels
    and Tracks 296
Editorial Work Products 297
Unfinished versus Finished Results 297
OPERATIONAL CONCERNS 297
Computer Capability 298
Editing Methods 298
Operating the System 298
Training Programs 299
Benefits 299
THE HUMAN SIDE OF THE NONLINEAR
    EDITING EXPERIENCE 299
BENEFITS FOR THE EDITOR AND CLIENT
    AND PREDICTIONS FOR THE FUTURE 304
TALKING WITH EDITORS 305
How Do Digital Nonlinear Systems Contribute
    to the Production Process and Benefit the
    Client and Editor? 305

18 The Future of Nonlinear Editing 307

EVOLUTION OF THE THIRD WAVE 307
Boundaries 308
Incorporation of Digital Video Compression
    into Existing Product Lines 309
Metamorphosis 310
TRANSFORMATION OF OFFLINE AND
    ONLINE EDITING 310

ALTERNATIVE APPLICATIONS 311
Broadcasting 312
Being in Two Places at Once 312
AFFECT OF DIGITAL VIDEO COMPRESSION
    AND DIGITAL NONLINEAR EDITING ON
    THE SHOOTING PROCESS 312
Changing the Work Flow 313
WIDESPREAD SYSTEM AVAILABILITY
    AND USE 314
DEFINING THE THIRD, FOURTH, AND FIFTH
    WAVES 314
The Third Wave: Digital Offline 314
The Fourth Wave: Digital Online 314
The Fifth Wave: Digital Uncompressed Media
    Management 315

19 Electronic Nonlinear Systems 317

THE VIDEOTAPE-BASED WAVE 317
BHP TouchVision 317
Ediflex I and II 318
Montage I & II 318
THE LASERDISC-BASED WAVE 319
CMX 6000 319
EditDroid I & II 320
Epix 321
Laser Edit 321
THE DIGITAL-BASED WAVE 322
Avid Media Composer 322
DVision 324
EMC2 325
Lightworks 326
Montage III 327
DEVELOPMENTS IN THE DIGITAL-BASED
    WAVE 328
QuickTime 328
DVI and JPEG Frame-Based Edit Systems 330

Glossary 331

Bibliography 337

Index 339

Preface

The first job I had in the film and television industry was to cut narration audio tracks for training films. The audio was given to me on large audio tape reels. I was given a splicing block, a razor blade, and a script. My job was to take out the pauses, the microphone pops, and the clicks and to make the narration sound as natural as possible. It taught me a lot, including patience and a sense of pace and rhythm, and gave rise to a quest for better ways of putting together pictures and sounds.

The process of editing film or videotape or putting together a presentation consisting of 35mm slides and an audio cassette requires creative and technical decisions. It is extremely rare that the final project is in the same form as the first edited attempt. Projects require some time to evolve, and the film and videotape editor needs time to try ideas.

Nonlinear editing techniques and systems allow the editor to try different ways of putting together the pieces. No longer is it acceptable to not try an idea simply because there isn't enough time to do so. The emergence of digital nonlinear editing techniques will fundamentally change the manner in which pictures and sounds are combined, rearranged, viewed, and distributed.

I have since put my splicing block in storage.

Acknowledgments

A number of individuals have contributed their time and counsel to this effort. This book would not be complete without acknowledging them: The managerial staff and employees of Avid Technology, Inc., for their dedication and commitment to building creative editing and communications tools; Curt A. Rawley, William J. Warner (who knew it was possible), Basil Pappas, Paul Dougherty, Tony Black, Peter Cohen, Alan Miller, Howard A. Phillips, Ryan Murphy, Patrick Quinn, Christine Cataudella, Steve Tomich, Rick Cramer, Tom Werner, Margaret Chevian, Eric Ridley, Jim Ricotta, Steven Goldsmith, Debbie Newell, Mark Newell, Steve Tymon, Jo-Ann Tymon, Briar Mitchell, and Vivian Craig; and Aware, Inc., C-Cube Microsystems, Inc., Digidesign, Inc., Evertz Microsystems, Inc. (Carter Lancaster), IBM, Imagenda in Natick, MA, Intel Corp., La Casa de Tres Hombres con Avid, Inc., Quantel, Inc. (Kelly Murphy), 3M Corp., Szabo Tohtz Editing and Skyview Film & Video in Chicago, The Edit House Boston, Truevision, Inc. (Tom Ransom, Celia Booher), and ViewPoint Computer Animation in Needham, MA (Carlo DiPersio, Glenn Robbins).

Particular thanks to Peter J. Fasciano and Eric C. Peters. A special mention must be made to three individuals who contributed to the visuals in this book: Jeffrey Krebs, Rob Gonsalves, and Michael E. Phillips.

Grateful thanks to my family and friends for their support.

1

Word Processing, Linear and Nonlinear Editing, and the Promise of Digital Nonlinear Editing

The editing of film and video images is influenced by technical and aesthetic tasks and decisions.

Editing film and video images is a combination of aesthetic judgments and a technical mastery of the film and videotape crafts.

These two sentences deliver essentially the same message. What is important is that an attempt was made to try the original sentence in a new way. The tool used to write these sentences, a word processor, allowed the writer to try out variations easily. One could argue that if the tool were too cumbersome and not as conducive to experimentation, then there would only have been one attempt made at the sentence.

Having the ability to add and delete words or sentences easily allows you to try different ways of writing and expressing your thoughts. Film editors, videotape editors, producers, directors, writers, and those who work with moving pictures and sounds always find themselves wanting to try things in different places and in different ways, and usually, these individuals will exercise this propensity right up to the last possible minute!

Linear editing means that ideas must proceed in a sequential order: Idea one is followed by idea two, and so on. Nonlinear editing means that ideas (shots) can be tried in any order and can easily be rearranged in the same way that words can easily be rearranged when using a word processor. As architect Frank Lloyd Wright wrote,

"Conceive the piece in the imagination, let it live there, gradually taking more definite form. When the thing lives for you start to plan it with tools. . . . Complete the harmonious adjustment of its parts."

Being able to try things, to play "what if" scenarios, not with words but with moving pictures and sounds, is what nonlinear

1

editing is all about. Trying things and not having to make creative compromises is at the heart of nonlinear editing. Being able to feel that the one edit that should be shortened, the one thing in a program that is just not quite right, can be changed without sending the project over budget or off schedule, is what all nonlinear editing systems promise.

Digital nonlinear editing is an alternative to analog nonlinear editing, which until 1990 was largely confined to specialized sections of post-production in New York and Los Angeles. The promise of digital nonlinear editing is that it will bring together not only film, video, and audio, but also a variety of other media that have never had one common environment in which to coexist.

Digital nonlinear editing offers the following benefits:

1. Creative flexibility. Technology becomes more transparent, freeing up the individual to concentrate on the needs of the presentation without regard to mastering the technical details of how the system operates.

2. Ability to integrate different media easily. Whether a program consists solely of video or a combination of video, film, 35mm slides, and so on, digital nonlinear systems allow the user to combine these different media easily into a completed program.

3. Savings in time and money. A digital nonlinear editing system offers a savings in time, resulting either in the ability to explore additional ideas or in decreased costs.

4. Preparation for digital integration. With an undeniable movement to all things digital, when to introduce and retrain staff to the concepts of digital media manipulation becomes an important question. Digital nonlinear editing systems provide a cornerstone of that educational process.

WORD PROCESSING: NONLINEAR EDITING WITHOUT PICTURES AND SOUNDS

It hasn't been that long since we were working almost entirely with typewriters. Although IBM introduced an electric typewriter in 1935, through the mid-1960's, manual typewriters were the norm. If we were creating documents, we were doing it by hand, and for business, we were using manual typewriters. You can extrapolate the difficulties: If you typed something and made a mistake, you used one of several, usually messy, methods of erasing the mistake. For important presentations, you had to retype the document.

By the early 1970's, electric typewriters were outselling manual typewriters. At the same time, dedicated word processors began to arrive in the business marketplace. With the lowered cost and increased availability of microcomputers, the idea of creating machines that would be used for manipulating text began to flourish. By the early 1980's, word processing was big business.

Companies such as Xerox, Digital, Royal, Wang, and Data General were all making dedicated word processors. These systems were designed specifically to handle the repetitive tasks required when creating documents.

It wasn't until the mid-1980's that word processing progressed to its next stage of development. This involved the movement away from dedicated word processing systems that consisted of hardware and software. This movement of software from one computer hardware system to another represents a significant step. It is at this stage that software developed for one computer is rewritten in order to run on a different computer system. The movement was afoot. Now you could start to buy software instead of having to buy the whole machine. Early word processing software programs were WordStar, EasyWriter, and VolksWriter.

Has this technology made a difference in what and how people write? Does the student become a better writer because she has an opportunity to go back and rework a paragraph? Does an established author write differently because he has the opportunity to try writing in many different ways? Does the film screenwriter write a better film because she can offer several different endings to the director? Overall, the answer to these questions is a resounding yes! The writers mentioned benefit from being able to hone and develop the initial idea. Similar benefits are realized when the digital nonlinear editing system is used.

THE EVOLUTION OF DIGITAL MEDIA PROCESSORS

The manner in which text is manipulated and documents are created has changed dramatically! The evolution and availability of inexpensive word processing software has been achieved in a short period of time. The manner in which film and video images are combined and edited will inevitably repeat in their development path in the same manner as the typewriter evolved into the word processor. We will see the same evolutionary path. At first, the machines will be dedicated systems that are used to edit film and video into professional presentations and are used by professionals trained in the art and craft of making presentations. Next we will see the availability of less sophisticated systems that will be affordable and within the grasp of more individuals who are less classically trained both aesthetically and technically in the art of making presentations. Finally, these new machines will evolve in the same manner in which word processing software evolved to the point where the computer's hardware platform is not important. This movement will lead to the availability of powerful software that is affordable on the mass market. We will most likely come to know these new series of machines and the software that directs these machines as *digital media processors* directed by *digital media managers*.

The evolution of these machines will mean that the machine is not always going to be something that you can see and touch. Instead, these media processors will be machines that are artificially created through the use of software. When we make a phone call from our computer and call up another computer, can we point to the telephone being used? Can we say, "Ah, there is the physical telephone; it looks like the telephone I use everyday"? No. Instead, we rationalize and understand that the computer is using a modem to send and receive signals. But the undeniable fact is that a machine, one that we all know in appearance, has changed. A piece of equipment, a machine, has been supplanted by a different technology.

The facsimile (FAX) machine has become extremely popular. Now, however, it is possible to send and receive faxes via modem because there is computer software that can direct a computer to act like a fax machine. So, if you already own a computer, it may not be necessary to purchase a fax machine. Instead, purchase the software that will accomplish the same task. While we now use dedicated machines to perform tasks and these machines were designed to accomplish only one task, the unceasing growth of technology will bring these capabilities within the realm of software programs.

The Turing Machine

Originally a mathematical concept, the term *Turing Machine* describes a general-purpose machine that can do anything if given the right software instructions. Every computer is a potential Turing Machine, and in theory, the computer can become any existing machine if given a set of instructions that emulate how the original machine operates.

Digital video, digital audio, and digital media can all be viewed as mathematical problems to be solved. The Turing Machine concept of general-purpose computers that are able to adopt the appearance and function of any given analog environment is especially important in the evolution of digital nonlinear editing. The digital nonlinear editing system will begin to offer more of the functions that heretofore were confined to dedicated analog hardware. As a result, the computer will draw upon its capabilities of being a Turing Machine, offering characteristics of the traditional machines used for editing film and videotape. This evolution is going to affect the process of editing and will change how presentations in the professional, institutional, and consumer arenas are made.

THE INCREASING COMPLEXITY OF THE EDITING PROCESS

Ask any film editor what he does, and he will tell you he manipulates images and audio to tell a story. Ask any videotape editor what she does, and she will tell you she manipulates images and audio to tell a story. Even though film and videotape editors work in different environments and with different equipment, their jobs are essentially the same. They both use

Figure 1–1 The silent movie art card and background music were at one time the most complex information presented to the person watching a film. Illustration by Jeffrey Krebs.

Figure 1–2 Multilayered graphic treatments often contain several visual themes within one frame, presenting the viewer with a great deal of information.

their skill and craft to give a scene its rhythm and an edited sequence its drive. Finding, defining, shaping, and delivering the message are the tasks that both editors have to accomplish. Balancing the creative aspects of craft with the technical details of film and videotape editing can be a formidable juggling act.

The elements that make up a presentation can be many in number and can be quite disparate, but it wasn't always that way. Consider the early silent movie. Usually, it consisted of just three elements: live action of actors and actresses, still art cards, and music. Music provided the mood, and the art cards provided the dialogue (Figure 1–1).

Now consider the images that appear every day in graphic treatments for television commercials (Figure 1–2). The number of visual layers, each providing one additional statement to the intended message, is quite far from the days of the simple silent movie art card. Today, more ingredients are being added to the process of making presentations.

There used to be only two key ingredients to editing: knowing which shots to include and knowing which shots to leave out. Today, the editing process has come to require many more tasks: slow motion, repositioning shots, electronic painting, and so on. The amount of work that is sometimes done to pictures before they are even edited into a program has become quite a significant portion of the post-production process. Most viewers are completely unaware that these "fixes" have been made. All are attributed to "the magic of movies."

Consider the 35mm motion picture. Until only recently, most feature films were shot using 35mm film. The 35mm film negative was loaded into the camera, exposed, and processed. Then a print was made from the processed negative, and this print was the film that the editor cut together to form the movie. There just weren't that many additional visual elements used in the presentation. Films, for decades and decades, were made in this fashion: Shoot the live action, edit the film, finish the sound for the film, put the titles and credits on the film, and show the film.

In the late 1980's, the integration of many types of media into the feature film began. Today, it is not at all unusual, depending upon the look of the feature film that the director is trying to establish, for portions of the film to include material shot on 35mm film, 16mm, Super 8, professional videotape formats (such as D2 and 1"), and consumer videotape formats (such as Hi 8, and VHS). These various formats can further be enhanced or degraded through electronic means. Since these different formats will record images in different ways depending upon how their limitations are stretched, integrating many formats into one feature film can provide exciting visual results; it all depends upon what the director is trying to express visually.

Technological manipulation of film and video has increased dramatically. For example, if a period piece is being filmed and it simply wouldn't be correct to see telephone poles in the frame, technology is used to fix the shot after it has been filmed. Spending time and money by using electronic painting systems to erase the telephone poles and replace these sections with a sky background is the only viable alternative. It would be unlikely that the poles could be physically removed!

Using computer-generated objects and 3D animation to create environments and characters has also become very popular. As the viewing audience, not only of feature films, but also of television, corporate, business, and educational videos, continues to grow in sophistication and expectation, there has been an increased pressure on program makers to deliver messages in newer fashions. Now more than ever, there is a reliance on new technologies to provide these functions. The use of improvements in digital technology not only to enhance, but also to repair visual and aural material is clearly on the increase. Today, using technology to make pictures and sounds "larger than life" is a reality.

THE COMING TOGETHER OF FILM AND VIDEO EDITING

The art of editing film was long considered to be resistant to technology because film editing requires a modicum of equipment. This attitude is changing as users of film and video struggle with ways of bringing the two forms together and determine what is required for both to coexist. Does film need video, and does video need film? The answer is yes.

The current manner in which the film and video worlds are being brought together revolves around the user of computers. Does film editing have anything to offer to video editing, and does video editing have anything to offer to film editing? Again, the answer is yes. The use of computers will bring the two art forms together because computers and digital technology can reduce both film and video into a common element: the digital bit.

What does this marriage mean for the person putting together a presentation? Whether this person is a film editor, a videotape editor, or a creator of multimedia presentations, being able to command and balance a number of different technologies, some old, some new, and seamlessly integrate their by-products into the completed program can take much more concentration than simply trying to find out what shots should and should not be used.

It is important to note that what will converge are forms of media manipulation and combining media. There are misconceptions about the media of film and video and whether they will combine or whether they will both be superseded by some new form of imaging, whether it is high-definition television or some future technology. Shooting film and shooting videotape will continue to be with us for some time to come. However, the standard, traditional, and known methods of editing these media forms will change dramatically, and they will change far more quickly than the span of time that film editing and videotape editing have been in existence.

2

The Editing Process: Film and Videotape Post-Production Procedures

If we charted the development of different editing techniques, we would find the following:

Film Editing	c. 1900
Analog Audio Editing	c. 1945
Videotape Editing	1956
Videotape Editing with Timecode	1970
Digital Disk-Based Audio Editing	1985
Digital Disk-Based Picture Editing	c. 1989

Film editing, as both a craft and a means of story telling, has been in use longer than other methods. For decades and decades, and in the face of the growth of computers and digital technology in the 1980's, film has been considered the quintessential "low-tech" method of making presentations. But, as a craft and as a method of practice throughout the great centers of film production—Hollywood, Europe, and the biggest center of film production of all, Bombay!—film editing has persisted and remains the most reliable method.

FORMATS AND STANDARDS

In discussing whether a project should be edited on film or on videotape, more than once the following words have been uttered, "We'll do it on tape because we've got a better chance at making the deadline, but if anything goes wrong, we'll fall back on the film edit." Editing film is a process that has had the benefit of time and experience to develop work methods that are understandable and shared throughout the world. Film is a unique standard in a world in which standards are difficult to achieve. A 35mm film running at 24 frames per second (fps) can be run on any 35mm film projector anywhere in the world.

A videotape, however, may or may not play back properly depending upon where you are in the world. A videotape program edited in the United States cannot be played on equipment in Europe because the technical standards differ. In the U.S., video plays back at 30 fps, while in Europe the standard is 25 fps. How can you be sure that your program can be seen anywhere in the world? Shoot it on film, and have a complete and edited film at the end of the post-production process.

Audio signals are easily played back from country to country. Audio has a certain amount of information (resolution) and can be interpreted at various speeds. There are different speeds at which audio can be recorded and played back (15 inches per second [ips] and 7.5 ips are common speeds), but the important aspect is that the information does not have a territorial stamp; it is not native to a particular country's recording or transmission method. A 15 ips recording on 1/4" reel-to-reel audio tape can be taken and played on any reel-to-reel machine capable of this playback speed.

The issue of being able to hear and see a program regardless of the technical makeup of the format being used is critical. While film and analog audio can each be viewed as an international medium, videotape is a limited medium in its "interchange" from country to country. This is especially true in the 1990's, with new videotape formats appearing with dizzying frequency. It seems that what everyone agrees on is that standardization is necessary but that standardization "right here, right now" is a bit further off than anyone would like. The economic importance of foreign markets to the earning power of a film requires that the medium itself be easily transportable and transmittable from country to country. Film, not videotape, offers that flexibility.

Film editing, although it may be viewed as "low tech," is highly standardized. Not only has film editing developed work methods that translate well throughout the world, but film editing has developed a hierarchy of personnel that is remarkably similar from place to place. Apprentice editors, assistant editors, and editors share similar tasks from country to country in the world of film editing. Although each knows his or her responsibilities, this extremely mature approach extends to the other participants in the process. For example, the film lab that is doing the processing and printing and the optical house that is creating the opticals for the program both know and perform their roles in the post-production process.

This hierarchy has not yet developed to any degree of maturity in the world of videotape editing. The diversity of equipment available in the electronic post-production process of videotape editing makes such standardized methods more difficult to achieve. From one editing facility to another, methods and approaches can vary greatly. Film editing, on the other hand, does not have such a diversity of equipment available. This limited variety of equipment is yet another reason why film editing methods are much more standardized than video editing methods.

FILM EDITING PROCEDURES

To really understand what digital nonlinear editing provides for the user, it's necessary to trace its roots. Nonlinear editing begins with film editing, and to judge any system that you may be using or evaluating, an examination of how film is edited is appropriate.

Editing a film, whether it is a short industrial film designed to explain how a machine works or a two-hour motion picture for theatrical release, involves the basic steps outlined below. The amount of film shot and the amount of time allotted to merge the diverse elements that make up different films will vary, but the post-production process will be quite similar. The steps below involve traditional methods; later we will identify some areas in which new technologies are being used.

Shooting and Preparing for Editing

The most common type of film used for feature motion pictures is 35mm four-perf (perforations per frame). A film roll is loaded into a camera in what is usually a ten-minute load. Sound for the film is recorded on a separate 1/4" reel-to-reel audio tape. The most common audio tape recorder for film is the Nagra. Because of these two systems, one for picture and one for sound, film shooting is often called a *dual-system approach.*

At the end of each shooting day, the exposed camera rolls and sound rolls are gathered together. The exposed film negative is processed by a film lab. This negative is then used to create a positive, or print. The takes that the director wants printed are usually referred to as *circled takes.* These are noted on a log sheet and given to the lab so that the correct portions of the negative are printed. Excessive time is not taken with regard to the quality of this print since it will be viewed and physically cut during the editing process. Later, more time will be taken to ensure that the final film print is of the highest quality.

The sound rolls are on a different format: 1/4" reel-to-reel audio tape. Thirty-five millimeter film editing requires that the picture and sound ultimately be edited together. To prepare for this, the audio reels are transferred to 35mm magnetic track (mag track). The circled takes for audio are transferred to provide the accompanying sound for the circled picture takes (Figure 2–1). The audio takes that are transferred to mag track are noted on the sound transfer report, which is a log sheet.

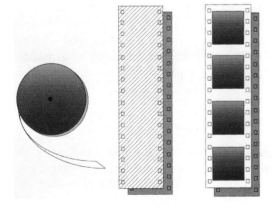

Figure 2–1 The quarter-inch sound roll is first transferred to 35mm magnetic track. This magnetic sound track is then combined with the 35mm picture track. Picture and sound now exist in the same 35mm format, and editing can begin. Illustration by Jeffrey Krebs.

Synchronizing Dailies

While the film is being processed, the mag track is marked at the sound of the clapsticks striking together at the beginning of each take. When the printed takes of film arrive, these *dailies* (also called *one-lites*) are marked with an *X* at the point where the clapsticks (also known as *slate, clapper, sticks*) first join together during the slating process.

The next step involves synchronizing, or syncing, the soundtrack to the picture. By lining up the reference point for the picture and the reference point for the audio, the 35mm film and the 35mm mag track are now in sync. If we look at an actress deliver her lines, we will see and hear perfect synchronization.

These synced dailies are reviewed by the members of the team: director, editor, producer, possibly cast and crew. It is necessary to see how new material will affect previously shot material both from a contextual point of view as well as from a performance perspective. This continues as more film is shot.

Thirty-five millimeter film has unique identifying codes spaced at specific intervals. These are *edge numbers*, and they are essential in being able to describe, numerically, what frames are being used in the project (Figure 2–2). If we use only a few feet of film printed from the camera negative, we must have a way of matching the work print to the original camera negative. Edge numbers let us do this.

However, the mag track does not have any identifying edge numbers. The process of *coding* (also known as *inking*) places ink numbers along the edge of the mag track. These new numbers are coded on both the film and the mag track instead of trying to match the original edge codes. In this way, both picture and sound share identical reference numbers. This is one of the reasons why the editor can easily put picture and sound back in synchronization if sync is lost through an editing operation. Do the numbers agree? If so, the material is in sync. Are the numbers different? Then the material is out of sync.

At this point, the footage is almost ready for editing. One additional step is to prepare the dailies for editing by separating them according to the routine that the editor prefers. Most often, dailies are broken down into one roll that combines the film and mag track for one take. This roll is wound and labeled with the scene and take number (Sc 101, Tk 3). In addition, some editors prefer to have a short description about the scene on the label.

The Nonlinear Mind Set

Editing a project can begin when the shooting process is completely over, but often, editing occurs simultaneously with shooting. There are scheduling reasons for this, of course, and

Figure 2–2 Film negative has edge numbers that identify the film frames. When a print is made from the negative, the same numbers are printed through from the negative to the positive. In this way, the same frames have the same numbers. Illustration by Jeffrey Krebs.

beginning the editing process earlier can help in meeting the deadline for program delivery. As the cost of making films increases, the goal is to minimize the amount of time between the initiation of a project and the release of the project. Costs cannot be recouped until there is a film that can be shown to an audience.

An author of a book can write the story from beginning to end or perhaps from end to beginning. An author also can write chapter 8 first, then chapter 3, then chapter 1, until all the chapters have been written. Then, the process of honing the different chapters into a finished form proceeds until the book is finished.

Writing a book thus can be a linear process(1, 2, 3, . . .) or a nonlinear process (8, 3, 1, . . .). Linear simply means that if there are eight steps in a process, you must proceed through step 1, 2, 3, and so on. Nonlinear means that the steps can be followed in any order. The essential point is that the needs of the story and perhaps the working style of the author or film editor, rather than the technology of the medium, should ideally dictate the methods used. If the author can write a better book by first writing the ending and then the beginning and then the middle, she will choose that method. If another author needs to build up to a conclusion by methodically working from the beginning of the book to the middle and then on to the end, he will choose that method.

There are, of course, different types of authors: those who are adept at novels, those who master the short story, and those who work in only one genre. There are also different types of editors who excel at certain genres. Some editors are masters of the feature film and work exclusively in that discipline. Some editors excel at music videos; others work only on television commercials. It is important to realize that a genre does not equate to whether it is the domain of the film editor or the videotape editor. It would be incorrect to think that the feature film is the sole province of the film editor or that music videos are the sole bailiwick of the videotape editor. The final delivery method often becomes confused with the type of editor associated with the project. Within these specialties are subspecialties. There are editors who excel at sound work, while others excel in special visual effects editing. Editing, it should be clear by now, is a general term that covers quite a range of work.

Editing the Film

Although certain technical tasks must be mastered, the process of film editing is an artistic craft that, above all else, allows the film editor to work in much the same fashion as the various authors mentioned earlier. A scene can be constructed by working backward from how the editor wants the scene to conclude. A scene can be edited from its logical starting point. The necessity of editing the best possible piece is that the editor must have the flexibility to decide how to approach the story telling of that scene, and the editor must have enough opportunity to reedit!

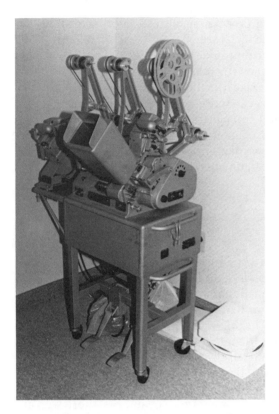

Figure 2–3 The Moviola, an early mechanical film editing system. Photo by Michael E. Phillips.

The editing procedure can vary depending upon the type of editing machine being used. There are Moviolas, Steenbecks, Kems, and so on (Figure 2–3). If a Steenbeck is being used, the editor begins by threading a segment of film and its corresponding mag track. In general, there will be at least one picture head and at least two sound heads. One film segment can be viewed while two mag tracks are heard. Different flatbeds offer more picture and sound heads. If it is absolutely necessary that the editor be able to see more than one picture playing back, a system with more than one picture head must be used.

Once picture and track are aligned using the code numbers to ensure synchronization, the film editor can run the footage at normal, slow, or fast speed. The film editor works from left (feed reel) to right (take-up reel). When the editor reaches the frame where a shot should start, the frame is marked with a Chinagraph marker. The end frame is also marked. Now it is time to cut the film. The film and mag track are placed in a sync block (synchronizer), and a film splicer (butt splicer) is used to cut the film on the line between frames and to cut the mag track. The same occurs for the end frame.

This is the first shot that will be used. The material from this take that will not be used is called the *trim*. The frames before the start frame that was marked are called *head trim*. The frames after the end frame that was marked are called the *tail trim*. Trims are important because they may be needed again. Head and tail trims are hung off a hook in the *trim bin* (Figure 2–4).

This process of selection and cutting continues as the editor juggles through the many different source reels of film to pull the appropriate material. Eventually, the editor has chosen all the shots that will be used in the scene. Now it is time to edit. Because film is physically cut into segments called *film clips*, the individual pieces of film can be rearranged easily. Each film splice represents a cut, and a splice can be made anywhere in the roll

Figure 2–4 In the film editing room, strips of film representing work in progress are hung off metal hooks in the film bin. Photo by Michael E. Phillips.

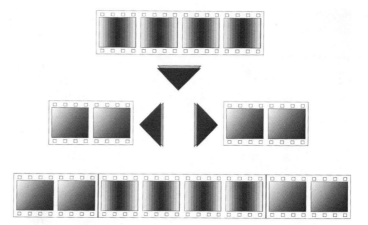

Figure 2–5 A single piece of film can easily be added between two pieces of film by physically cutting the film apart and splicing additional material between the sections. This is nonlinear editing. Illustration by Jeffrey Krebs.

of film. If we have three pieces of film and want to change their order, we simply remove the segment we want to move, put it in the new position, and resplice it. Conversely, if a shot is added that simply doesn't work in its place, it can just as easily be removed (Figure 2–5).

As editing continues, there will be a number of strips hanging in the trim bin, and there will be a roll of film on the take-up reel. The latter represents the rough assembly of shots that make up the scene. Next, the editor plays back the scene and decides where changes should be made. In the event that shots should be rearranged, splices can be removed, and the shot order can be changed.

Trimming shots is a different matter. If a shot needs to be shortened, the editor can move back and forth over the splice point to determine how many frames need to be removed from the shot. Then, the film and mag track are placed in the splicer, and those frames are cut out. They will go into the trim bin. The film is spliced together again, and the cut is judged. If a few more frames need to be removed, the trimming process is repeated. Obviously, as these trims get more and more exact, the trims become smaller and smaller. This is one reason why film editors often joke about having to "find my trim in the bottom of the bin." (Actually, small trims such as these are usually placed in a box rather than in the bin.)

If the editor decides that a shot should be longer, the trim for that section is located in the bin, and the desired number of frames are added to the shot. Editing continues. The film editing process is limited to picture cuts and picture and sound overlaps. An overlap occurs when both picture and sound do not "transition," or cut, together. For example, if we see two people having a conversation, we cut from one person talking to the other person talking. This is called a *straight cut* since both the picture and the sound cut together as we switch from person to person. However, if we want to continue to see the first person for a few frames while we hear the second person talking, this is an overlap. We overlap the picture of person 1 onto the audio of person 2. Eventually, we cut to person 2's picture.

Optical effects, such as dissolves, wipes, and fades, are not immediately available to the film editor. Instead, when a transition

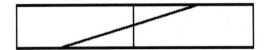

Figure 2–6 By specifying where a dissolve should occur, the film editor directs the optical laboratory to create a new piece of film for the optical effect.

other than a cut is desired between two shots, the editor provides directions for the optical laboratory. These directions are indicated by marking the film with a Chinagraph marker (Figure 2–6). During the editing process, optical effects are not seen. Instead, the editor will use the markings to judge where the transitions will be.

Editing continues on this work print until the *rough cut* stage is reached. Depending upon the complexity of the project, the following steps can occur sequentially or concurrently. The rough cut stage (also called the *first cut*) is the first complete viewing of the entire program. At this stage, changes are noted, scenes may have to be reworked, and reediting can be extensive or minimal. This process of review, reedit, and change continues until the *final cut* is achieved. The final cut is the point where the picture portion of editing is complete, or "the picture is locked."

Work begins on all the sound elements that will be used in the film. Again, depending upon the project, many of these activities could be happening simultaneously, but traditionally, audio work begins after it can be ensured that the picture portion is finished. Sound effects editing, composition of music, and dialogue replacement are just a few of these items. After the optical effects are delivered by the laboratory, they are spliced into their correct position, and the entire film, usually broken down into ten-minute reels, becomes available for final audio dubbing. Each reel is projected while the different sound elements are mixed together for that reel of film.

While dubbing is occurring, the original camera negative is being *conformed* based on the work print. The most original element (the earliest generation) in the entire process will always be the first item we started with: the original film that was loaded into the camera. This original negative will be cut in the exact manner that the work print was cut. This is the process of conforming the film negative. Negative cutting is as final as the process gets since, once the negative is cut, it cannot be replaced (though there have been countless examples where unique fixes have been used in dire emergencies, such as replacing a torn frame with another very similar but not exact frame).

The first version of the film to be viewed with optical effects and the complete and mixed optical soundtrack in place is called the *answer print.* Here is a last-minute opportunity to check everything involved in the presentation, with particular attention to the quality of the soundtrack mix and to the scene-by-scene color correction done to the film. Scene-by-scene correction is a process of adjusting color values to match from one scene to the next.

The last stage is the *release print.* Additional film copies will be made from a copy of the original negative to distribute to theaters or from which videotape copies will be made.

The film editing process is based on years and years of refinement, study, and practice by film apprentices, assistants, and editors, and it is supported by the film industry infrastructure: the writers, directors, producers, film laboratories, sound houses, optical houses, and dubbing houses. Most extraordinary is that, while there may be variations from location to location, the process is remarkably consistent from country to country.

While the film editor's job and the craft of film editing are laborious, they are also very flexible. At any point, the editor can simply rearrange a shot. Audio from one take can be used to replace audio from another take. The editor can construct backward, forward, or from the middle outward. Film is a nonlinear process. Making changes does not require paying a penalty for making the change other than the time that it takes to cut and splice!

THE DEVELOPMENT OF VIDEOTAPE

Prior to the existence of videotape to record and edit images, a process called *electronic transcriptions* was used. These were electrical signals that were recorded on glass discs, somewhat similar to the methods used to create phonograph records. These discs represented the only method to record visual signals before magnetic tape. Electronic transcriptions are not to be confused with Kinescopes, film recordings of televised signals, which appeared and became quite popular in the 1950's.

The first videotape recorder was developed by the Ampex Corporation in 1956 and used 2" recording tape. Videotape offers one attribute that film does not: It is not necessary to wait for processing to see if an image has been recorded. Early videotape recorders were large machines with a great deal of electronics, and they were expensive. Still, the ability to record an event and then broadcast it at a later time brought news and happenings to greater groups of people and was invaluable in fostering worldwide communication.

The recording method of videotape is known as *single-system recording*. Image and sound are simultaneously recorded to the same piece of videotape. Film production is based on dual-system recording, in which the image is recorded on film, and the sound is recorded on audio tape.

While the normal play speed of film is 24 fps, the number of frames per second for videotape will vary depending upon where the recording is made. In NTSC (National Television Standards Committee) countries, the frame rate is 30 fps; in PAL (phase alternate line) countries, it is 25 fps. These three rates—24, 30, and 25 fps— are the most commonly encountered "normal" play rates. The earliest videotape machine from Ampex ran at 30 fps.

Since videotape is a magnetic medium, unlike film, the frames are not visible to the eye. There was no way to edit videotape until 1958, when Ampex introduced a videotape splicer and special magnetized particles that were spread on the tape (Figure 2–7). Control track pulses, signals recorded onto the videotape and used to identify videotape frames, became visible through the use of these particles. The editor would then cut between the frames. When the desired frames were aligned, the videotape would be joined with special cement.

It is interesting to note that videotape editing started out as a completely nonlinear form of editing, just like film. How ironic, upon retrospect, that so much time and effort would be spent

Figure 2–7 The Ampex videotape splicer was used to join the ends of two-inch videotape in the same fashion that film is spliced together. Courtesy Peter J. Fasciano. Photo by Michael E. Phillips.

trying to emulate film's nonlinearity by electronic nonlinear editing systems, given videotape's history! Editing videotape by actually cutting it lasted for quite a while. These methods were used throughout the 1960's and even into the early 1970's. Although the process was successfully used on many programs, it was quite laborious.

Several attempts were made to use machines to edit videotape electronically. Why was there an interest in moving away from the manual process of cutting the videotape? Computer and video technologies were advancing, but a more important factor was that the schedules and deadlines of television began to require that programs be put together in less time. Developing a way to view an edit without actually having to commit the edit became an important goal. Early attempts at bringing the videotape editing process under more control included recording cue tones on the 2" videotape to signify where an edit should be made. When the cue tones were sensed, an edit would be made. This was akin to having an automatic method of making a cut (switching) from one source reel to another. This became a way of being able to repeat edits in a reliable fashion.

In 1967, a method of consistently identifying videotape frames was developed. Film, with its edge numbers, offers a way of identifying each frame on the film roll. Videotape had no such identifying scheme. *Timecode*, in the form of hours, minutes, seconds, and frames, such as 01:05:12:20, is a signal that is recorded onto videotape and that identifies each and every video frame. Timecode was developed by the EECO Company, and in 1972, the Society of Motion Picture and Television Engineers (SMPTE) and the European Broadcasting Union (EBU) standardized the code, and it became known as *SMPTE timecode*.

Timecode was revolutionary. It not only meant that frames could be identified, but that they could be identified quickly. Timecode is a series of numbers, and computers can find numbers fairly quickly, certainly faster than a human searching for a particular frame in a reel of videotape. The precise location of timecode points and repeatable edits became a reality. If we are at the end of a videotape reel and want to cue to a particular frame, we enter the timecode number, and the cueing is easily and automatically accomplished. If we fast forward the tape several minutes into the reel and want to go back to the same point, it is easy to do so.

VIDEOTAPE EDITING PROCEDURES

Videotape editing and the equipment associated with it can vary greatly, depending upon the type of project being edited. Walk into any film editing room anywhere in the world, and the equipment will bear great similarity; there will be some type of editing table and film bins. Walk into a video editing room (called *suites* or *bays*) anywhere in the world, and there will be very little similarity. Videotape editing can be done in a very basic fashion using a modicum of equipment or in a very complicated fashion using a great deal of equipment.

The two common forms of videotape editing are known as *cuts-only systems* and *A/B roll systems*. The cuts-only system most closely emulates the film editing environment. In this system, there are two machines: one source and one record. The only transition that can be made is a cut. An A/B roll system comprises three machines: two source machines and one record machine. The A/B roll system allows the editor to make transitions, such as dissolves and wipes, and involves the use of a video switcher. The tape machines receive instructions from the edit controller; the common language is the timecode. The edit controller usually consists of a computer that directs one or more source machines and the record machine. The edit controller coordinates these machines, keeps track of the timecodes for each shot used in the program, and generates an edit decision list.

Videotape editing is based on selectively recording material from a source videotape to a destination videotape called the master tape (Figure 2–8). This copying process proceeds in a linear fashion. That is, the first segment is copied to the master tape, then the second segment, and so on. Because videotape is no longer physically cut, reordering segments is not as easily accomplished as it is on film. Instead, if a change needs to be made, it is necessary to begin recording again on the master at the point where the change is being made.

Because it is difficult to reorder shots, videotape editing relies on the use of a *preview* function. The editor chooses a point on the master tape where the next edit will occur. This is marked as an *in point*, and the timecode for this point is entered into the edit controller. Then the source machine is played, and another in point is chosen for the new material to be copied to the master tape. Next, the editor can either directly edit the material onto the master, or more likely, the preview function will be used.

Whether working on a cuts-only system or an A/B roll system, the editor hits the preview button on the edit controller, and the machines rewind to a *preroll point* (usually three to five seconds before the edit point) and then roll forward to display the edit. Preroll enables all devices to gain full operating speed before recording. Since the timecode is a perfect reference for each frame, previews can be done over and over with the same result. There will be no change regardless of how many times the preview button is hit.

Figure 2–8 In videotape editing, selected pieces of footage are copied from a source videotape to a master videotape. Illustration by Jeffrey Krebs.

Trimming Shots

Trimming shots in film editing involves either chopping out frames or putting them back. In videotape editing, trimming involves "plusing" or "minusing" frames to be copied to the master. If the editor wants to see more frames before the original point chosen on the source, she moves the footage back by minusing from the timecode in point. For example, if the timecode in point is 01:09:14:12 and the editor wants two more seconds of the shot, the editor types in -2:00, and the in point changes to 01:09:12:12. By moving the edit point back, two more seconds of material will be copied to the master.

After the timecode in point is altered, the editor can preview the edit, and if it is satisfactory, the edit is committed. The recording process is a copying process, and it occurs in real time. If we are editing film and we have a 12-second take, we splice this on and continue. If we are working in tape, the source and record machine will synchronize together, preroll, and then a 12-second copying process records the source material to the record master. All linear videotape editing is a real-time process.

In an A/B roll system, transitions such as dissolves and wipes become available through the use of a *switcher*. Two source machines are synchronized while one shot dissolves into another. The result is recorded onto the master in the third machine. If the length of the dissolve needs to be changed, a new duration is typed in, and the edit is remade. Because timecode allows this repetition and because there is a preview function, the editor can try different transition effects and transition lengths. But, at some point, the edit must be committed, and then the next shot must be placed in a successive position on the master.

The editing process continues until the picture portion is completed. At one time, final sound was edited simultaneously. However, as the level of sophistication of analog multitrack audio machines grew and as digital audio workstations appeared in 1985, the videotape editing process began to emulate the film editing process: Picture editing is completed, and then audio editing and mixing are undertaken as a separate stage.

Completing Audio for Video as a Separate Stage

If the final audio will be posted separately from the picture editing stage, once the video picture is locked, the post facility that is handling the audio editing tasks on the project will receive a copy of the program. If the basic dialogue or ambient tracks that were included during the picture cut are acceptable, these will be preserved and recorded directly from the master to a multitrack audio tape machine (analog or digital). Timecode from the original master tape is used as the matching reference, and this timecode will also be transferred to the multitrack audio machine. This process is called *audio layup* or *audio layover*. From here, the audio designer adds additional sound effects, music is recorded and edited into position, and additional audio work is done.

When all the tracks have been laid out on the multitrack, the process of mixing the audio to the master begins. Mixing simply means that the various tracks will be at different volume levels. These different levels are mixed together so that when we listen to the program, the items that are intended to be softer and louder than other elements are arrayed accordingly. As sections of the program are played on the video monitor, the final audio mix is done. If there is any reason to interrupt this process, resuming the mix is easily accomplished. In film, this entire process is called the *dubbing stage*. In video, this process is called *mix to pix*: mixing audio to the locked picture.

The process is completed by taking the final audio mix and recording it back to the original video master. This process is called *audio layback*. The master videotape is then copied, or dubbed, for distribution. The term *dubbing* means something quite different in the video world than in the film world!

Integrating and Orchestrating Equipment via Timecode

Editing videotape with timecode ensures that the editor can orchestrate the efforts of not only a series of videotape machines, but also a wide variety of other devices, such as digital video effects units (DVEs), character generators (CGs), and audio consoles. Triggering any of these devices can be accomplished from the edit controller based on timecode and *general-purpose interface* (GPI) switches, which instruct a machine to turn on and turn off.

If the editor is previewing an edit in which many different devices are being used, the repeatability and precision of timecode are a critical advantage. For example, let's say that the editor is working on an edit where one source will dissolve into another, and as the dissolve finalizes, a character-generated title will be keyed (superimposed) over the picture and then moved completely off the picture with a digital effects unit. In this very common type of edit, a fair amount of orchestration must occur. Looking at the editing console display, we would see something like this:

	Record Master	Source 01	Source 02
IN	1:09:03:12	1:01:45:12	1:02:12:09
OUT	Open Ended	1:01:48:12	
VIDEO			
Delayed Effect:	1:09:06:12		
Diss. 030			
GPI 001:	1:09:08:12		
GPI 002:	1:09:14:12		

Although it may appear as if there are only numbers here, to the trained eye, there is a complete story in those numbers. Remember, these are timecode numbers, and every machine in this editing process will perform its role when the proper instruction is issued, based on the reference common to all the machines: the timecode. Everything that the editor wants to happen for this edit is shown on this screen. Here's how it works.

The first thing we want to see is a cut to our first source. The edit begins at an in time of 1:09:03:12 on the master. The shot is on source reel 01 at an in time of 1:01:45:12. Shown on the screen is video, which is the *track* on which the cut will be edited. Each source videotape will be identified to the edit controller by giving the reel a number, in this case, 01.

This cut continues for three seconds. At 1:09:06:12 on the master, there is a delayed effect, a dissolve, from source 01 to source 02. Source 02 has an in point of 1:02:12:09. The dissolve rate is 030, or 30 frames. At this time, the video switcher, under computer control, makes an automatic transition from one tape machine to the other; the editor sees the dissolve take place.

A GPI trigger occurs next at 1:09:08:12 on the master. 001 is the GPI code for the particular device being triggered. In this example, device 001 is a keyer that will insert the character-generated title. At the precise timecode of 1:09:08:12, the title is inserted. Thus, we see the dissolve, and one second after the dissolve completes, the title is keyed over the picture of source 02.

Another GPI trigger occurs at 1:09:14:12 on the master. 002 is the GPI code for a digital video effects unit. Thus, after the title is keyed over the picture, it remains there for six seconds, from 1:09:08:12 to 1:09:14:12, at which point the digital effects unit moves the graphic off screen. At the precise timecode of 1:09:14:12, the digital effects unit is triggered, and an effect that was programmed into the DVE moves the graphic off screen.

All these commands occur under computer control and under the scope of just one edit. The benefit of this orchestration is that the editor can see how the elements of an edit are working together and can adjust each element separately. For example, if everything is working fine in terms of timing, but the dissolve needs to be just a bit shorter, the editor can change one parameter, the dissolve duration, from 030 to 025: five frames faster. When the edit is previewed, all the other parameters will remain intact, and a slightly faster dissolve transition will be made.

What is the alternative to using a computer and timecode to achieve such an edit? The editor would have to be extremely well coordinated. He would have to trigger manually all the different devices, operate the switcher, and be able to do so at exactly the right moment. And, the editor would have to be able to repeat this over and over until the desired result was achieved. This was the state of the art of videotape editing until timecode and computers brought coordination to the disparate machines used in the videotape editing process.

Are all of these commands, timecodes, and triggers helpful or intrusive to the editing process? There is certainly more orchestration involved here than on a film editing table, and there are also more items that the editor must keep track of, but the coordination of these different devices is quite an achievement. The ability to do all of these things at one time and repetitively cannot be overlooked or underestimated. It is an extremely powerful combination of capabilities tied together by a computer that allows the editor to change even the smallest parameter.

LINEAR VERSUS NONLINEAR EDITING

Consider the following situation. Let's say that you have video-taped a series of three weekly company meetings, and you need to make one videocassette copy from the three separate original tapes for your supervisor. You connect two videocassette machines and play tapes in one and make the recording in the other. The order you choose is chronological:

1. Meeting 1 on March 14.
2. Meeting 2 on March 21.
3. Meeting 3 on March 28.

You finish copying the tapes, and your supervisor tells you that the finished tape should start out with the last meeting and end with the first meeting. You have made the compilation tape exactly backward to the request. There is only one way to make the change. The entire recording process must be repeated. Since videotape editing has migrated away from actually cutting the videotape, the order of the meetings cannot be physically rearranged. All the work that was done is of no use now. Instead, the lengthy process of making the copies, in real time, must be undertaken until each of the three segments is recorded onto the finished tape in the final order:

1. Meeting 3 on March 28.
2. Meeting 2 on March 21.
3. Meeting 1 on March 14.

If you were to look up the word *linear* in a dictionary, you would find that it is an adjective that means of, relating to, or consisting of a line: straight. Linear editing means that you must adhere to the principle of assembling your program from beginning to end, and once you put the second shot in, the first shot cannot easily be replaced or altered, even to add or delete just one frame. Everything downstream of the change must be rerecorded. The physical nature of the medium enforces a method by which the material placed on that medium must be ordered.

Compare this with an example of nonlinear editing. The editor has three shots (Figure 2–9). The editor first decides that the shots work in the order of clock, couch, and lamp. After viewing the sequence, the editor wants to change the order to couch, clock, and lamp. This rearrangement is easily accomplished. The editor removes the film splices, reorders the shots, and then splices the film together again (Figure 2–10).

Film editing is nonlinear editing. Splices can be made anywhere, and material can be added or deleted anywhere. The entire program is in a very fluid state and can be changed at any point. This nonlinearity of film editing has always existed, but it has never been discussed as an attribute; it was always taken for granted.

If you looked up the word *nonlinear* in a dictionary, you would not find it. Nonlinear refers to the concept that the physical nature of the medium and the technical process of manipulating

Figure 2–9 The first editing attempt results in the images being ordered clock, couch, and lamp.

Figure 2–10 After the editor has respliced the images, the new order is couch, clock, and lamp.

that medium do not enforce or dictate a method by which the material must be physically ordered.

Although film lets us edit nonlinearly, film does not give us *random access*. Random access refers to speed, an inherent part of electronic, nonlinear editing. Consider a roll of film. If the roll is positioned at the beginning, and we want to get to a shot that is in the middle of the roll, we must wind the film through all the intervening material to get to the desired location. We cannot simply point to the wanted shot and immediately go to it. Nor does videotape offer random access. To get from one shot to the next on a reel of videotape, we must wind from point to point.

> Nonlinear + Random Access = Free Form
> Free Form = The Way We Think?

While editing, ideas will often surface at random. Someone will suggest something, or a direction will become apparent as two shots are juxtaposed. Free-form editing, like free-form thinking, must be encouraged, and the editing system must provide easy access to these free-form thoughts.

The Linear Process of Videotape Editing

Linear editing and sequential access cause the user to think ahead and plan how a sequence will be edited. When editing linearly, it's difficult, as many editors will remark, "to let the cut evolve." With film, a scene can be edited, left for a period, returned to, and easily reedited. During this time, the editor can be working on other scenes and thinking about how these scenes affect previously edited scenes. Since there is no associated "penalty" in going back and reworking a scene when editing nonlinearly, editing can be more unstructured, allowing the editor to try ideas.

Consider the following general statistics about videotape editing:

1. While there are many types of video projects, a well-paced video will have edits every two to six seconds. Therefore, there are about ten to 30 edits per minute.

2. Based on studies done in professional video editing facilities, each edit will take about six minutes, including shuttle time, video and audio setup, and preview time.

3. As a result, a range of one to three hours of editing will be required for each finished minute of program.

Of course, there will be exceptions. If the footage is particularly well organized, less time will be spent searching for the shots to use. If there are ten source reels instead of one hundred source reels, less time will be spent loading and unloading tapes.

In general, what follows are the realities of linear editing and how the issue of making changes is addressed.

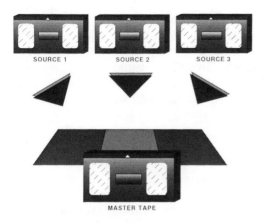

Figure 2–11 Material from different source reels is recorded in proper sequence to the master tape. Illustration by Jeffrey Krebs.

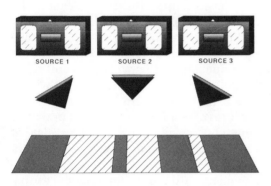

Figure 2–12 As additional edits are made to the sequence, the result is a structure which is difficult to re-order. Illustration by Jeffrey Krebs.

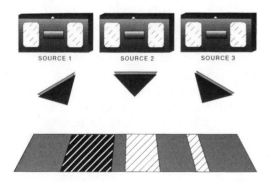

Figure 2–13 If we remove the second shot, everything that follows will be in the wrong place. Illustration by Jeffrey Krebs.

Step 1: Building the Master Tape (Figure 2–11)

In video, a sequence is created by copying sections from the different source tapes to a master tape. The editor needs time to review the log sheet and determine a timecode from which to start. The machines must shuttle to their points, and the edit must be previewed and, finally, recorded. The different patterns shown in Figure 2–11 represent material taken from the three source videotapes.

Step 2: Locking in a Structure (Figure 2–12)

As editing continues, additional material is taken from the source reels. Each edit requires more shuttle and preview time. After many edits have been made in the sequence, the structure has been established.

Step 3: Making a Change (Figure 2–13)

While editing, we decide that the second shot should be removed. Since we are editing on videotape, we cannot physically cut the material; we can only erase the shot. If we do this, all material following will be in the wrong place. This means that we will see the first shot, and then we will see black where the erased shot was previously. Then we will see the remaining material. We cannot easily "pull up" the material to abut the first shot.

Step 4 (Option A): Reedit from the Point of Change

After erasing the shot, there are two ways to proceed. The first option is to reassemble all the shots that follow the change. This reediting procedure can be very time consuming, but it ensures that the best quality is preserved because the original tapes are recopied to the master tape.

Step 4 (Option B): Copy to Submaster (Figure 2–14)

The second option to reediting involves recording the material downstream of the change to another piece of videotape. If the material represents 30 minutes worth of edits, it will take 30 minutes to copy the section to a submaster. The submaster represents all the material that was recorded to the master tape but is now in the wrong place. Then, this section will be recopied to its new location, which will take another 30 minutes to execute.

Shown at the top of Figure 2–14 is the downstream material that has been recorded to a submaster tape. This submaster is then recorded back to the master tape (middle). The result is that the material that was in the wrong place is now in the correct location on the revised master tape (bottom).

Essentially, we have "block moved" the remaining portion of the sequence to abut the change. This process took 60 minutes, and a tape generation was lost. Losing a generation means that we are no longer editing from the original source videotapes. Instead, we are editing from a copy. Depending upon the tape format, severe signal artifacts can be seen whenever tape is "taken down a generation."

Whether we are editing a program for worldwide distribution or making a compilation tape of three company meetings, linear

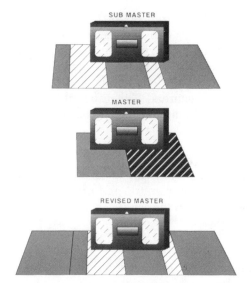

Figure 2–14 Copying footage to a submaster allows material downstream of a change to be rerecorded to its correct location. Illustration by Jeffrey Krebs.

editing requires that we decide on the best possible plan of action before we edit. We can preview as many times as needed to choose the correct frames, but at some point, we must commit. While there are options available if we want to make changes, both options have their drawbacks. Going back and making changes is not a scenario that anyone editing linearly likes to consider.

Understand the Needs of the Program before Choosing the Editing System

Both film and videotape editing offer substantial tools to the person who is putting together a presentation. Understanding what the program requires will generate a series of questions that will help us assess the capabilities of the editing system.

Is the project scripted, or will putting the material together occur without benefit of the structure of a script? What is the estimate on the amount of footage that will be shot? Can the material be edited in just one or two ways, or could the footage be put together in many different ways? If shot on film, will the program finish on film or videotape? How will audio editing be addressed? What is the likely procedure for changes that will be requested, and how much time will there be to provide the changes? Does the project involve a fast editing pace or a slow editing pace (e.g., commercials vs. dialogue)? If a great deal of footage has been shot, will all of the footage have to be instantly accessible, or can the editing process be broken down into segments? What other elements will the program require during the editing stage? Will there be titles and graphics? Will slow or fast motion be needed? Will regular review copies of the project be needed or will it be viewed at the first cut stage?

Most of these questions (with the exception of cutting pace and footage accessibility), have always been asked whether a program is being planned for a film edit or a videotape edit. This is true regardless of whether the systems are linear or nonlinear. If a very large amount of footage is being shot for a project to be edited on film, more cutting tables and more assistant editors are required. If scheduled for a videotape edit, we would consider whether the project should be edited directly, or whether there should be a form of "preediting." Editing a project directly is usually referred to as *online*, while preediting the project is referred to as *offline*.

3

Defining the Electronic Nonlinear Random-Access Editing System

When the film editor cuts a frame, he's undeniably working with images. They're right there in the palm of his hand. When he rocks a mag track back and forth across a sound head, the sound is at his fingertips. The tape editor can't enjoy these sensations. She touches neither the pictures nor the sounds. She doesn't directly manipulate images or audio. Instead, she manipulates numbers—timecode numbers—that place images and sounds in their proper positions.

When a film editor views a sequence and feels that his cuts are long, he goes to the shots and removes frames. He uses a splicing block and tape. The videotape editor accomplishes the same result by "positive trimming" the shots. She uses a computer—in effect a computerized splicing block—which acts like a sophisticated calculator.

Why do many videotape editors want the flexibility that film editing affords? Why do many film editors want some of the tools that videotape editing has to offer, but not at the expense of having to use that calculator? The answers to these questions become obvious when we consider the difference between linear and nonlinear editing and what is meant by an electronic nonlinear random-access editing system. A basic definition of this system has four elements:

Electronic The system is driven by a computer that provides speed, data management, and graphical interfaces between the user and the system hardware.

Nonlinear The physical nature of the medium does not impose constraints on how the material must be ordered. Shots can be tried in a different order, and a series of shots can be easily moved around as a group.

Random-Access The system allows the user to seek a particular section of material without having to proceed sequentially through other material to reach that location.

Editing System Hardware and software allow the editor to combine picture and audio tracks and, at the very minimum, yield an edit decision list (EDL). For some programs, the result of the nonlinear session is a finished product. For other programs, the EDL is used to recreate the sequence during final program assembly.

Although there are nonlinear systems that have some combination of these attributes, all four characteristics are important to a fully flexible system. Having nonlinear capabilities without random access means that cueing problems are encountered. Each tool takes its part in the overall process, and the absence of just one of these tools becomes a very large liability in the system's flexibility.

An implicit characteristic of the nonlinear editing system is its ability to let the editor create multiple versions of a sequence. One very common characteristic that both film and video share is that they are both inherently incapable of providing the flexibility of multiple versions. When editing film, it is possible to rearrange shots in a scene until we are satisfied with the result, but it is not immediately possible to create another version of that scene. Instead, we have one version of the scene comprised of shots in a certain order.

In film, if we wanted to keep our original version of the scene, create an entirely new one in which we change the order of a few shots, and then view each version one after the other, we would have to place a special order to the film lab. We would order a dupe (copy) of the scenes used in the first version and then make the changes using the dupe footage. This would become our second version. Ordering a dupe work print is the only way we would be able to preserve the first version. (However, enterprising editors have been known to videotape the sequence directly off the flatbed's picture head and use the videotape as a temporary version!) With videotape, to create the second version we would either edit a completely new version or we would use a combination of reediting and taking material down a generation to create the second version.

Any nonlinear editing system must provide multiple versions without forcing the editor to order additional copies of footage or to degrade the signal by losing tape generations. Working in an electronic nonlinear editing environment automatically means that several versions of a sequence can be made.

THE RISE OF DIGITAL SUPPORT, DIGITAL MANIPULATION, AND DIGITAL STORAGE OF VIDEO

The trend of video and audio being manipulated by computer began in the mid-1970's as timecode and computer-controlled video editing systems began to appear. These editing systems offered more control over the various devices in the editing room.

This assignment of tape machines as peripheral devices to the computer was a natural progression of the assignment of unique identification numbers (timecode) to each video frame.

More important, however, is the trend of analog video being represented as digital data. There are two aspects: the recording of video as a digital, not analog, signal and the digital control and manipulation of that video signal. The migration of analog video and audio signals toward digital processing and the ultimate origination of video and audio as digital data will continue as the all-digital production system and the "digital media processor" stages begin to appear and dislodge what has, up to now, been an analog-dominated world of production and post-production. The introduction of a number of significant products represents the continuing development of the digital processing and manipulation of pictures and sounds.

Timebase Correctors Manipulate a Signal's Waveform Characteristics

The timebase corrector (TBC) is a device that can manipulate the analog waveform of a video signal. A timebase corrector is an electronic device that is used to correct and stabilize the playback of the video signal. Prior to TBCs, it was often the case that a tape made on one machine would not play on another machine. A number of factors can cause this. Problems sometimes occur when a tape is copied. Many elements make up the video signal in addition to the video image. Vertical and horizontal sync pulses ensure that the video image is displayed in a stable fashion. Making copies of a videotape without a TBC can cause signal degradation not only of the picture, but also of these stabilizing signals.

Another source of trouble is a mechanical intolerance from machine to machine. If a videotape recorder's tape guide path is worn enough to affect the manner in which the tape is wrapped around the recording head, this tape may exhibit problems when played on another machine.

Narrow-window analog TBCs were developed to process and replace unstable sync signals. *Narrow window* refers to the amount of time distortion that the TBC is capable of correcting, specifically, two microseconds. By reprocessing these sync signals, there is a better chance of playing back a tape with distorted sync information.

Wide-window digital TBCs began to appear in the late 1970's and provided one to 12 lines of error correction. This was a breakthrough and allowed a number of the emerging tape formats such as 3/4" to be used for broadcast situations. As 3/4" tape began to be used for fast-breaking news stories, it was necessary to broadcast these original tapes. The use of TBCs to digitally correct the error between the studio playback machines and the field recorder was necessary to ensure that the tape could be broadcast.

Digital Framestores Process Video Information

Digital TBCs paved the way for digital framestores, devices that can store from one to several frames of video. One major reason why framestores were developed was for broadcast transmissions. In a television broadcast facility, all sources that will be linked together and will be part of a news broadcast need to be synchronized to one another. If they are not in synchronization, it is likely that when different video sources (e.g., tape or camera) are punched up on the video switcher, some of the sources will not display stable pictures.

Genlocking is the name of the process that provides this common synchronization. When sources are genlocked together, a master sync generator provides outputs for horizontal and vertical sync signals and the subcarrier, the basic stabilizing signal. In this case, the master represents the television station's switching and routing system. Each source to be genlocked will be matched to the master's horizontal and vertical sync and to the subcarrier. When this is done, internal sync generators in each source are running at the same frequency and phase as the master.

When a reporter is on location and a live camera source is sending a signal back to a television station via microwave transmission, there is no way to genlock that camera. Instead, the camera signal is sent into a digital framestore. This framestore processes and genlocks the signal. It can then be put on the air. If the transmission feed from the camera to the framestore suddenly becomes faulty, the framestore will usually freeze and display the last complete frame that it processed.

If you have ever watched a live sporting event that was being broadcast via satellite, you may have seen the live action freeze occasionally. You were watching the digital framestore display the last complete frame while trying to lock onto the next series of incoming frames.

Digital Video Effects Devices Manipulate Moving Video in Real Time

Digital framestores, in turn, gave rise to a number of new and revolutionary devices in the mid-1970's, such as digital video effects systems that could move and reposition moving video signals. As a video source was sent into the DVE, the framestore units would process frame after frame in rapid fashion.

Each DVE offered new features that changed the picture in some way. Whether the DVE was used to move pictures across the screen or to shrink or expand the picture, DVEs soon became yet another tool in the editing process. Other DVEs could break apart one picture into several smaller pictures. Visual effects that formerly were impossible or too costly to create became routinely available. Manufacturers of these early systems included Vital (SqueeZoom) and NEC (E-Flex). By the mid-1980's, a variety of digital effects systems were available, including Ampex's Digital

Optics System (ADO), the Abekas A53D, Pinnacle's DVEator, and Grass Valley Group's Kaleidoscope. Regardless of the system used, the essential point was this: The picture that came out of these devices usually bore no resemblance to the picture that went in.

Film editors have access to special optical effects other than dissolves. Repositioning a shot (repositions), making a shot larger (blow-ups), changing the screen direction of a shot (flops), and stopping action (freeze frames) are all types of optical effects that are specially ordered from the laboratory. The time involved to create these effects before being able to view them can be significant, depending upon the complexity of the effect.

Trying a visual effect and changing its movement path by altering positions is easily accomplished with a DVE. An effect can be viewed, judged, changed, saved to computer disk, and recalled weeks later. The ability to receive instant feedback and to try things in this manner promoted some crossover between film and video. While a film is being edited, tools from the electronic post-production can be used to create visual effects with these digital effects systems. In this way, the filmmaker is able to get an idea of how the eventual film optical will work.

Digital Still Store Devices Store and Recall Frames

A further breakthrough was the introduction of digital still store devices in 1983. The Abekas A42 was an early still store. Although the framestore unit can process frames without manipulation and DVEs can both process and manipulate frames, these devices are not designed to store and recall frames; they are intended to be pass-through devices for frames.

The digital still store, on the other hand, can be thought of as a digital version of a slide machine that projects 35mm photographic slides. The slides are advanced either randomly or, if part of an audio-visual presentation, based on cues on the accompanying audio cassette.

Television stations and networks in the early 1980's were frequent users of the 35mm slide. While the news anchor spoke, some type of graphic was almost always placed over the anchor's shoulder. The graphic was most often a slide being cycled through a slide machine.

Digital still store devices allowed the storage of images that could be recalled upon demand. Full-resolution video could be played from a tape machine into a still store device, and one video frame would be captured and stored at its original quality with little observable degradation. The use of computer disks to store the material provided rapid retrieval and rapid searching. The number of frames that could be stored varied according to system configuration. Each frame had a unique address on the disk, and when this number was typed into a small console, the frame was recalled.

The digital still store made using slide machines to recall graphics an outdated technology. Although there is an access time of a few seconds in recalling from the computer disk, it is nothing compared to the problem of moving linearly through the slide tray. Today, it is common to use digital still stores to store and recall images. Still stores have provided a digital, nonlinear solution to the cumbersome alternative of sequentially recalling slides.

Although the cost of such a system is far above that of the basic slide machine, the benefits of being able to store an image instantly without having to first photograph, process, mount to slide, and then project the image in an environment where schedules and deadlines are tight are a main reason why the sales of digital still store devices jumped dramatically after their introduction. These sales were also bolstered by the advent of electronic paint systems, whose images needed to be stored and recalled quickly for display and editing purposes.

Images Are Stored to Computer Disk

The digital still store brought a digital solution to the analog world of the slide machine, and it did so in a relatively inexpensive fashion, given the benefits received. There was precedent for such a solution. The actual storage of analog video signals on computer disks had been in regular use through the 1970's. One of the earliest reasons for storing frames of video on some type of fast-access medium was to address the need for slow motion and replays during sporting events. Before the introduction in 1978 of a standardized 1" videotape format that could provide slow motion, all such effects were accomplished through the use of computer disks and storing frames to disk.

The Ampex 100 used large-capacity disks. Analog video signals were recorded as FM signals onto these disks. Once the material was on the disk, the disk could be played back at speeds other than normal play. The disks could record directly and could hold approximately 30 seconds of motion video. When the sports director wanted to replay a shot or play it in slow motion, the disk was commanded to the start point, jumped back to that point (which would take several seconds), and then a linear playback would occur.

Digital Disk Recorders Provide Simultaneous Digital Playback and Recording Capabilities

By the mid-1980's, digital still stores and DVEs were in wide use. Combining several images to create a "layered" fabric of elements became a common technique. For example, let's say that we had to create an optical effect in which a mechanical plane is flying over a background of clouds while a boy in the foreground of the frame looks over his shoulder to see the plane.

The way that this was accomplished prior to the appearance of digital disk recorders was to put the piece of tape that had the clouds on one tape machine, the plane on another, and the boy on another. The proper start points for each would be chosen, and with all machines rolling together in synchronization, the signals would be processed through a video switcher, and a variety of keying techniques would be used to combine the three elements into one seamless element. Achieving these effects depended upon having many tape machines working together. When that was not possible, material had to be taken down a generation to continue to layer the elements.

Digital systems merged in the mid-1980's. There were devices that could record and recall still images (still stores), and there were devices that could store limited amounts of motion video (disk playback). Combining the ability to play back images, add additional information, and rerecord the result without signal loss prompted the creation of a new product.

The Abekas A62 digital disk recorder was developed in 1985. It offers the ability to simultaneously play back and record images. The A62 uses magnetic computer disks to accomplish these functions. It offers a way of recording and rerecording element upon element without tying up many tape machines or experiencing generation loss.

At first, the device was solely a simultaneous playback and record device. However, the manufacturer realized that including a digital keyer within the system would provide increased functioning: It would be possible to combine and modify images without signal degradation. If an external keying system were used, the signal loss would be apparent after a few generations. This new product introduction was highly successful. Those who were in the business of routinely providing optical effects and who required multiple generations to achieve those effects simply had to have this system to maintain the quality and integrity of the original signals.

Prior to having this capability, combining many visual layers was routinely achieved by using a large number of 1" tape machines, many video keyers, and digital video effects units. With the digital disk recorder, fewer overall machines are involved in the process. Increasing overall functioning and doing it from within one singular system is the ongoing goal of digitally transforming, storing, and manipulating the analog video signal.

Digital disk recorders, although they are primarily used for combining images, can also be used to perform editing functions. Some editors experimented with using disk recorders to do editing. By cycling from selected frame to selected frame, digital disk editing was being done. In essence, disk-based, nonlinear editing could be tried. This technique never came into common use, but it was tried, and it was an option. The main reason that digital disk recorders have never been commonly used in editing situations is due to their limited storage time; approximately 50 to 100 seconds of full-resolution images.

Still, digital nonlinear editing at full resolution had never been available before, and some users of digital disk recorders were more than willing to experiment with the possibilities. For example, if a 30-second commercial is being edited, and it is too long at 50 seconds, this material could be loaded into the digital disk recorder. Existing edits could be "trimmed" by cycling from one point to another. This was digital nonlinear editing, albeit to a very limited degree, but it provided a sense of things to come!

While film and video have had limited reasons to coexist, the benefits of digitally manipulating and storing images have served to bring film and video closer. Today, it is routine to electronically previsualize in video what will eventually be complex film opticals to determine direction and possible cost. It is also routine to transfer film elements to video, create the optical effect in an entirely digital environment, and then transfer the results from this digital environment back to film. The electronic tools of digital imaging, having advanced steadily through the last ten years, are now in a position where work done in the electronic digital system can be directly used in the film world.

Although the increased interdependence of film and video will continue to develop and prosper, the editing processes of film and video will also continue to evolve and become similar, with a sharing of technique and technology. The entire image management process began a route toward digital manipulation. We now examine one small portion of that phenomenon by exploring the further migration toward electronic editing and a movement and incorporation of the nonlinear film culture and work methods into the electronic post-production methods.

TALKING WITH EDITORS

What Do You Try to Do for Directors or Clients When You Review Their Footage?

Tony Black, ACE, Washington, DC:

Editing is story telling, whether feature, documentary, or commercial. I help a director or producer tell the story in an interesting and appropriate way. It's unlikely that you would use the same techniques for a historical documentary as you would for a slice-of-life commercial. I'm trying to grasp the overall content of the footage: Are there enough visual alternatives to offer a variety of suggestions for a cutting style? The client will have his own ideas of what he expects the finished product to look like. It's up to the editor to tell him whether or not those expectations are realistic.

Paul Dougherty, editor, New York:

Invariably, the client shapes my performance. The editor has to be a bit of a chameleon adapting to each client. Some are very forceful and want little more than a button pusher. Then there are others who don't have either the time, ability, or inclination to get involved and want you to take total control and run with it. The trick is to "read" the client as to whether they want you to be servile, opinionated, or somewhere in between. I hope that I'm bringing a sensibility to the process beyond grading footage by a "craft report card." I'm looking for magic moments that might be lost on others and will fight to keep those moments in. It's gratifying when you succeed; then you feel you've put your stamp on the piece.

Basil Pappas, editor, New York:

When I am seeing material for the first time, I am trying to form my own opinion of the footage, independent of the script. Much of the evaluation will depend upon the type of project. If it is a commercial, there certainly can be storytelling, but advertising the product is of prime importance. Editing is a lot about following your instincts: What does the project mean? What does the client need it to do? You are also evaluating technical aspects: Is the picture as good as it can be? How can we improve the look and sound?

Many people will let the editor make the first cut and come back and comment. The comments depend on who these people are. A director will be looking at specific things, such as performances. And nonlinear is great because if the director wants to slow things down a bit and let performances breathe, it's very easy to do it on a nonlinear system. There is also the economic restraint that you may have which causes you to think: How much time do I have to present the best possible edit I can accomplish? Because of all these considerations, what the project should be, what technical things need to be done, and how much time and money are available, it is important to establish a good relationship with the client and to become a partner in the vision.

Peter Cohen, editor, Los Angeles:

No matter what kind of project I am editing, my first responsibility is to bring alive the vision of the people involved in the project. Although there are usually many parties involved (director, producer, writer), the first vision I work towards is the director's since the material was shot with that vision in mind. When I am working on the recuts and the client is with me, I am working with direct input, and I add my interpretation of the material as appropriate.

When I review the client's source material, I make notes on several levels, among them, performance, technical aspects, pacing, and continuity. The priority of these levels changes with each different style of editing. The same amount of story that happens in a 30-second commercial may take an hour in a feature film, or two minutes in a music video, or ten minutes in a sitcom. The style determines the pacing and rhythm.

Alan Miller, editor, New York:

If a shoot is in progress, the first thing I try to do is determine if the coverage is sufficient. Usually, I try to let the director know of any problems that I see that could be corrected on set. If the shoot is already complete, I try to see how the story plays, and where there are problems. Identifying problems early helps the director–editor relationship. It also gives me a head start on my own viewpoint of the piece. It's very important for an editor to have a viewpoint of his own. Often the editor is the only objective person in the editing room. Everyone else is seeing what he thinks he shot. The editor has to see what's really there.

4

Offline and Online Videotape Editing

The 1970's saw editing evolve from the 2" videotape format used by professional broadcasters to the 1/2" open-reel EIAJ (Electronics Industries Association of Japan) format used in industrial and educational applications. The decade also saw the eventual introduction of additional videotape formats, most notably, 3/4" videotape. Less expensive electronic field systems were developed, and the use of film for everyday events such as news and sports programs began to decline. Videotape editing, now with SMPTE timecode as a standard referencing method, began to be used for many types of programming.

No one really agrees on when the terms *offline* and *online* editing began to appear. It's quite plausible that the terms were influenced by the broadcasting industry. When using a video switcher to "live switch" an event, the act of taking something "online" means that the source should be put on the program bank of the switcher, then fed to the broadcasting transmitter. Taking something online meant that the final decision was made: "Send it over the airwaves." Putting a source on preview or taking a source that was online and moving it off the program bank was taking it offline: "It's not ready yet."

In offline editing, the resulting program is not finished; it is in a form of preview. In online editing, the program is finished; it is ready for final viewing. Traditionally, offline systems and online systems differed greatly in their capabilities, and thus, the distinction between the two was clear. That distinction has become very blurred.

In 1974, the 3/4" videotape format began to be used widely as affordable 3/4" players and recorders became available. Three-quarter-inch tape machines that could be edited on manually could also receive instructions from edit controllers based on the language that they shared: SMPTE timecode. At this time, 3/4" videotape was of lesser quality than 2" videotape. Half-inch open-reel videotape was the equivalent of today's VHS videotape.

Editing systems that could control these machines appeared in various models. Each model had a series of features and capabilities that served to distinguish one edit controller from another. Another distinguishing factor was the method by which the editing system controlled the connected tape machines.

Some systems could make edits based on SMPTE timecode, and others could only make edits based on control track timecode. The difference between the two is that control track is not a unique frame identification as is SMPTE timecode. Control track editing is akin to putting a tape into a machine, zeroing the tape counter, advancing and counting frames to the edit point, and editing. If another tape is put in the machine and the counter is zeroed, there is no way of putting the original tape in and using that counter to determine where an edit should be made (or remade).

THE OFFLINE–ONLINE LINK

The process of editing videotape was rapidly divided into two camps: offline systems and online systems. At the time, all finished online editing was taking place on higher quality 2" videotape. Due to the cost of these machines and the typical scheduling constraints, it was necessary to develop methods that would ensure that the time required to edit a program was spent efficiently and economically. The solution was to use these new 3/4" machines to develop a form of preediting. This stage would ultimately turn into the method of offline editing.

The edit systems that began to appear were unique in that their purpose was to divide the work between the more capable editing controllers and the less capable ones. The CMX 50 editing system (the CMX Corporation was a joint venture of CBS and Memorex) was an A/B roll edit controller that could control three 3/4" tapes and permitted dissolves. One purpose of the CMX 50 was to provide a less expensive alternative to the more functional CMX 300 editing controller, which was being used to edit 2" videotape. A more important purpose was to use the CMX 50 to preedit a program.

If a program was destined to be edited on 2" videotape, editing decisions had to be final decisions due to the linear nature of tape editing as well as the fact that the editing cycle for television was short. As a result, there was little time to experiment with the order that shots should take. What was necessary was to bring footage to the edit, to know exactly how shots would be used, to edit, and to make the editing room available for the next project.

Experimentation and preparation for the 2" online session was the only goal of the offline session. It was achieved through the use of the CMX 50. The compatible link between the offline and online systems was SMPTE timecode. The original 2" videotapes were copied to 3/4" tape, and the timecode was regenerated from the originals as the copies were made. The same frame on the 2" and 3/4" videotape shared the same timecode number.

Editing was done offline. Because the machines cost less and there usually was a modicum of other hardware in the offline editing room, usually more time was allotted to allow the editor and client to "rough out" ideas and edits for the program. Although it was linear, there was more time to try things and to experiment since the goal at this stage was to make all the decisions that would then be carried out during the online session.

Once all the edit decisions had been made, a list of these edit points, an edit decision list (EDL), was then transferred into the online edit controller, and the program was edited into its final form from the original source tapes. This offline–online process reduced the time spent experimenting in the online room and provided an opportunity to create better programs by virtue of having more time to try ideas during the offline stage. This has always been the profile of offline: less functional machines, but more time available to try ideas. Ultimately, the program is not finished at this stage, but work products are generated from this process that should make the online session proceed more efficiently. The promise of video offline is a better result and a more efficient and less expensive online session.

THE REASONS FOR OFFLINE EDITING

Offline editing cannot solely be judged as a step used to streamline the efficiency of the online suite. In addition to trying to make more creative decisions while not under the pressure of having to vacate the online suite, there are a number of factors that necessitate the process of editing before the final program is achieved.

It is important to realize that a number of different motives are at work via a number of individuals in any edit, whether it is offline or online. While *motives* may bring with it negative connotations, it is simply necessary to remember that each person may have a personal agenda. Whether it is the director, producer, writer, advertising agency, or another editor, each person requires a certain amount of input to the program. When the effort is a collaborative one, the additional input is most welcome. When the agendas clash, as they often may, it is left to the editor to try to maintain the integrity of the program while allowing each person to continue to participate in the process.

Ultimate authority regarding how a program will turn out rests with different individuals, depending upon the project. If the project is a commercial, the advertising agency ultimately reigns since it is responsible to the client. For a feature film, final authority might rest with the director or perhaps with the studio financing the picture, depending upon contractual agreements. What is the editor's role in all this? The editor can shape and hone, but the editor must take all the concerns and suggestions of the various individuals and attempt to incorporate them into the material. The editor must know when to speak, when to listen, and when to chastise if the integrity of the piece is beginning to be compromised and diluted. Offline editing allows these individuals an opportunity to realize their responsibilities to the material.

Offline editing is also necessary because the people for whom the program is being created often cannot take part in the editing process, and as the program content increases in complexity, it may not be realistic to have all the individuals present who must deem the offline ready for online. Any number of concerns can arise during the offline stage; legal approval may be necessary to use certain material or a content expert may need to approve the information presented before the program can be finished.

These experts can be very far away from the editing process, and yet their input is critical. The editor may find, while searching through footage, that a shot that was not logged is perfect for a section of the program. If the client believes that it is improper to use the material or if the client does not have permission to use the footage and if the program is not reviewed before release, a costly recall of the program could occur.

As a result of client feedback, changes may be required. Most often, when changes are made to the offline version, the program will be taken down a generation, and changes will be made to the new copy. Loss of picture or sound quality is meaningless since the program will go to online anyway. Last-minute changes are also made during the online session, and although this can be time consuming as new material is sought, this happens all too often. While the theoretical goal of an online session is no experimentation and just an automatic conform of the EDL, it is rarely the case that an edit is not changed in some small or grand manner.

The "attitude" found in the offline editing suite is characterized by its purpose: to achieve the best possible plan of action for the footage. Offline, and not online, should be the place where ideas and edits are tried and discarded or kept. The creation of an EDL that can be used to create the final show is the prime work product from this stage.

THE PROCESS OF OFFLINE EDITING

Offline editing did not begin in earnest until 1980. At the time, 3/4" machines were in abundance, they were in use by many types of clients, and they were affordable. It was necessary to find an alternative to creating programs from scratch in the online suite. Until 1980, 2" three-machine A/B roll editing cost approximately $350 to $500 per hour, while offline cost only $50 to $75 per hour. It made economic sense to work in offline prior to online.

Most offline editing suites have two machines where cuts-only editing is done. Other offline rooms have A/B roll capability. Regardless of the type of equipment, the process of offline editing has remained remarkably consistent over the years, although many of these various steps have been made easier through the use of computers. The usual steps are as follows:

1. "Window" dubs
2. Logging footage
3. Paper edit list
4. Offline editing
5. List cleaning

Figure 4–1 When window dubs are created, timecode information is placed as an overlay window over copies of source footage as a step in the offline process.

Step 1: Window Dubs

Window dubs are also referred to as *burn in* copies. The first step in the offline process is to copy either all original source material or selected source material to a lower format videotape, such as VHS. VHS is a common choice because the tapes can be taken home and loaded into an inexpensive consumer machine. During the copying process, the timecode from the original footage is read by a timecode generator, and a visual overlay is displayed over the footage (Figure 4–1). Window dubs are these burn in copies.

Step 2: Logging Footage

The process of logging the footage should have begun during the shooting stage. If a log is not available, one is created by viewing the window dubs. The tapes are scanned for approximate start and end timecodes for each take. These times are written down, and any further description is noted (Figure 4–2). Some form of preference is given to the various takes, either by circling one of them (circled takes) or giving them some scale of usability.

The logging process can certainly be time consuming, but to be sure that the best possible footage is used in the program, we must know what footage is available. Also, if a take we use in the program needs to be changed or if the director wants to try a different take, it is important to be able to have a good log that will allow the alternative material to be located quickly.

Step 3: Paper Edit List

Once the logging process has been completed, the actual editing can be done or the editor can further organize the edit by creating a paper edit list. This list outlines how the program will be put

PRODUCTION: Dragonfly   DATE: 6/8   Page 5 of 8

Reel	Scene	Take	Timecode Start	Timecode End	Description
5	43B	1	08:15:39:29	08:16:04:05	MS David
	43B	2	08:16:24:02	08:16:47:00	" Great
	41	1	08:16:55:00	08:17:32:00	Enters Room
	41	2	08:17:42:00	08:18:18:00	" OK
	41	3	08:18:28:00	08:19:00:00	" KEEPER
	44	1	08:19:25:00	08:19:31:00	Hand on Doorknob
	44	2	08:19:42:00	08:19:51:00	" Best
	44A	1	08:20:12:00	08:20:19:00	Low Angle Knob
	44A	2	08:20:29:00	08:20:42:00	"
	44A	3	08:20:53:00	08:21:05:00	" KEEPER
	41C	1	08:21:15:00	08:21:25:00	Door Opens
	41C	2	08:21:37:00	08:21:52:00	" Good
	40	1	08:22:15:00	08:22:32:00	Looks Out
	40	2	08:22:44:00	08:22:59:00	" OK
	40	3	08:23:05:00	08:23:22:00	" NG
	40	4	08:23:33:00	08:23:52:00	" Best
	40A	1	08:24:10:00	08:24:15:00	Nods Head
	40A	2	08:24:22:00	08:24:32:00	" Best

Figure 4–2 Using window dubs, footage is logged according to timecode points and is described further.

together before any videotape is placed in a machine. It is a helpful stage in that it furthers the logging process by requiring more specific decisions to be made regarding what shots will be used (Figure 4–3).

During this stage, the log sheets are consulted, and the different source tapes are reviewed for appropriate in and out points for each shot. These points are written down as actual edits that will be made when the tape editing process begins. The paper edit list has the source reel number, the timecode points for the shot, the tracks that are used, and the type of transition. This editing process is entirely conceptual. Since no tape-to-tape editing is being done, it is left to the person creating the paper edit list to imagine how everything will work together. It is not an easy task by any means, and it requires a good understanding of how editing is done. Offline editing can begin with a minimum of equipment. Much of the preparatory work involves cassette machine, paper and pencil, and an imaginative mind.

PRODUCTION: Sc. 40-44 Reel 08 DATE: 7/14 Page 1 of 1

Event	Effect	Source In	Source Out	Rec In	Rec Out
1	Fade 30	08:16:27:02	08:16:37:00	1:13:02:16	1:13:12:14
2	Cut	08:18:32:04	08:18:53:01	1:13:12:14	1:13:33:11
3	Cut	08:19:45:16	08:19:49:10	1:13:33:11	1:13:37:05
4	Cut	08:20:55:12	08:20:58:01	1:13:37:05	1:13:39:24
5	Cut	08:21:38:06	08:21:44:02	1:13:39:24	1:13:45:21
6	Cut	08:24:24:10	08:24:27:02	1:13:45:21	1:13:48:13
		End of	Scene to 45		

Figure 4–3 The paper edit list represents how the different source material will be ordered during the offline process.

Step 4: Offline Editing

The two most tangible results of the offline process are the creation of an EDL and a videotape copy of the work in progress. If we create a list of editing decisions based on timecode during this offline process, we can reproduce the program at any time by running this list during the online stage.

The majority of offline editing suites are capable of using timecode. Offline rooms that offer only control track editing have become obsolete because no EDL is created during the editing process. Without the EDL, there is no available record of what timecode points should be used. The equipment in an offline editing room will vary; it can range from simple two-machine editing to A/B roll editing with additional devices. A minimal offline room consists of one source and one record machine and an edit controller that uses SMPTE timecode to coordinate the editing process.

The offline process consists of copying source material to the record master based on the instructions from the paper edit list. Minimal attention is given to the quality of the audio and to the quality of the picture. Instead, it is somewhat accepted that the audio and video levels should be adequate enough to form a coherent opinion of how the program will eventually be conformed. More time, care, and attention will be paid to these issues during the online edit. For now, the sole intent is to see what works and what doesn't work in terms of editing pictures and sounds together.

If there are only two machines and an edit controller, only cuts will be available. If there are more machines and a video switcher, dissolves, wipes, and keys can be visualized. If a character generator is available, titles can be created and placed over images. Traditionally, offline editing was designed simply to exercise the editing options that the footage offered and to produce an EDL. The integration of other devices into the offline suite occurred for two reasons: to offer additional tools so that the editor can visualize more aspects of the program and to justify raising the hourly price of the suite.

When the offline process is complete, the EDL will contain all the directives to create the finished program during the online assembly. Table 4–1 is an example of how edits appear in list form.

Step 5: List Cleaning

Most EDLs created in the offline process are called *dirty lists*. Because more material is recorded than will eventually be used and because edits are made and then superseded by other edits, there will be extraneous information in the edit list. In the process of list cleaning, the editor identifies the extra footage and unnecessary edits and removes them. If an EDL emerges from the offline session and is not cleaned, the online session will not proceed as efficiently due to unnecessary edits being performed. Several computer programs can be used to automatically perform this function. After cleaning, the EDL and the offline videotape master are brought to the online session.

Table 4–1

TITLE:	SEQUENCE 1						
FCM:	NON-DROP FRAME						
FCM:	DROP FRAME						
001	051	V	C	01:02:49:15	01:02:58:06	01:00:00:00	01:00:08:21
002	051	AA	C	01:02:49:15	01:02:57:06	01:00:00:00	01:00:07:21
SPLIT:	VIDEO DELAY= 00:00:01:00						
003	051	AA	C	01:11:01:11	01:1:12:04	01:00:07:21	01:00:18:14
003	051	V	C	01:11:02:11	01:1:12:04	01:00:08:21	01:00:18:14
004	051	AA/V	C	01:11:42:24	01:11:51:10	01:00:18:14	01:00:27:00

THE PROCESS OF ONLINE EDITING

Although 2" videotape editing was the standard method of finishing programs through the 1970's, 1978 saw the introduction, development, and eventual adoption of standardized 1" type C videotape as the online mastering format of choice. Outfitted with timecode, online editing on 1" became the standard for mastering throughout the 1980's.

The online edit suite represents a very large capital investment. It has a variety of videotape machines, a video switcher, and audio mixing board. There may be a variety of peripheral devices, each of which performs a specific task. There may be a DVE, a character generator, a camera for shooting title cards, a digital disk recorder, and a variety of monitors for each device. Videotape machines may be near, but they are usually behind glass partitions to reduce the operating noise. A great deal of cabling is required to route the signals from machines to consoles (Figure 4–4).

The control panels that the editor operates are usually remote control units. If the editor is using a DVE, only the remote panel will be in the suite. The actual processing unit of the device will be in a central machine room or a central switching location. Tying into these machines can be a very straightforward affair, depending upon how the facility manages its signals. If the editor is working with three videotape machines and needs a fourth, it may be very easy to simply select a machine from the central location, choose a destination, and have the machine appear as a source on the edit controller screen.

Because online is characterized by using many different types of equipment, it is necessary for all equipment in the online suite to operate in synchronization with one another. The cost of the different machines and the engineering time and costs involved

Figure 4–4 In the online edit suite, final editing takes place. Courtesy Szabo Tohtz Editing and Skyview Film & Video, Chicago.

in making the room operational have made the building of the online room an expensive undertaking. Depending upon the equipment, the cost of building an online room in the early to mid-1990's ranges from $500,000 to $1 million. The linear offline room, depending upon complexity, might cost $20,000 to $40,000.

To justify the capital expenditure, the rates of renting an online room are much higher than those for the offline room. Depending upon the capabilities of the room, online ranges from $300 to $1,000 per hour. The breakdown of costs is based on how many hours each device is used. A sample work sheet is shown in Table 4–2.

The costs involved can be quite extensive depending upon the number of hours required to finish the program, and the rates given in Table 4–2 are very conservative. Is it any wonder, then, why offline exists? If the price difference and the capabilities difference between offline and online rooms were not so great, how could the ratio of hourly charges of 5:1 or 6:1 be justified? For these reasons, it makes sense to try to go to the online suite as prepared as possible.

The Steps of Online Editing

The usual steps of an online editing session are not as easily defined as those of the film editing process or the offline editing process. This is due to the fact that there are many different types of equipment in the online room. For example, rather than start off with the tape-to-tape editing, if a project is being done that requires a great number of character-generated titles, the editor may elect to create these first and store them to computer disk so that they will be ready when they are needed.

Since online is a linear building process, the editor must be able to look ahead and consider the next series of edits and how what is happening now will affect the program later. Since the hourly rates of online can be significant, there is an emphasis on being able to use time efficiently so that the client's editing budget is not exceeded.

Table 4–2

	Hourly Rate ($)	No. of hours	Total ($)
3-VTR edit (standard)	300	8	2400
Digital effects	200	3	600
Character generator	75	4	300
Title camera	15	3	45
Digital disk recorder	200	4	800
Additional VTR	100	1	100
			$4,245

Online editing is the creation of the finished program shot by shot. The finished program is being made as each edit is recorded. The usual flow is as follows:

1. If offline was done, the offline EDL is loaded into the online edit controller.
2. If there is no offline list, editing begins with little or no materials from the preedit stage.
3. The technical quality of video and audio is maintained.
4. Additional devices are used if appropriate.
5. The program is either mixed or sent for audio sweetening.
6. The finished picture and sound are married together.
7. The program is duplicated and distributed.

If an offline edit was done, the purpose of online is straightforward. The online editing stage becomes a conform stage, sharing the same task as the film negative cutter who must conform the negative to the work print. The creative task is finished, and online is the step-by-step process of replicating what was done during offline editing. First, the offline EDL is loaded into the online edit controller. The list will usually be on 8", 5 1/4", or 3 1/2" computer floppy disk.

Once the list is loaded, the editor will determine if edits should be done individually or if edits can be grouped together. For example, if there are 25 edits that use the same source tape, rather than unloading and reloading that source tape, the editor may be able to make many or all of the edits using that tape before moving on to another. Various auto-assembly modes are designed to make the most efficient use of tape shuttling time.

Next, the editor will find the source tapes that are needed for the series of edits and will put these tapes into machines. Online edit suites offer different formats for the program master. The most common mastering formats are D2, 1", and Betacam.

Online suites have at least three videotape machines, so two source tapes can be loaded while the master tape is on the third machine. Once the source tapes are loaded, they must be set up to play back correctly. There will usually be reference signals for picture and sound at the beginning of each tape reel. These signals comprise color bars and tones and are standardized so that setting a machine to play back the reference signals ensures that recorded picture and sound are played back at their intended levels.

The color test pattern comprises several bars, beginning with white and continuing in descending order of brightness: yellow, cyan, green, magenta, red, and blue. White is measured at 77 IRE (International Radio Engineers) units, and black is measured at 7.5 IRE. All of the remaining color bars are measured at 75%. The white level is set to 75% of full level to accommodate the high levels of chroma. Audio tone is generally at 0 db.

Once the tapes have been properly set up, editing can begin. Since the decisions have already been made during the offline stage, the online stage is concerned not with changing any edits, but with ensuring that the offline list is run properly. If the offline editor indicated that there should be a title from one timecode to another, the online editor will place these keyed titles into the show as it is being edited. If the offline editor did not have access to a DVE but noted in the EDL that a shot needs to be flopped from right to left, the online editor will maintain the edit points specified in the offline list but will process the shot through a DVE, flip it on the x-axis, and edit the flopped shot into its correct position.

Careful attention is paid to the quality of video and audio signals while the show is being edited. If an actor's facial tones are slightly green or if an audio passage is too low, these things are corrected now. There won't be another time when these issues can be addressed; this is the final stage.

An efficient online process is dependent upon knowing as many things as possible about the incoming project. Will the offline EDL be clean, will all the tapes be of similar format, or will the online suite require different playback machines? A simple question that sometimes is completely forgotten and yields great embarrassment is how the EDL will be delivered. If the client brings a 3 1/2" floppy disk and the online edit controller accepts only 8" disks, time will be spent in making the file transfer if the appropriate equipment is available. Most of these concerns are expressed at the outset, and most facilities have equipment on hand that can accomplish these tasks. However, it is important to note that online is not a place where surprises should be introduced.

Despite the best planning, it may be necessary to make slight changes to the offline list. Often, slight audio imperfections are heard, and these are usually fixed by changing the edit points by one or two frames. Why do such imperfections exist? Some of this has to do with the integrity of the offline equipment and the discipline of the offline editor. If we cut very close to the beginning of a word, we may clip it (inadvertently shorten it) by one frame, and this will require some adjustment during online.

Despite the presence of the offline EDL, it is very unusual that all directions in the EDL will be preserved without some changes. Some amount of experimentation may be required. For example, if the offline editor could not see titles keyed over an image but made notes regarding when the titles should be inserted, slight adjustment of these times and durations may be required. As the editor sees the titles inserted for the first time, it may be necessary to adjust the timings until the desired result is achieved.

The online scenario just described began with an EDL and a rough cut from the offline edit. What are the steps of an online session when an offline has not been done? The steps are similar, but the actual choice of editing points is an ongoing process. Clearly, more time will be spent this way. The online editor is now able to function in a much more creative role by guiding the choice of these edit points and the flow of images as the piece is edited. Even so, online, as a linear editing process, will not allow the editor to easily make changes.

Audio Editing in the Online Room

Until the late 1980's, audio editing in the online editing room went hand in hand with video editing. As picture was being finished, audio was being finished. If a section of audio needed to be equalized or changed in any way from the original recording, it was done during online.

The limitation of two program audio tracks on 1" type C videotape (four tracks on other formats such as D2 and Betacam) meant that any additional tracks of sound required additional steps. For example, if editing on 1" videotape, the picture was first locked, and the two program audio tracks were used for sync sound. This program would then be copied to acquire additional tracks of sound. For example, if three copies of the show were made, six tracks of sound could be added, such as stereo music, narration, two sound effects tracks, and an ambient noise track.

These copies and the original master would be rolled together (four videotape machines playing while a fifth machine was recording), eight tracks of sound would be running simultaneously, and the editor would then operate the audio mixing board to combine the eight soundtracks to create a stereo audio mix. This finished mixed track would then be laid back to the original video master. The program would then be complete and ready for duplication.

Whether to finish sound in the online editing suite today depends on the number of tracks needed and the complexity of the sound mix. One of the immediate problems of the method described above is that care must be paid to signal degradation via the copying process, although this will be less likely if digital tape formats are used. Additionally, how intensive will the audio portion of the edit become? Does original music need to be composed and edited to the piece? Will Foley work be done (creating and editing sound effects that were not recorded during shooting)? Will the audio requirements far outweigh the capabilities of the audio mixing console in the online suite? Most importantly, is the craft of audio editing within the capabilities of the online editor? If the person is not capable of providing the same artistic and technical attention to the program's audio requirements that was paid to the program's picture requirements, then the correct person for the job must be found.

Because of these concerns, audio finishing is being done less frequently during the online session today. It depends upon the budget. The audio portion for television commercials is not done in most online rooms. Rather, picture is locked and then sent on to the audio sweetening stage. Audio for lower budget programming is mostly done during the online stage because the budget will not include any provision for taking the project to audio sweetening. During the 1970's and 1980's, audio for video programs was always regarded as less important than the pictures themselves. This has improved considerably, especially throughout the late 1980's when, with the introduction of digital audio workstations, audio came to be regarded as just as important to the presentation as the picture.

THE REASONS FOR ONLINE EDITING

The online edit suite is where all the ingredients that make up a program are combined to finish the program for final delivery. If an offline session has been done, when the project enters the online stage, the editor has a clear idea as to how the material will be ordered. If editing is taking place from the beginning stages, many of the same motives and concerns that various individuals will bring to the editing process will remain. If the online is a conform of the offline EDL, there will be fewer participants since most of the decisions will have already been made once the rough cut is approved.

However, if the program is being created during online and changes are later deemed necessary, assessing how the changes should be made will depend upon where the changes occur, how many there are, and what tape format the show was mastered on. For example, if the show was mastered on a digital tape format such as D2 and the changes affect the length of the show, the choice will be made to copy the master and make the changes. Because of the digital format, generation loss won't be as big a factor as if the master was on an analog tape format. Making the required changes while preserving the best possible quality at the least cost will always influence what method is chosen.

The "attitude" found in the online editing suite is divided into two domains. First, if the online session serves as a conform process to the offline EDL, the editor's job is to recreate the original intent while making sure that all technical considerations are met. While this sounds like a task that offers no creativity, this is not true. Even though the edit points may have already been chosen, it is still left to the online editor to accomplish everything that could not be done during offline, such as the perfect audio crossfade that makes a picture transition so much better or the choice of a slow motion speed that could not be visualized during offline due to machine limitations. Second, if the project being edited comes to online with no shape or focus given to the footage, the online editor's job is no different than the offline editor's job: to provide some direction as to how the material should be ordered based on what the project is meant to accomplish.

It is also important to note that not every project requires the offline stage. Some projects are simple enough to go from shooting right to online. Consider the 30-second commercial that consists of one continuous shot with a title keyed over the image during the last five seconds or the political advertisement that consists of the candidate speaking to the camera for 25 seconds and a dissolve to the campaign slogan for the last five seconds. Picking the right take and making the dissolve in the right location are the tasks to be done. The audio requirements for this type of project could far exceed the audio capabilities of the offline room, so bringing the material directly to online is a better choice. If the piece needs to be on the air quickly, offline editing will be superfluous.

THE UNFULFILLED PROMISE OF LINEAR OFFLINE EDITING

The majority of film editing is an offline process. The shaping of the entire project, whether it is a commercial or a feature film, is a process of selecting and discarding material. Film enters its online stage at the moment that the film negative is cut. When the negative is cut, the film is in online.

Further, the online for film is not left up to interpretation. All decisions have been made and finalized. Unlike videotape online where changes can be made even though they may be time consuming, once a film is in its online stage, changes may be impossible to make. When the film negative is cut, at least one adjacent frame will be unusable because two film perforations will be lost for that frame. This is because when film negative is spliced together, two perforations beyond the frame are used to make the splice.

Videotape's offline and online stages simply and logically mimic the natural process of film editing. In film, editing is done on positive prints. In video, offline editing is done on copies of the original footage, mostly as a safeguard against harming the original material. Then, in film, the negative is cut, and in video, the original videotapes are used during the online edit.

Videotape offline during the 1980's was used with varying degrees of success. However, linear offline has been viewed as falling short of achieving its promise to save time and money during the online process for two basic reasons: the discipline required and offline limitations. To be successful, offline editing, whether done in the world of film or videotape, relies on proper record keeping. Knowing what pictures were used, what reel they came from, what audio was used, and the source for this material requires good logging procedures. Proper list management is critical to a decreased assembly time during online. If record keeping is not good or if the EDLs that are generated have flaws, unnecessary time will be spent during online editing to duplicate work already done during the offline stage.

The first major reason why linear offline editing has not been embraced more fully is the discipline that the record keeping process requires. This aspect cannot be overemphasized. All it takes is for one source reel to be edited in the control track mode instead of in the timecode mode, and all edits for that reel will be useless during online. While this may sound like an egregious error that would not likely happen, the fact is that it can happen, and avoiding this possibility requires discipline and awareness.

The organizational aspects go hand in hand with the creative aspects. You can edit a stunning sequence, but without absolutely accurate records or an EDL that can be used to recreate the piece, you will be forced to recreate that sequence during online, using the rough cut as a blueprint. Shots will have to be found on source reels, and points will have to be found manually.

The second reason why linear offline has not been as successful at decreasing the amount of time spent in online has been due to the creative limitations of the offline room. Because it rents at a far lower price per hour than the online room, what sense would it make to install a great deal of image-manipulation hardware into the offline room? The tools in that room will most likely be some mélange of older equipment left over from the consolidation of other edit rooms.

What are the consequences of creative limitations during offline? If you are using a basic cuts-only offline system with two tracks of audio, the best you can hope for is to cut dialogue tracks to get simple content approval from the client. Anything more than that will require more previsualization tools. If the client needs to see dissolves to decide if the piece is working or not, this is a limitation. Without being able to lay up additional audio tracks, sound effects and music have to be imagined.

Often, an editor will go through these initial stages of cutting picture and dialogue and will then play back the sequence while music is played off an audio tape or compact disc just to get an idea of how everything could work during online. Depending upon the job, the personalities involved, and the degree of envisioning that has to take place, offline can yield many creative limitations. If directors or clients cannot get a good indication of how the project will appear as it exits the offline stage, how can they be expected to agree to the online session without reserving the right to make changes?

THE EVOLUTION OF OFFLINE VIDEOTAPE EDITING

To provide additional tools that can reduce these creative limitations, offline rooms in the late 1980's began to move from cuts-only to A/B roll systems. More than two audio tracks were not usually available, but editing with Betacam or MII tape formats did provide four audio channels. In addition, lower cost digital effects units found their way into the offline suite. Character generators based on more affordable personal computers found their way into the offline room.

By the end of 1989, the offline video room could differ radically from facility to facility. Offline at one facility could consist of two machines with a timecode controller, while another facility's idea of offline would be an A/B roll system with DVE, character generator, and a video switcher. If the definition of *offline* is that the program is not finished when it leaves the editing suite, how do the capabilities afforded by all these peripheral devices affect that definition? At what point does offline begin to have the capabilities necessary to finish a project?

What was accomplished by outfitting the video offline room with all the additional hardware? Throwing hardware at a problem has never resulted in success. The problem was not that the offline process was suspect, it was that the technology of offline did not allow for the creative job that had to be done during the process of offline. The reason linear offline has failed to deliver its promise is that it is a linear solution to a nonlinear task. Film editing is an inherently offline solution that is successful because it is a nonlinear solution. Offline videotape editing is where the creative aspects of a program's creation are delegated, but the linear methods that must be used are not in keeping with the requirement of being able to try any idea and make changes as necessary.

The solution was to try and bring nonlinear editing to the video offline room. This was the essential problem that offline video editing faced until nonlinear editing systems began to be used for offline purposes. This was occurring in limited areas by 1984, but the explosion in nonlinear offline approaches and usage did not occur until 1990.

EDITING MODES: DIFFERENT FORMS OF THE EDIT LIST

When an EDL is transferred from the offline edit system to the online edit system, the online editor may be able to save a considerable amount of time during the auto-assembly process by working with specific groups of edits. There are different ways of ordering an edit list to increase efficiency.

An A mode list orders the events based on the record in times. As shown in Table 4–3, the list is sequential as the event numbers increase in increments from 1 to 11, and for each edit that is done, the master tape will be recorded from start to end in a linear fashion. Although this method does not make the most efficient use of shuttling time for the source reel and does not decrease the number of reel changes, if there are a number of transitions such as dissolves or wipes, these edits must be approached from start to end. An A mode list takes the most time to execute.

A B mode list introduces the concept of "checkerboard" auto-assembly (Table 4–4). In a B mode list, the order is determined by

Table 4–3

```
TITLE: SEQUENCE 1, A MODE
FCM: NON-DROP FRAME
001 001    AA/V  C          22:26:45:20 22:26:51:09  01:00:00:00 01:00:05:19
002 002    AA    C          02:08:27:05 02:08:46:29  01:00:05:19 01:00:25:13
003 002    V     C          02:08:27:05 02:08:46:27  01:00:05:19 01:00:25:11
004 002    V     C          02:08:46:27 02:08:46:27  01:00:25:11 01:00:25:11
004 003    V     D    005   00:03:16:24 00:03:23:28  01:00:25:11 01:00:32:15
005 003    AA    C          00:03:16:26 00:03:23:28  01:00:25:13 01:00:32:15
006 003    V     C          00:03:23:20 00:03:33:07  01:00:32:15 01:00:42:02
FCM: DROP FRAME
007 003    AA    C          00:03:02:23 00:03:09:19  01:00:32:15 01:00:39:11
FCM: NON-DROP FRAME
008 001    AA    C          22:26:38:13 22:26:41:06  01:00:39:11 01:00:42:04
009 003    V     C          00:03:33:07 00:03:33:07  01:00:42:02 01:00:42:02
009 002    V     D    005   02:08:52:00 02:09:03:08  01:00:42:02 01:00:53:10
FCM: DROP FRAME
010 002    AA    C          02:08:52:02 02:09:03:08  01:00:42:04 01:00:53:10
011 001    AA/V  C          22:26:47:10 22:26:59:00  01:00:53:10 01:01:05:00
```

| | Reel # | | Effect | Duration | Source In | Source Out | Record Out | Record In |
| Event # | | Tracks | | | | | | |

Table 4–4

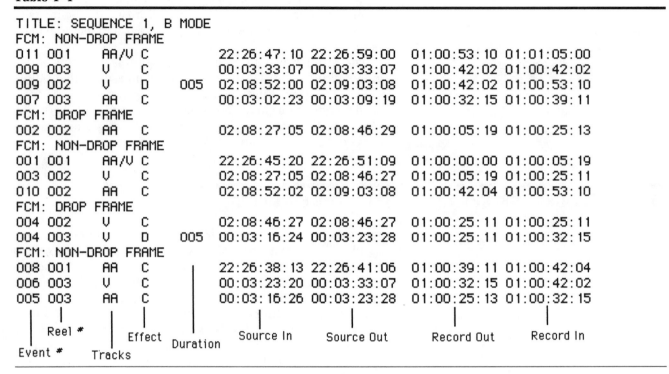

Event #	Reel #	Tracks	Effect	Duration	Source In	Source Out	Record Out	Record In
TITLE: SEQUENCE 1, B MODE								
FCM: NON-DROP FRAME								
011	001	AA/V	C		22:26:47:10	22:26:59:00	01:00:53:10	01:01:05:00
009	003	V	C		00:03:33:07	00:03:33:07	01:00:42:02	01:00:42:02
009	002	V	D	005	02:08:52:00	02:09:03:08	01:00:42:02	01:00:53:10
007	003	AA	C		00:03:02:23	00:03:09:19	01:00:32:15	01:00:39:11
FCM: DROP FRAME								
002	002	AA	C		02:08:27:05	02:08:46:29	01:00:05:19	01:00:25:13
FCM: NON-DROP FRAME								
001	001	AA/V	C		22:26:45:20	22:26:51:09	01:00:00:00	01:00:05:19
003	002	V	C		02:08:27:05	02:08:46:27	01:00:05:19	01:00:25:11
010	002	AA	C		02:08:52:02	02:09:03:08	01:00:42:04	01:00:53:10
FCM: DROP FRAME								
004	002	V	C		02:08:46:27	02:08:46:27	01:00:25:11	01:00:25:11
004	003	V	D	005	00:03:16:24	00:03:23:28	01:00:25:11	01:00:32:15
FCM: NON-DROP FRAME								
008	001	AA	C		22:26:38:13	22:26:41:06	01:00:39:11	01:00:42:04
006	003	V	C		00:03:23:20	00:03:33:07	01:00:32:15	01:00:42:02
005	003	AA	C		00:03:16:26	00:03:23:28	01:00:25:13	01:00:32:15

the source tape and the record in time. The event column shows that the events are no longer ordered sequentially. This results in the edits being determined by the record master time. The advantage is that the edits that require a specific source reel will be done at one time, use of that reel will be finished, and the next reel will be similarly ordered.

Checkerboarding refers to the fact that the master tape is assembled in a nonlinear fashion. Edits are made at different points on the master while other sections are left black. Figuratively, the master tape resembles a checkerboard, with some sections filled in and some left blank until the appropriate edits are made to fill the entire master tape.

A B mode list does not minimize source shuttle time, but it does minimize record shuttle time. Therefore, a B mode list is used when the source reel is short, and the record master is long.

A C mode list is also a checkerboard list (Table 4–5). A C mode list orders the edits based on source reel and source in times. Unlike B mode lists, where the record in time determines the order of edits, a C mode list orders the source times based on the order in which they appear on the source reel. If you compare events 7 and 5 from the B mode list to the placement of events 7 and 5 in the C mode list, the more efficient use of source shuttle time will be evident.

A C mode list therefore minimizes source shuttle time, while the record shuttle time is not minimized. A C mode list is used when the source reel is long, and the record master is short.

A, B, and C mode lists are fairly standard features on most edit controllers. D and E mode lists are a bit more unique and may not

Table 4–5

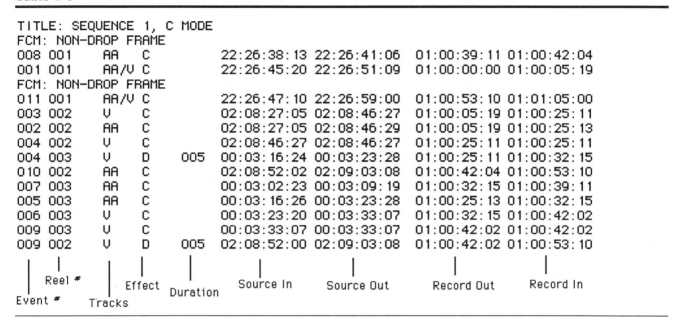

```
TITLE: SEQUENCE 1, C MODE
FCM: NON-DROP FRAME
008 001    AA   C            22:26:38:13 22:26:41:06  01:00:39:11 01:00:42:04
001 001    AA/V C            22:26:45:20 22:26:51:09  01:00:00:00 01:00:05:19
FCM: NON-DROP FRAME
011 001    AA/V C            22:26:47:10 22:26:59:00  01:00:53:10 01:01:05:00
003 002    V    C            02:08:27:05 02:08:46:27  01:00:05:19 01:00:25:11
002 002    AA   C            02:08:27:05 02:08:46:29  01:00:05:19 01:00:25:13
004 002    V    C            02:08:46:27 02:08:46:27  01:00:25:11 01:00:25:11
004 003    V    D    005     00:03:16:24 00:03:23:28  01:00:25:11 01:00:32:15
010 002    AA   C            02:08:52:02 02:09:03:08  01:00:42:04 01:00:53:10
007 003    AA   C            00:03:02:23 00:03:09:19  01:00:32:15 01:00:39:11
005 003    AA   C            00:03:16:26 00:03:23:28  01:00:25:13 01:00:32:15
006 003    V    C            00:03:23:20 00:03:33:07  01:00:32:15 01:00:42:02
009 003    V    C            00:03:33:07 00:03:33:07  01:00:42:02 01:00:42:02
009 002    V    D    005     02:08:52:00 02:09:03:08  01:00:42:02 01:00:53:10
```

Event #	Reel #	Tracks	Effect	Duration	Source In	Source Out	Record Out	Record In

be available. A D mode list is similar to an A mode list, but all transitions other than cuts are placed at the end of the list (Table 4–6). Here, the two dissolves in the sequence are placed at the end of the list. An E mode list is similar to a C mode list, but all transitions other than cuts are placed at the end of the list (Table 4–7). Unlike a D mode list, the remainder of the list is in C mode.

Table 4–6

```
TITLE: SEQUENCE 1, D MODE
FCM: NON-DROP FRAME
010 002    AA   C            02:08:52:02 02:09:03:08  01:00:42:04 01:00:53:10
002 002    AA   C            02:08:27:05 02:08:46:29  01:00:05:19 01:00:25:13
006 003    V    C            00:03:23:20 00:03:33:07  01:00:32:15 01:00:42:02
003 002    V    C            02:08:27:05 02:08:46:27  01:00:05:19 01:00:25:11
007 003    AA   C            00:03:02:23 00:03:09:19  01:00:32:15 01:00:39:11
FCM: DROP FRAME
005 003    AA   C            00:03:16:26 00:03:23:28  01:00:25:13 01:00:32:15
011 001    AA/V C            22:26:47:10 22:26:59:00  01:00:53:10 01:01:05:00
008 001    AA   C            22:26:38:13 22:26:41:06  01:00:39:11 01:00:42:04
FCM: NON-DROP FRAME
001 001    AA/V C            22:26:45:20 22:26:51:09  01:00:00:00 01:00:05:19
004 002    V    C            02:08:46:27 02:08:46:27  01:00:25:11 01:00:25:11
004 003    V    D    005     00:03:16:24 00:03:23:28  01:00:25:11 01:00:32:15
009 003    V    C            00:03:33:07 00:03:33:07  01:00:42:02 01:00:42:02
009 002    V    D    005     02:08:52:00 02:09:03:08  01:00:42:02 01:00:53:10
```

Event #	Reel #	Tracks	Effect	Duration	Source In	Source Out	Record Out	Record In

Table 4–7

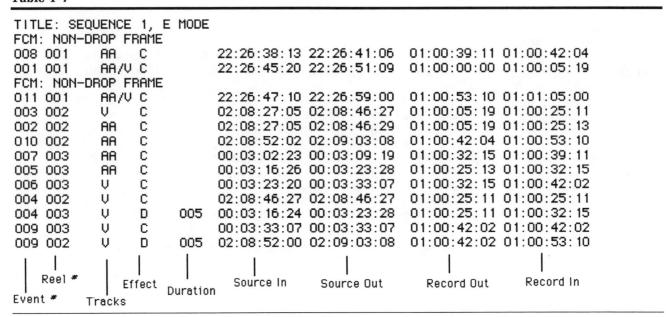

```
TITLE: SEQUENCE 1, E MODE
FCM: NON-DROP FRAME
008 001    AA   C        22:26:38:13 22:26:41:06  01:00:39:11 01:00:42:04
001 001    AA/V C        22:26:45:20 22:26:51:09  01:00:00:00 01:00:05:19
FCM: NON-DROP FRAME
011 001    AA/V C        22:26:47:10 22:26:59:00  01:00:53:10 01:01:05:00
003 002    V    C        02:08:27:05 02:08:46:27  01:00:05:19 01:00:25:11
002 002    AA   C        02:08:27:05 02:08:46:29  01:00:05:19 01:00:25:13
010 002    AA   C        02:08:52:02 02:09:03:08  01:00:42:04 01:00:53:10
007 003    AA   C        00:03:02:23 00:03:09:19  01:00:32:15 01:00:39:11
005 003    AA   C        00:03:16:26 00:03:23:28  01:00:25:13 01:00:32:15
006 003    V    C        00:03:23:20 00:03:33:07  01:00:32:15 01:00:42:02
004 002    V    C        02:08:46:27 02:08:46:27  01:00:25:11 01:00:25:11
004 003    V    D   005  00:03:16:24 00:03:23:28  01:00:25:11 01:00:32:15
009 003    V    C        00:03:33:07 00:03:33:07  01:00:42:02 01:00:42:02
009 002    V    D   005  02:08:52:00 02:09:03:08  01:00:42:02 01:00:53:10
```

| Event # | Reel # | Tracks | Effect | Duration | Source In | Source Out | Record Out | Record In |

5

The Operational Aspects of Film and Videotape Editing

Editing film is a unique situation where constant attention can be paid to the picture and sound elements while a modicum of attention is paid to the equipment that is used. Online videotape editing combines the creative elements of editing with the significant requirement that the technical aspects of videotape editing be understood and mastered while the creative decisions are being made. Before an edit session begins in the online videotape room, a considerable amount of time may be spent in preparing the room for the project.

The condition of the editing room must be examined: Are enough videotape machines assigned based on the videotape formats used in the program? Will character generation be needed? Whenever a variety of machines are utilized, will they find a common junction at the video switcher? Are all the sources in synchronization with each other, and is this the responsibility of the editor or the engineer?

The online editor must constantly balance these technical issues while still maintaining the focus of the editing session, which is, after all, to be conscious of the program being edited and the integrity of achieving the program's aims. When you consider all the disparate and intricate pieces of equipment that can exist in the online room, from digital tape machines to digital video effects units to digital disk recorders, a great deal of technical mastery must be learned and balanced along with the artistic side of the editing process.

In addition, the training of the video editor must include not only an understanding of the technical aspects of hardware operation, but also a background on troubleshooting the technical things that can go wrong in the editing process.

The learning process for the videotape editor is further complicated because each of the peripheral devices usually operates in a different manner. A character generator requires a certain amount of operational knowledge, while a digital video effects device requires different knowledge. One system may be more intuitive in its operation but provide simpler features, while another system may be more difficult to learn but offers more complex features. It is left to the editor to learn and master all the different devices that may be utilized as part of the online session.

THE CREATIVE PROCESS OF FILM VERSUS VIDEOTAPE EDITING

The creative process of editing is equally difficult in film and video. Choices have to be made, and options have to be tried. How much effort is expended on manipulating machines or the medium itself is an issue. If time to delivery is a factor, variations of a cut may not be tried. Making the editorial process easier on the person putting together the presentation means removing barriers to the manipulation of the material. If you were told that you could make only 18 film splices for a scene, you would think very carefully about what to do and what not to do in approaching the editing of that scene. If you were told that you had only one hour to cut a scene and you were working on a linear videotape editing system, you would try to plan as carefully as possible. Opening up the possibilities and making these constraints more palatable can be achieved with the use of nonlinear editing systems.

Videotape editors have traditionally been trained as online editors. Less emphasis is placed on the offline editing process, and most online videotape editors have not been trained to adopt the discipline necessary in achieving a successful offline edit. A majority of the training that the editor undergoes involves technology: maintaining standards of video signal ranges and learning what can and cannot be broadcast. They learn to be proficient at list management and in a range of activities that can best be described as "taming the technology." In many ways, the workload for the online editor can be very complicated, especially when a project comes into online without any preediting work having been done. When this happens, the videotape editor must perform the offline and online edits at the same time.

Film editors, working in the inherently offline process of film editing, can devote more time to developing the creative and relational skills that serve to move a client through the online process. Both the film and videotape editors must always know how to work with and balance the personalities of their clients.

In the early 1980's, when offline videotape editing was emerging but online editing was fully entrenched, there was a noticeable trend. Film editors were sought by clients for the initial editing stages, whether on film or on videotape. This was especially true for advertising agencies, which often requested film editors and eschewed online editors, who were, in a generic sense, perceived as "button pushers," offering little to the creative task at hand.

Although this perception was quite evident in the 1980's, it bears noting that when film editors were sought out, many additional aspects proved to bear on the decision. Whereas videotape editors must also concentrate on many other items, the film editor can concentrate on the issues of taste and execution.

The issue was not whether to go to the film editor versus the videotape editor. Instead, being able to exhaust the possibilities contained in the footage and being able to enter the film editing environment where images and sounds dominate was very important.

Most likely, and more fairly, clients went to film editors because the medium was nonlinear, and nonlinear editing affords possibilities. Clients were not seeking out the film editorial process with the notion that it was "nonlinear." The only thing that mattered was that the film editing process allowed scenes to be edited in different ways, and there was less likelihood that the film editor would be reluctant to make changes since film can be rearranged easily. Instead of hearing what needed to be done to make a change and the time involved when editing in the videotape world, the client could enjoy more creative freedom by editing in the film world. To some extent, this attitude influenced the production methods that were chosen: Should we shoot on film or on videotape?

Extending the possibilities of nonlinear editing into electronic offline videotape editing began in earnest in 1984. The main purpose was to offer the same creative flexibility that the film editor and the client accustomed to editing projects on film were realizing. For the editor, nonlinear videotape editing offered a chance to try more ideas and make changes without worrying about generation loss or list management. For the client, it offered a chance to make changes easily without as much concern about the difficulty or additional cost of making the changes.

Can Film and Videotape Editing Be Improved?

The process of editing film can be viewed as something of a cloistered activity. Shut away from the world of technology, the film editor can put together a 30-second television commercial or a two-hour documentary from tens of hours of footage. But, even with the flexibility that film editing offers, can the film editing process be improved? In what areas can technology add to the creative process, and can time be saved without requiring a complete change of working methods? What problem areas of film editing would editors like to see addressed? Although any ideas can be exercised in the film editing room, at what point are ideas not tried because the tasks become too intricate or because they simply take too long to try?

For videotape editing, the most significant area for improvement is to be able to edit in a nonsequential manner. Bringing nonlinear editing into the offline editing room serves two purposes. First, the editor is able to try more creative options during the critical time of editing: shot selection and positioning the shots. Second, the offline editing concept, always treated as an inferior stage in the videotape world, is a natural stage in the film world. Moving videotape editing toward the film editing structure means that there must be an emphasis on offline first and online second.

ELECTRONIC NONLINEAR EDITING SYSTEMS

The development of electronic nonlinear editing systems began in 1970. There have been three approaches, or waves, of these systems. The first wave is based on videotape, the second on laserdisc, and the third and most recent wave, on digital. While each system is different and the technology used in each wave differs significantly, an examination of the goals of the electronic nonlinear system is warranted.

There are four basic theories of operation for electronic nonlinear systems:

1. Achieve random and rapid access to material
2. Edit nonsequentially
3. Make changes easily
4. Output work products that serve to reduce the time spent conforming

Common Stages of the Electronic Nonlinear Editing Process

Some basic stages and characteristics common to all electronic nonlinear editing systems:

1. The original footage is transferred to an identical medium or a different medium.
2. If transferred to an identical medium, additional playback devices are employed.
3. Computer methods are used to log and track footage.
4. The editing process replicates the film editing model.
5. The work product stage is a video EDL, a film cut list, a viewing copy of program, a final auto-assembled version of the program, or direct output from the system.

Editing on any electronic nonlinear editing system begins with a transfer of the original footage. Transferring the material to the same medium or to a different medium improves random access to the material. If you are editing on film, you have to wind from the beginning of the film roll to the end to get to a shot toward the end of the reel; the same is true for videotape. Random access means that you can reach the needed material in less time and in a nonsequential fashion. A random-access system allows the user to move through material in a nonlinear fashion.

Random access is a requisite of the nonlinear system. If we have a videotape that has many shots that we need to get to quickly, one solution is to make several copies of the videotape. By transferring the footage from one tape to many and then playing back the tapes with several machines, the necessary shots can be accessed much more quickly than if only one tape and one tape machine were used.

Transferring footage from one medium to a different medium is merely a different approach to improving random access. If the original footage is on videotape and we transfer the footage to the same medium and use multiple machines to play back the footage, we will improve random access, but the material must still be searched sequentially; videotape must be shuttled backward and forward.

Transferring material to a different medium can dramatically improve random accessibility. If we have a 35mm motion film camera roll of approximately ten minutes and we need to wind from the beginning to the end, a powerful winding motor on the film flatbed could perform the operation in approximately two minutes. If the ten minutes of material is on videotape, winding can occur at top speeds of about 50 times normal play. However, if the ten minutes of footage is transferred to laserdisc, any frame can be accessed in about 1.5 seconds. Finally, if the material is transferred to computer disk, access time can be as low as 12/1000 of a second.

When original footage is copied to an identical medium, multiple playback machines are used. While one machine is playing material, other machines are searching for upcoming material. If we have eight shots that we need to get to quickly, eight machines, each searching for one shot, will provide faster access to the material we need. The tape-based wave of nonlinear systems, such as Montage and Ediflex, operate in this fashion.

One of the most time-consuming jobs of the editing process is keeping track of footage as it originates and as it passes through the post-production process. Describing the footage that was shot, noting what camera roll or videotape reel the shot is on, knowing which take the director preferred, and describing a variety of other characteristics, are a critical part of the editing process.

Logging is primarily a manual process. Hand-written reports, log sheets, the script supervisor's notes, and camera reports are all designed to allow the editor to more efficiently view footage. More importantly, good logs tell the editor how to find related footage. Rifling through the logs to find the desired footage is necessary; after all, how can the best possible performance be ensured if the editor is not looking at the best possible material that was shot?

Most electronic nonlinear systems seek to preserve the level of detailed information that is entered into these various logs. However, instead of looking at multiple pieces of paper, the editor usually finds a computer log, which can be an extremely powerful database. Logging shots, cross-referencing material, and searching and sifting for material based on criteria can all be accomplished quite easily once a computer is being used to rifle through the descriptions for footage.

Electronic nonlinear systems seek to preserve the basic film editing model, which allows shots to be positioned, easily trimmed, and easily rearranged. These operations are equally easy on both the picture and audio tracks.

Work Products of Nonlinear Editing

Electronic nonlinear systems differ in the types of work products that they provide. This is a critical distinction among the various systems. When one analyzes which work products the nonlinear system offers, a clear indication will emerge of the system's purpose. Based on the work products, we can judge whether the nonlinear system is intended to be solely a previsualization device for the program being created or whether the system offers features that make it a finishing device.

Video EDL

Some systems generate an edit list that indicates how the program was put together. This is the most basic work product, but it is still important to note that systems that purport to be nonlinear editing systems may not offer even this most basic feature! There are different EDL formats, each specific to a particular online edit controller. This list can be interpreted by the online editing system and allows the conforming of the original source material in the order necessary to complete the program. Without an EDL, all the work done during the editing stage will have to be painstakingly recreated. With an EDL, there is no doubt as to which frames were used in the editing process.

Film Cut List

Some systems generate a different form of EDL. Instead of numbers that relate to the timecodes of the source videotapes, the film cut list (or *negative cut list* as it is called) indicates which film edge numbers should be cut from the original film negative. These original strips of negative are then hot-spliced together to deliver a finished film of the program.

As with the case of the video EDL, these lists of how the original material is to be conformed cannot have any errors. If there is an error, the incorrect source material will be used. Although this can be remedied in the case of a videotape conform, the results for a film conform could be disastrous since the film negative can be cut only once.

Viewing Copy of Program

During the editing process, it is usually necessary to have presentation copies of the program as it is being edited. To view edited sequences as they are completed, the various shots will be played back in their assembled order and recorded to a videocassette. This can then be reviewed and critiqued, eliciting changes to be made in the sequence. The record machine will usually be of U-matic or Betacam videotape format. This output tape represents the visual blueprint of the editing session and can then be used as a reference when the online edit stage begins.

Final Auto-Assembled Version of Program

It is with the auto-assembled version that nonlinear editing systems begin to distinguish themselves. If the nonlinear editing system is capable of running an auto-assembly routine, the system is in effect conforming the edit list and finishing the program.

Auto-assembly options use the original source material and record the appropriate sections to a master tape. When this is complete, either the program will be finished entirely or additional items will be added to the master tape. For example, the nonlinear system could first be used to offline the program and create an EDL. Second, the system could be used to perform an auto-assembly. Third, this auto-assembled tape could be taken to an online suite so that additional effects such as character generation can be keyed over the video.

Nonlinear systems that offer auto-assembly options are usually associated with third-wave digital nonlinear systems. Not only do these systems serve as the offline system, but they also act as an edit controller, directing multiple machines and additional devices. The ability of the digital nonlinear editing system to be used first as the offline system and then as the online edit controller is important and serves notice that the digital nonlinear system is not destined strictly to offline functioning.

Direct Finished Output From System

Until 1992, nonlinear systems were never concerned with being the final step in the post-production process. Editing on a nonlinear system, whether it belongs to the tape-, laserdisc-, or digital-based wave, was always primarily focused on providing work products that were designed to be taken to a conforming session.

Direct finished output from a nonlinear system means that the playback of the program from the system becomes the final product. The system is both the offline and the final conform station. Once the original footage has been transferred into the system, editing is done, including the incorporation of titling, optical effects, and audio mixing. The completed program is then finished and recorded to videotape.

This focus has begun with the digital-based wave of systems, and it represents a new direction for the entire field of editing and program creation. The main issues raised by suggesting that the direct output from a system can be used as the final output are concerned with quality. Offline nonlinear systems have always resulted in a great portion of the work being duplicated. Little attention is paid to picture or sound quality because it is accepted that attention to quality control will happen later during the conform session.

Can a nonlinear system become a "nonlinear online system"? Can the quality of picture and sound that have been transferred be akin enough to the original material so that a conform session using the original source material becomes unnecessary? This is the primary concern of a system that purports to be able to provide finished results. If there is no discernible loss of picture and sound quality or if the output of the system is of sufficient quality to meet the needs of the program, why pursue an online session? If the required editing tools and features—such as dissolves, audio crossfades, and titling—are available, for many types of programs direct output from such a system can indeed be the final conformed program. As the final step to finishing a program, the online suite is not always the inevitable path that a program must take on the road to completion.

TALKING WITH EDITORS

How Are the Film and Video Editing Cultures Alike and Different?

Paul Dougherty, editor, New York:

Generally, film editors have a better reputation, and there is some reason for this. They tend to be more engaged in content selection and have a better feeling about the edited work as a whole. Due to the fact that many tape editors learn their craft at video facilities, they tend to be most involved on an edit-by-edit level, but do not necessarily have the big overview. Due to the facility influence, tape editors often have an "hourly" mentality.

One defense of tape editors is that, unlike film editors, they are used to working with several versions of a cut. Usually with film editors, earlier versions are just a memory. Tape editors have and must keep track of several versions because that's the way tape editing works, and the final cut is a hybrid of the many versions. As tape editors are generally more technical, they probably have an edge when it comes to complex nonlinear systems. The other thing is that before nonlinear, tape editors probably would prepare a more formal paper list before starting their first cut. You'd like to think this discipline is good for something.

Tony Black, ACE, Washington, DC:

I find more differences than similarities. I have worked in both and find that film editing devotes more time and energy into the visual aspects of cutting the project. All video editing I have done has been governed, to a large extent, by the clock. How much time will this take? Time is money!

Basil Pappas, editor, New York:

Film editors are more used to the idea of swapping ideas and changing scenes. Video editors need to scrutinize the assembly process because so much is involved in changing ideas. Any tape editor is going to think twice about list management. All of these things weigh heavily. One difference I have found is that a majority of film editors are a lot keener on using sounds to overlap and layer dialogue, and they will usually stop right then and there in order to make a sound cut work, instead of doing it later.

Peter Cohen, editor, Los Angeles:

They are alike only in their desire to reach the same finished product. The approaches are entirely alien to each other. Linear videotape editing requires a constant, ever-present consciousness about the entire project because of its inflexibility towards making changes.

For the opposite reason, film editing allows you to tunnel into the specific piece at hand with little regard to the final outcome. You can always fix it later.

Alan Miller, editor, New York:

The biggest difference is time. Film editors are used to working at a more leisurely pace than tape editors. The cost-per-hour of video edit rooms makes a slow pace unacceptable. Most tape editors are used to working faster and making quicker decisions. Unfortunately, this often leads to compromises because of time and money. Film editors have the luxury and the medium to constantly fine-tune material. Film editors work for months on a 90-minute show while tape editors might get two weeks. So the initial mind set is different at the start of a project.

Is There a Bias Against Videotape Editors?

Paul Dougherty, editor, New York:

Because I started doing online editing in the late 70's, I can remember the roots of this bias. Video editors in the 60's and 70's often learned television in the army and did not attend college. That was true of the people who taught me, but those who "made it" in the big markets were skilled editors. Any bias against such editors would be unfair class-based snobbery. In the early 80's, you got a new generation of editors who were a somewhat different breed. First, they attended college and probably studied communications, but frankly I'm not sure how big a difference this makes in their editing skills. Mostly this meant they were more like their clients. But most importantly, they were more likely to be aspiring film or video makers who had ambitious designs and "pet" projects on the side.

Basil Pappas, editor, New York:

I don't see much of a bias; occasionally you'll hear clients specifically ask for a film editor, but this is far less common now. In any case, the really inquisitive film and video editors will want to learn more about the whole electronic nonlinear process. Other editors will concentrate more on graphic layering, compositing, and conforming.

Tony Black, ACE, Washington, DC:

When video editing emerged, it was so technically complex and difficult that only an engineer could do it. As a result, video editing was dominated by engineers. Even though it has become much easier to video edit, the online suite, especially, is not a good creative atmosphere with all its technical gadgets and distractions. A film cutting room, by contrast, is so simple as to be almost laughable; but nothing stands in the way of your editing.

Peter Cohen, editor, Los Angeles:

I believe that, unfortunately, there is still a bias against online video editors. It requires a great technical aptitude to be an online editor, and creativity is sometimes secondary. The creativity is left to the offline editor and the clients. It is rare to see an online editor working without a client sitting over the editor's shoulder. There is certainly a tremendous amount of talent required in translating the client's concepts, but because of the expense of online editing time, speed is often the most important factor for the online editor.

Editing requires a combination of talents, basically a certain amount of artistic talent and a certain amount of technical talent. Different styles of editing require various levels of competency within these two basic talents. Because of my familiarity with the concepts of both styles of editing, I have little frustration with either. A good editor is the person who knows how to get the most out of whatever system is being used, regardless of the system's limitations.

Alan Miller, editor, New York:

Absolutely. The bias against tape editors as being engineers and not creative people is a myth that has its roots in the very earliest days of video editing. I started as a tape editor, and it took years before people would consider trusting creative decisions to me. I think the confusion still exists because there are so many levels of tape editors. The complexity of video edit rooms plays into people's basic fear of technology. I constantly hear film editors say they shiver at the thought of using a computer like the one my 11-year-old daughter uses. Also, while film editing is basically story editing, video editing runs the gamut from story editing to special effect editing. In film, this would be a separate group of people.

6

Videotape-Based Systems

Through the early 1970's, film was still the predominant medium for programs as diverse as commercials and segments for nightly news shows, many of which were shot on 16mm film. As 2" videotape machines became affordable, more programming was shot and edited on tape. Two questions arose: How could the benefits of film editing be preserved while the labor and time involved is reduced? Is there a way to combine the strengths of film editing with the strengths of videotape editing?

The first electronic nonlinear editing system was a hybrid that utilized videotape, computer, and analog recording techniques; it was unique, and it was quite revolutionary. It paved the way for the first major wave of nonlinear systems, those based in videotape. In the late 1960's and early 1970's, technologies were beginning to emerge that could be used in attempting to bring together film editing tools with the electronic environment of videotape. The earliest attempt occurred in 1970.

THE CMX 600

The CMX Company was a joint venture by CBS and Memorex. CBS, the broadcasting entity, and Memorex, known for its expertise in analog recording techniques, decided that technology existed that could be combined to create a new method of editing.

This nonlinear editing system, the CMX 600, was developed in 1970 and saw regular, though limited, use by 1972 (Figure 6–1). The basic operation of the CMX 600 involved using magnetic computer disks that were modified to store analog video. Film was transferred to videotape, or the original 2" videotape was transferred onto six to 12 39-megabyte (MB) computer disk drives. At the time, 39 MB disk drives were considered to be very large!

Instead of using videotape to record signals from a camera, computer disks were used to record analog signals from videotape. It is important to remember that the format of the transfer was an analog FM recording; it cannot be considered digital, or digitized, video. The format of the video was still analog in nature. Not until the third wave of nonlinear editing did the

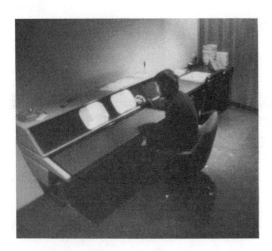

Figure 6–1 The CMX 600 is an electronic nonlinear editing system. Courtesy of Art Schneider and CMX Corporation.

intrinsic nature of the material being fed into a nonlinear system change from analog to digital. This is a critical point, and it cannot be overlooked. While the CMX 600 was the first electronic nonlinear system, it cannot be considered a digital system because the stored material was not digital in nature. The very reason that this system is called a hybrid is because it does not fall easily into any of the three waves of nonlinear editing. Nondigital editing systems still meant limited processing capabilities and loss of quality through successive generations.

Approximately five minutes of video could be stored on each disk drive for a total of approximately one hour on 12 disks. Because of the limited storage time, the system was used primarily for commercials as they had less overall footage than television programs, although several television comedy shows were also edited on the system. The pictures were often described as exhibiting a "sparkling effect" in playback, referring to the granularity of the images. The CMX 600, despite some reservations regarding the quality of the pictures, was quite an innovation and offered instant random access to all the frames stored on the magnetic disks.

The software program, written in Assembler language and based on a PDP-11 processor, had limitations as far as editors were concerned. Tom Werner, who had extensive experience with the CMX 600 at One Pass Film and Video in San Francisco, notes, "It gave you random access, and we used it a lot in the late 70's and early 80's, but it never did dissolves well, and edit lists were displayed in a manner foreign to most editors. We decided to change the software ourselves and develop a different capability for the machine. By using it to just store digital audio, we began to edit audio in a digital environment, which was quite innovative at the time."

Despite its innovation, the CMX 600 eventually faced difficulties. Not all of the computer software was complete, and reliability was a factor. Priced at over $200,000, it was an expensive system. While there weren't many 600s, the concept certainly intrigued post-production professionals. The 600 proved that different technologies could be used to create new methods of accomplishing old tasks. Most importantly, it made enough people in the film and videotape editing communities think about how their current working methods could possibly evolve.

Were other people also experimenting with modifications that could make this early nonlinear system do things that it wasn't originally intended to do? If it could be used for digitally storing audio at a time when editing audio was still largely accomplished by using razor blades to cut 1/4" reel-to-reel audio tape, what lay in store not only as technology continued to bring together these different areas of editing techniques, but as people began to customize the features of the system to their own tastes and requirements? This was heady stuff back then, at a time when 2" videotape editing involved machines that took up most of the space in a large room.

Although the CMX 600 was in operation for several years, its use was concentrated mostly in Los Angeles, though a few systems made their way to New York. Despite the introduction and promise that this electronic nonlinear system seemed to hold, it eventually ceased being used since, by the early 1980's, the use of videotape and the sophistication of computerized videotape editing systems had grown considerably. Timecode meant repeatability, and repeatability meant that other machines could be brought into the editing process, all orchestrated by cue points dictated by the presence of timecode.

THE FIRST WAVE

For almost 12 years, electronic nonlinear was not revived and no new systems were introduced. Film editing, a known entity and a standard and accepted way of putting together all types of programs, was considered a "safe" method; whatever could go wrong had been experienced before. Electronic linear videotape editing, however, was evolving at a rapid pace, and with the introduction of devices that began to change the analog nature of video by digital means, a blend of both analog and digital technologies began to emerge. But, as is usual with such a rapidly emerging blend, users were often reluctant to be the first to try a new system, especially when faced with the pressure of delivering a program under a tight deadline.

From this merging came a series of systems that sought to combine the nonlinear aspects of film editing with the repeatability of videotape editing. These systems formed the first wave of electronic nonlinear editing systems: those based in videotape. These systems were nonlinear, but they were not random access. Because videotape was involved, access to the material was still sequential. The editor still had to cue the point on the videotape and wait for this to occur.

What qualities of film and videotape editing did this first wave of nonlinear editing systems seek to preserve and enhance? Editing in a nonlinear system means preserving the techniques of film: the ability to splice a shot anywhere into a sequence while the sequence expands accordingly and being able to remove a shot and have the sequence contract accordingly. Picture and sound had to be treated separately, as they are on film. From video editing, the first wave adopted the ability to see optical effects, such as dissolves. The system also needed to be *dual-finish cognizant*. This means that the system creates a negative cut list as well as a video EDL that is used to conform the source videotapes for a videotape release. Finally, the system had to create an interim version of the show: a screening copy.

Beginning in 1984 and lasting until 1989, true electronic nonlinear editing systems began to appear and prosper. In 1984, the first system to appear was the Montage Picture Processor™. The name carried positive references to the methods of word processing over to pictures and sounds. Following the Montage was the Ediflex system in 1985 and TouchVision in late 1986. Each system offered a different user environment and work

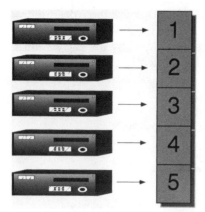

Figure 6–2 In first-wave nonlinear editing systems, multiple machines are used to seek out the desired sections on videotapes. After reaching their cue points, the machines play back the sequence in its intended order, shots 1, 2, 3, 4, 5. Illustration by Jeffrey Krebs.

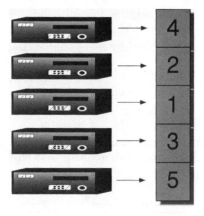

Figure 6–3 By directing the machines to cue to different sections of videotape, the sequence being edited appears to have been reordered. Here, the original sequence of shots 1, 2, 3, 4, 5 is changed by merely directing the source machines to play back material in the new order of 4, 2, 1, 3, 5. Illustration by Jeffrey Krebs.

method, but all shared the same basic method of bringing nonlinearity to the editing task.

The underlying concept of the tape-based systems is achieving nonlinear editing by using multiple videotape machines. A sequence is created by cueing each machine to a different shot and then playing back the shots in their intended order (Figure 6–2). In this way, editing can proceed in a nonsequential manner. Nothing has to be committed to a final order until the editor is content with the way the shots work together (Figure 6–3).

Although each system differed in its design and characteristics, some general observations can be made. All the systems used a series of videotape machines that varied in videocassette format. Some systems, such as Montage and Ediflex, first used VHS tapes, while TouchVision initially used VHS and also offered 3/4" videotape. Later, the Montage incorporated 1/2" Betamax tape.

The editing process begins with a transfer of the original material shot on videotape or a series of film rolls to videotape. The material on this "master" videotape is then copied to several videotape cassettes. The result of this real-time transfer process (dubbing 30 minutes essentially took 30 minutes plus preparation time) is that each videocassette in a tape "load" has identical footage with identical timecode.

Using VHS tapes, a complete load can offer approximately 4.5 hours of source footage at one time. If more material is associated with a project, the first tape load is removed, and the next load inserted. The number of cassette copies is directly related to the number of playback machines being used. Some systems offer from six to 27 playback machines. A common combination is 17 machines, at 4.5 hours per load.

The theory of operation is simple: By having identical footage on all the tape machines, the editor has a nonlinear editing system. Instead of editing on film where the splices would first have to be removed and the order of the film pieces changed and then respliced, here just the directions to the tape machines are altered.

For example, let's assume that a system is configured with 12 VHS machines. The editor is creating a sequence with ten shots. Since the same source material is on the duplicate cassettes, the system can easily find the first shot in machine 1, the second shot in machine 2, and so on. After the editor chooses appropriate in points for each of the shots, ten machines will work in tandem, cueing to their in points. Each shot is then played back at the correct time, and the sequence can be viewed in its entirety. By virtue of having material spread over numerous playback machines, the editor is able to move shots around and experiment with different ways of putting a scene together.

THE PLAYLIST

When we talk about electronic nonlinear editing systems, it is important to realize that usually no actual recordings are being made. Instead of recording material from one source to a record master, the order of shots is changed to represent a different playback order. Nothing gets recorded to a master when you edit nonlinearly until the final assembly or output stage. Instead, electronic nonlinear editing systems, regardless of which wave they represent, all share the concepts of a *playlist* and of *virtual recording*. Although material appears to be recorded, the sequence being edited is not actually being recorded anywhere. Viewing an edit or a sequence of edits *gives the impression* that the entire sequence has been recorded and is now being played back. This distinction is important.

We can see the result of moving shots around without committing anything to tape. This type of visualization is geared toward the concept of a playlist. A playlist determines which shots will be played and in which order. If we want to create a sequence where we see a shot of the sun, then a shot of a boy on a beach, and finally a shot of a girl swimming, the playlist, represented by timecode numbers, would look like this:

Segment	Play	From	To
1	Sun	05:00:02:01	05:00:05:12
2	Boy on beach	05:05:08:09	05:05:18:09
3	Girl swimming	05:02:04:01	05:02:09:01

In essence, these are the directions that the tape-based nonlinear editing system is issuing to the various machines. When the editor begins to put a sequence together, whether editing in film or videotape, the from and to points (start and end points from the source videotapes) are being chosen. In film, these points are chosen visually. With videotape, the points are chosen visually but are determined by choosing appropriate timecode numbers.

In this example, the sequence has three segments; we see the sun, a boy on the beach, and a girl swimming. Three of our 12 tape machines are involved in showing this sequence. The first tape machine plays the shot of the sun, the second machine plays back the boy, the third machine plays back the girl. Once the machines reach their cue points, the sequence will be shown in the proper order of shots and with the desired from and to points which the editor chose.

If we could vocalize the commands of the playlist, it would be as follows:

1. Tape machine 1, cue to here.
2. Tape machine 2, cue to here.
3. Tape machine 3, cue to here.
4. Wait.
5. Tape machine 1, play from here to here.
6. Tape machine 2, when 1 reaches its out point, start playing from your in point.
7. Tape machine 3, when 2 finishes, begin to play.

Remember, you can make changes at any time to the playlist without penalty.

When the sequence needs reordering or when the in and out points for shots need to be trimmed, the playlist is easily altered.

Making changes to the playlist results in the sequence being displayed in the new order.

For example, if we want the sequence to start with the boy, then show the sun, then the girl, we alter the playlist.

Segment	Play	From	To
1	Boy on beach	05:05:08:09	05:05:18:09
2	Sun	05:00:02:01	05:00:05:12
3	Girl swimming	05:02:04:01	05:02:09:01

By altering the playlist, the order in which the machines play back their segments is changed; as a result, the sequence is shown to us in a new order.

If in addition to changing the order of the shots we wanted to make some slight adjustments to the in and out points, we would alter the from and to points.

It is always interesting to note how terminology seems to creep in from different disciplines. It is not important if we refer to these points with the electronic term from/to, the film term head/tail, or the videotape term in/out. They all mean the same thing: the beginning and end of the shot.

If we now like the new order of the sequence, but we decide that the shot of the boy should be changed slightly, we make an adjustment.

Segment	Play	From	To
1	Boy on beach	05:05:08:09	05:05:18:09

becomes

1	Boy on beach	05:05:09:09	05:05:20:09

We alter the from point and start one second later than the original point; we alter the to point and continue for two seconds longer than the original point. The new playlist is as follows:

Segment	Play	From	To
1	Boy on beach	05:05:09:09	05:05:20:09
2	Sun	05:00:02:01	05:00:05:12
3	Girl swimming	05:02:04:01	05:02:09:01

The tape machines cue again, and the sequence reflects the new trims for the shot of the boy. Reordering shots in this manner borrows from the word processing activities of cutting, copying, and pasting text from one place to another. By altering the instructions that the playlist holds for the tape machines, nonlinear editing is achieved in a repeatable fashion.

The first-wave systems offer a choice as to how many machines are used as playback sources. Why is there a choice and not a fixed number? The reason has to do with the demands placed upon the machines by the playlist. If the demands of the playlist begin to stress the cueing limitations of the machines, there are a number of consequences.

Let's return to our sequence. We are working on a system that has 12 tape machines. We begin to edit our sequence, and the playlist shows 12 segments:

Segment	Play	From	To
1	Boy on beach	05:05:09:09	05:05:20:09
2	Sun	05:00:02:01	05:00:05:12
3	Girl swimming	05:02:04:01	05:02:09:01
4	Sailboat	05:04:12:11	05:04:16:21
5	Man fishing	05:02:10:09	05:02:16:03
6	Radio in sand	05:08:09:05	05:08:13:02
7	Fishing rod tip	05:02:17:02	05:02:19:05
8	Man looks to water	05:03:11:01	05:03:13:03
9	Bobber disappears	05:08:17:01	05:08:19:16
10	Reel spins	05:11:10:01	05:11:11:24
11	Water ripples quickly	05:17:18:09	05:17:21:07
12	Hand moves to reel	05:22:09:05	05:22:13:03

Each machine cues to its correct position. Time is required to do this, and most systems provide status messages informing the editor of the remaining cue period. When the machines are ready, the sequence is shown. Each machine plays its role in the process. We see the boy on the beach, then the sun, the girl swimming, and the sailboat. Then we begin to see the fisherman and the ensuing series of shots involving the fisherman. Since we have 12 shots in the sequence and since we have 12 playback machines, the sequence is shown to us in its entirety.

Now, let's add more shots to the sequence we are building:

Segment	Play	From	To
13	Tight on man's eyes	05:03:24:00	05:03:26:07
14	Reel spins quickly	05:11:22:04	05:11:24:01
15	Fingers grip rod	05:22:17:00	05:22:19:04

If we look at the entire playlist, we see the following:

Segment	Play	From	To
1	Boy on beach	05:05:09:09	05:05:20:09
2	Sun	05:00:02:01	05:00:05:12
3	Girl swimming	05:02:04:01	05:02:09:01
4	Sailboat	05:04:12:11	05:04:16:21
5	Man fishing	05:02:10:09	05:02:16:03
6	Radio in sand	05:08:09:05	05:08:13:02
7	Fishing rod tip	05:02:17:02	05:02:19:05
8	Man looks to water	05:03:11:01	05:03:13:03
9	Bobber disappears	05:08:17:01	05:08:19:16
10	Reel spins	05:11:10:01	05:11:11:24
11	Water ripples quickly	05:17:18:09	05:17:21:07
12	Hand moves to reel	05:22:09:05	05:22:13:03
13	Tight on man's eyes	05:03:24:00	05:03:26:07
14	Reel spins quickly	05:11:22:04	05:11:24:01
15	Fingers grip rod	05:22:17:00	05:22:19:04

Since we have 12 machines, three machines must search to a second tape location, cue to the desired from point, and be ready to play back the material required in segments 13, 14, and 15. When the rate of editing increases in this manner, tape-based nonlinear editing becomes an orchestrated effort of commanding machines to reach the needed location in time. If the three machines that must now cue to a second point have enough time to do so, the sequence can be viewed in its entirety.

TRAFFICKING

All nonlinear editing systems must cope with "trafficking" problems (Figure 6–4). These problems diminish as the access times (the amount of time that passes before material requested is available for use) of the different waves of nonlinear edit systems improve. If a playlist segment cannot be reached in time, that material cannot be displayed. As a result, the sequence cannot be viewed in its entirety. Most tape-based systems use computer software that looks ahead to the tape cues that are needed and then determines which machine is closest to the desired cue points to minimize shuttle time.

If our sequence requires that several machines cue from points at the beginning of the cassette to points in the middle or at the end of the cassette, there may be situations where the sequence cannot be viewed in its entirety. For example, in our sequence with 15 playlist entries, as the sequence plays down, it is incumbent upon three machines to get to the locations indicated for shots 13, 14, and 15. Whichever three machines can get to these points as the sequence continues must shuttle and cue accordingly. If successful, the sequence is shown in seamless fashion.

If a trafficking conflict arises, one possible result is that we will see black on the screen, which represents where the shot should have been but is not due to the conflict. This is obviously not very convenient. Many systems, sensing that a trafficking issue is imminent, will take some time to record several shots together at a location on a tape (often called a prebuild). This section represents the shots that would have been difficult or impossible to cue to in the allotted time. When these shots are required, they are played back from this prebuild tape, and there are no trafficking problems. It should be noted that it will take time to make the prebuild, and if any of the shots within the prebuild must be changed, it will have to be reconstructed.

This is precisely why these first-wave systems offer a choice of the number of playback machines. More machines do not increase the amount of footage that can be stored since it will continue to be identical footage with an overall limit of about 4.5 hours. However, adding machines decreases the frequency of trafficking problems. With 27 playback machines, as some systems offer, there is an increase in the likelihood that we will be able to watch our sequence uninterrupted.

VIRTUAL RECORDING

Each wave of nonlinear editing systems adheres to the concept of *virtual recording*. Unlike videotape editing, where material is copied from a source machine to a record machine, with nonlinear

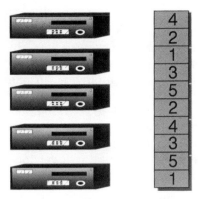

Figure 6-4 As the number of cuts increases, trafficking problems can arise if the number of tape playback machines remains constant. Will the additional cue points be reached in time? Illustration by Jeffrey Krebs.

editing, actual video and audio signals are not recorded to a master tape. Because nothing is recorded to tape (with the sole exception of the prebuild), it is far easier to adjust a sequence by altering the playlist than by rerecording everything as you would have to do with linear videotape editing.

Instead of copying video and audio from one tape to another, the "recording" process for nonlinear editing systems is actually a record-keeping process for the edits that will eventually be made. The ordering of the shots, the creation of sequences from these shots, the creation of scenes from these sequences, the creation of acts from these scenes, and the creation of programs from these acts are a process of keeping records about how playlists are created.

The playlist is the most important item created during the nonlinear edit session. It is used to create the edit decision list. Playlists do not take up much space in the computer's memory because they are just lists of numbers rather than the data-intensive files of video and audio. Virtual recording is often referred to as using "pointers" to the original material. When we created our playlist, the sequence we were shown was based on a playlist that points to certain material on different cassettes.

Without the concepts of virtual recording and playlists that merely point to how the original footage should be ordered, nonlinear editing would become much more complicated. If we had to record the material as we were editing, we have suddenly committed material to tape, and as soon as we involve tape, any change will involve rerecording. This defeats the entire purpose of being in a nonlinear environment. As soon as something gets committed to tape, changes become difficult.

MULTIPLE VERSIONS

All nonlinear editing systems should offer the ability to provide multiple versions of a sequence. This lets the editor experiment with alternative versions for a scene, and different creative views can be exercised. For example, after a scene is cut, we know that a playlist exists for that scene. This playlist can be duplicated. By duplicating the playlist, we receive a duplicate of the scene. Then, the new copy of the scene can be altered. As a result, there are now two versions of the scene. Even if only one frame is altered in the second version, it is different from the first version. This flexibility and power to change just one thing must be available in all nonlinear editing systems. The playlist allows entire edited sequences to be duplicated and then trimmed, rearranged, and so on.

When a project routinely requires the creation of multiple versions, nonlinear editing can be extremely helpful. For example, for television commercials that require different versions for foreign languages, creating the basic version and then duplicating it to change just the sections that are country-specific can be accomplished very quickly on the nonlinear editing system.

Another interesting development is the use of nonlinear editing systems to replace content but not intent. Take, for example, an actual case where nonlinear editing was used to edit different versions of a television trailer (promotional commercial) for a motion picture about to open nationwide. In one version of the commercial, the lead actor was showcased. In another version, the intent of the commercial–to provide information about the movie's premise–was not changed, but all sections featuring the lead actor were replaced by scenes featuring the lead actress. The replacement was easy, and two versions were available without a great deal of extra effort.

USER INTERFACE

Each tape-based nonlinear system offers different tools and different computer interfaces. What is important is how the first systems sought to blend the world of film editing techniques with the electronic world of videotape editing. The considerations that were made and the computer and manual controls that were offered were diverse.

All nonlinear editing systems use computers. At the heart of all of these devices is a computer that offers a software interface with which the user interacts. This computer issues directions to the hardware associated with the system. The environment that the software creates for the editor, if successful, gives the impression that the editor is in a cutting room and not in a computer software program. The distinction is how well the cutting room is emulated in the nonlinear editing system.

Tape-based systems, as we know, utilize a bank of tape machines with each machine having its own small monitor. A larger monitor, often 21", that displays the sequence being edited is usually positioned in front of the editor. These systems receive user input from a variety of devices: knobs and wheels, which provide the same tactile feedback as film editing systems, such as the Moviola and the Steenbeck, or a light pen, which is not unlike the film editor's grease pencil used to mark on film. Computer keyboards are included, but the use of the keyboard is confined to inputting text descriptions for shots.

Providing the editor with tactile devices was intended to give the editor the sensation of manipulating film through the fingers. This was an important step in the acceptance of these early systems. Imagine touching film for one's whole career and then being asked to touch a computer keyboard; reluctance was viewed as a natural inclination that needed to be addressed by providing system tools that recreated the working environment.

System designers of the first wave of nonlinear editing were quite successful in doing the best possible job at keeping the computer aspects in the background. Because personal computers were in their infancy in the early 1980's, exposure to computers by film and television editors was somewhat limited, and computers were viewed as not being thoroughly reliable. With the correct interface, being in a computer environment did not have to feel like being in foreign territory. In fact, it is quite fair to say that the designers of these first systems spent a great deal of time trying to create editing computers that did not look like computers.

THREE SYSTEMS OF THE VIDEOTAPE-BASED WAVE

In order to achieve a short learning period, creating systems that would work in a logical manner and that the editor could quickly relate to became very important for system designers. Of the three most popular systems in use, each recreates the editing environment in a unique way.

Montage

In the Montage system, the environment created and shown to the user duplicated the concept of the film editor's cutting room. Representations of footage (the pointers) were housed in electronic bins. As the inputting of material occurred, shots could be identified, or tagged, and a head frame and tail frame established for those shots.

The head/tail shots were displayed to the editor as black and white digitized frames. Digitizing is the conversion of an analog signal to a digital signal. In this way, the editor could see how a shot began and ended. The digitized frames were stored on the computer's magnetic disk while the actual footage was stored on the videotape cassettes. Digitizing the first and last frame of a shot was a unique approach that utilized the developing digital technology (Figure 6–5).

These digitized frames were stored in "work bins." Editing shots together and trimming, splicing, and rearranging the different shots were similar to the method of cutting film without the labor involved. As the process continued, the videotape machines cued and displayed the sequence accordingly.

The user interface was largely menu driven, offering a variety of tools that allowed the user to copy, trim, discard, splice, dissolve, and insert material. Instead of graphical icons that represented tools, text descriptions served as functions to be implemented. An electronic grease pencil let the editor write directly on the digitized frames to make notes and comments. The control mechanisms, two large wheels with paddle controls, were viewed by many users as excellent mechanisms. These were used to shuttle through footage, and they preserved some of the tactile sense of controlling the material "by hand."

In addition, optical effects were available in this first wave of nonlinear systems, and dissolves and wipes were features that were applauded. Suddenly, film editors could see a shot dissolve into another shot without having to wait for the optical effect to be readied by the lab. Film editors had not embraced videotape editing because it was linear, but here was a technology that

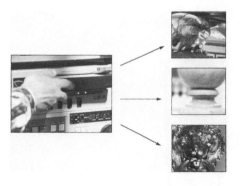

Figure 6–5 Footage contained on an original videocassette is digitized into the digital nonlinear editing system and displayed as representative frames. Illustration by Rob Gonsalves.

could provide nonlinear editing and offer something that film did not offer: the ability to make trims easily, to try different versions of an edit, and to see optical effects during the editing process.

With the Montage system, changes could be tried and edits reexamined. Without ever committing anything to tape, time could be spent honing a scene by trimming and adjusting until the editor was satisfied with the results.

At the completion stage, there are several work products. The first product is the transformation of the playlist into an edit decision list created for either a film finish (the negative cut list) or a videotape finish (the video EDL). The second product was somewhat ingenious: After the edit session, a printed storyboard was available, using the digitized pictures to show the edited sequence in visual form. This could be a helpful reference during the final conform stage. The final work product of this wave involves the creation of a viewing copy of the material being edited. Most frequently, a 3/4" videotape recorder is used. While each videocassette player cues to the appropriate material, the sequence is recorded in its edited form to the 3/4" machine. The result is a complete program that can then be evaluated.

Ediflex

In 1984, the Ediflex system underwent its testing, and it was introduced for regular use in 1985 (Figure 6–6). The inventor of this system was Adrian Ettlinger, who had previously invented the CMX 600. The Ediflex is based on multiple VHS machines and takes a somewhat different view of how footage is organized than does the Montage. Borrowing from the notion that footage could be further broken down and related to the shooting script, the logging process involves relating a shot directly to a section of the script.

Figure 6–6 The Ediflex nonlinear editing system. Photo courtesy of Ediflex Systems.

During the logging stage, computer input, via a light pen that is tapped against the computer screen, associates the videotape material with a certain section of the script. This marked script then becomes the main reference that the editor has in knowing what footage is available when cutting the sequence.

The editing environment that the Ediflex created for the user relied heavily on the script as the blueprint for how footage would be put together. All editing was based on the correlation of the footage contained on the videotapes to the script. This was something of a redefinition of the editing process. Editing was done, in effect, by joining pieces of picture and sound material together by appending the text descriptions to one another via instructions input by the light pen. It was certainly different from traditional editing techniques, but because the connection with the script was present, the Ediflex enjoyed great success with one-hour television episodic programming, which is entirely script driven.

In addition, a particularly useful feature, called *horizontal play* or *scan across* allowed the editor to view all footage associated with a specific section of the script. For example, if the editor wanted to see all the options for an actor's dialogue in a specific scene, several machines cued to the material and played option after option so that the editor could make a selection.

The development of the system relied heavily on input from film editors. Among the software features that users suggested are the ability to use Aaton™ code for negative matchback, optical effects in the form of dissolves, the use of a second audio track, and assembling a cut sequence to 3/4" tape to incorporate the cut material as a source. The playback of a scene from the various VHS machines is known as *real-time playback*.

Film editors who were first introduced to the Ediflex system were impressed by the concept behind its operation. Many of these editors had some computer experience, but other editors approached the system with some trepidation because it represented a departure from their normal work methods. Over time, this latter group came to appreciate that the system does not veer too much from the normal work process. The system software has changed to reflect the changing post-production market in Hollywood.

While still in prototype form, Ediflex was used on a cable series for a major motion picture studio in the United States. After a successful experience on this show, the system became available for rental. Since its introduction, the Ediflex has been used for many television series and feature films. The work products from an Ediflex session are similar to those of the Montage; the Ediflex provides both a negative cut list for film and a video EDL at the completion of the session. The process of creating screening copies is also similar.

BHP TouchVision

In mid-1986, Bell & Howell's Professional Division began testing its entry into the first wave of nonlinear editing. This product was created by a wing of the company, TouchVision, and it became

Figure 6–7 The BHP TouchVision nonlinear editing system. Courtesy of TouchVision Systems, Inc.

available for regular use in 1987 (Figure 6–7). Just as Montage was unique in offering an environment based on film work bins and Ediflex was unique in placing an emphasis on the script, TouchVision was unique in that its interface replaced text-driven commands with graphical icons.

The TouchVision system was based on six to 12 VHS tape machines and a 13" color monitor with an infrared touch-sensitive screen. If enhanced picture quality was required, eight 3/4" videotape machines replaced the VHS machines, and a large 21" color monitor displayed the sequence as the machines played it back for the editor.

The touch-sensitive screen was a unique approach in 1987, but more important was the adoption of a user interface that replaced text directions with graphical methods. This *graphical user interface* (GUI) provided icons that represented visual shortcuts in their emulation of editing tools. Although the icons are unsophisticated by today's standards, representing film rolls not as "Film Roll #3" but as a drawing of a film roll was a step toward the creation of a more visual instruction set for the user. Many editors felt that the icon-based format led to a shorter learning period and an easier transition from film and videotape to electronic nonlinear.

Editing on a TouchVision screen involved the display of one box on screen for each tape machine. Touching the tools "PIX" and "SND" would cause a cut to be made from material being played on the videotape machines. One of the unique aspects that TouchVision offered was the ability to slow down and, as the feature was called, "VIEW" the edited sequence. The other nonlinear systems in this wave provided a real-time preview but did not allow the user to crawl over the edit points. An edit-by-edit examination of the sequence was possible on TouchVision. The ability to "gang" several videotape machines together and jog them back and forth was a particularly important feature in the editing of programs involving multiple camera setups when the editor wants to simultaneously view as many angles as possible.

Multiple versions, film-style splicing, and all trimming options are available. The work products of the TouchVision system include the negative cut list, the video EDL, and a screening copy. TouchVision has been used for a variety of projects, but the majority have been feature films and episodic television. In 1990, a Macintosh computer platform was used in place of the original IBM-compatible computer. As a result, the sophistication of the graphical menus and the user interface grew considerably.

APPRAISING THE FIRST WAVE

The videotape-based wave of nonlinear editing continues today. Montage II, Ediflex, and TouchVision continue to be used primarily for episodic television, although feature films are also being edited on these systems. These longer forms of work do not involve the fast-paced cutting of commercials or music videos. Since there are access time delays involved in displaying the footage as the playlist demands grow, it is easy to see why the tape-based systems are so well suited to projects where the editing pace is less frantic.

The limitations of access time cannot be viewed as a liability as long as the correct system is chosen for the job at hand. If the editor is scheduled to edit a commercial that requires an extremely fast cutting pace and the job is scheduled on a six-machine tape-based nonlinear system, trafficking problems will most certainly occur. Again, proper evaluation of the correct system for the job is always required. Otherwise, frustrations inevitably are going to occur on the part of the editor and for the editor's clientele.

While systems based in videotape definitely qualify as being nonlinear, they do not function in a random-access environment. Since material must still be located via shuttling through the reels of videotape, these systems are considered to be direct-access systems. As long as one must wait while tapes cue, random access cannot be claimed as an attribute.

The tape-based wave represented many converging ideas and technologies. It was a first attempt by several companies to blend together the strengths of film editing and the faster access and immediacy of videotape editing, presented in an interface and working method that would not be completely alien to existing methods.

The major benefit for the editors and their clients was an increased ability to try out possibilities. "Should the shot be here, or should we put it later? Don't you think the shot should be a bit longer?" These are always questions that arise during a collaborative edit session. Prior to this wave of editing, there were only two ways to get answers to these questions. If editing on film, the film would have to be rearranged, and trims would have to be located, put back in place, and so on. If editing on videotape, the material would have to be rerecorded, or a generation would be lost.

For the editor who was intent on remaining current with technology and who was concerned that skills learned as a film editor would become defunct as the use of electronic technology grew and program delivery schedules shortened, the first wave of nonlinear editing systems provided an opportunity to move ahead both creatively and technologically. Each of the systems in the first wave had characteristics that helped it to garner its own loyal following. These systems were embraced by film and videotape editors, and they were eschewed by film and videotape editors. One thing was for certain. Unlike the CMX 600, which was in retrospect perhaps too early for its time, electronic nonlinear was here to stay.

ECONOMIC BENEFITS OF THE FIRST WAVE

There are many opinions as to whether nonlinear editing, regardless of whether the systems are based in videotape, laserdisc, or digital, save time over traditional methods of cutting a project. Many editors, directors, and producers feel that nonlinear is faster at every stage. Others feel that the initial cut, the first assembly, takes a similar amount of time as it would on film or on video.

However, one thing that both camps agree on is the overwhelming advantage that nonlinear holds over linear editing when it comes to changes. Whether the project is a commercial, an episodic television show, or a promotional videotape introducing a new product, changes will inevitably be required as others view the edited material. Nonlinear systems are extremely efficient at making the required changes and creating the new edit list. Only the new screening copy requires a linear assembly.

The adoption of tape-based nonlinear systems was not merely a decision made to provide more creativity. It is inconceivable that a new method of editing would be implemented unless the post-production personnel made some attempt to learn how the financial condition of the business would be affected. For the first wave of nonlinear editing to have done as well as it has, it must provide economic benefits to the users. These benefits are in the form of a decreased amount of time spent editing programs and a decreased cost of keeping individuals in the post-production chain involved in the process.

By being able to save time during the editing stage—and electronic editing has shown that, on average, more time is saved versus the traditional film or videotape editing of the same program—the overall schedule for a show from initiation to completion can be reduced, and money can be saved. The acceptance and practice of electronic editing by many of the production companies in Los Angeles that create programs in the episodic and movie-of-the-week (MOW) television categories are a testament to the economic benefits of editing on electronic nonlinear editing systems.

EVOLUTION OF THE FIRST WAVE

Product introductions in this first wave reached a peak in 1988. Although many of the systems have been popular and are still in current use, it is important to understand how the first wave of nonlinear systems is evolving. While it is never certain how manufacturers will fare with their future product plans, there is a definite movement by the designers of some of the first-wave systems to move beyond the technology of videotape machines.

The Montage III, for example, is based in the third wave of nonlinear editing, the digital wave. The Montage II continues to exist as a first-wave system. The Ediflex continues to exist as a first-wave system but plans are afoot to enter the digital domains. TouchVision, meanwhile, continues as a first-wave system, while its manufacturer has entered the digital domain with the introduction of DVision. Clearly, these system manufacturer are all intent on incorporating new technology if it makes sense in their business plans.

The videotape-based wave introduced and brought forth concepts that were replicated and improved upon in the ensuing waves of nonlinear editing. Trafficking, virtual recording, playlists, multiple versions, and graphical user interfaces could all be improved. How each of these basic operational and design concerns are addressed in the creation of new systems is examined in the next two chapters.

TALKING WITH EDITORS

How Did Film and Videotape Editors React to the First Appearance of Nonlinear Systems?

Paul Dougherty, editor, New York:

I knew nonlinear was the way of the future and absolutely essential for complex editing. I originally thought that it might be overkill for most jobs. Maybe "self-indulgent" might be a better way of putting it. After looking at the various systems in 1987, my reaction was simple. I did not like any of them, but I'd be obliged to learn whichever one came to dominate the market. Well, as everyone knows, the first generation of nonlinear systems never really caught on outside of Los Angeles. My other original reaction was that the design of these systems pandered too much to film editors. They went too far in burying timecode. No one likes columns of numbers on the screen, but you can't beat SMPTE as a precise labeling system for footage.

Basil Pappas, editor, New York:

Tape-based nonlinear was a difficult sell to both editors and clients in New York because the type of work was not conducive to this nonlinear approach. Since there was no episodic television show market here, there was little need for tape-based systems, since they were desired primarily for scripted show formats. Laserdisc was more capable of commercial spot editorial but suffered from disc turnaround and inability to handle late-arriving material. Some systems were tried, mostly for cutting commercials, but due to the problems with accessing materials for quick cut spots, there was little success. I found the first systems interesting, but because the type of work being done here in general did not require this approach to nonlinear, there was no reason for them to be pursued, either for creative or economic reasons.

Some manufacturers, especially Ediflex, put a very commendable effort into training editors and supporting people who were using the system. Even though the work in New York wasn't conducive to this type of nonlinear editing, Los Angeles being more appropriate, it was encouraging to see a manufacturer put effort into educating editors. Both Ediflex and TouchVision were very supportive with manuals and training.

7

Laserdisc-Based Systems

The second wave of nonlinear editing systems utilizes laser videodiscs (Figure 7–1). In creating such systems, the intent was simple. To avoid the cueing problems as well as the trafficking conflicts inherent in the videotape-based nonlinear systems, manufacturers decided to use the speed of laserdisc technology to afford improved random access to material.

The functioning of a nonlinear editing system that uses laserdiscs involves transferring picture and sound from a videotape onto a laserdisc. Once the material is on the disc, it can be accessed quite rapidly. Most laserdisc players are able to display information within 1.5 seconds after the request.

There are various systems within this category. The first system to be introduced was the EditDroid system, followed by the CMX 6000, Epix, and Laser Edit. All of these systems are still being used today.

TYPES OF LASER VIDEODISCS

Before we discuss the specifics of nonlinear laserdisc systems, a discussion of the technical characteristics of laserdiscs themselves will be helpful. There are two types of laser videodiscs: constant angular velocity (CAV) and constant linear velocity (CLV). The important differences for our purposes are as follows:

Type	12" CAV	12" CLV
Capacity	30 minutes per side; 54,000 frames	60 minutes per side; 108,000 frames
Channels of sound	2	2
Slow motion	Yes	No
Step frame	Yes	No
Still/freeze frame	Yes	No
Use	Industry	Consumer

Both types of videodiscs evolved during the late 1970's. Laserdiscs were sometimes referred to as discs for video and discs for data. Originally, tube lasers were used, and as a result,

Figure 7–1 The second-wave of nonlinear editing systems utilizes laser videodiscs which provide improved random access to material. Photo by Michael E. Phillips.

machines were quite large. Later, solid-state logic circuitry made videodisc players easier to manufacture and smaller in size. Initially, laserdiscs were used to permanently store video and sound for either presentation or archiving purposes, including displays for museum exhibits and interactive training programs. For these types of programming, videodiscs offered certain advantages. Unlike videotape, with laserdiscs, wear is much less a factor, and most importantly, discs offer improved random access to information.

TRANSFERRING VIDEO AND AUDIO TO LASERDISC

As with all nonlinear systems, material is transferred from one medium to another. In this case, either film or videotape is transferred to laserdisc. Laserdisc-based nonlinear systems use CAV-type discs, and therefore, the amount of footage per disc is limited to a total of 30 minutes.

The original footage is first transferred to a premaster tape that has continuous timecode and allows a total time of 30 minutes per disc side. This videotape contains the picture and sound information from which the laserdisc will be struck. A laserdisc master can be made of either polished glass or plastic. This smooth surface is covered with a photosensitive layer that is washed with photochemical materials.

Next, this premaster tape is played back into a laserdisc mastering machine. The video and audio signals excite a laser, and the resulting light beam creates a corresponding value as a frequency modulation (FM) carrier. Laserdiscs use the same FM standard as 1" type C videotape. This FM carrier is recorded onto the disc.

Sometimes people are confused about the type of signals that are recorded to the disc. The most often asked question is whether the video and audio are digital or analog in nature. The video signal is analog in nature. The two audio tracks that both CAV and CLV discs can store are almost always analog since the manner in which audio is stored on the disc is the same as for storing video. Audio modulations from the premaster videotape excite the laser beam and are modulated to two FM carriers. The audio resolution of laserdisc is approximately 15 kilohertz (kHz). Thus, recording both video and audio signals to laserdisc is based on FM modulation of these signals, a process identical to 2" and 1" videotape recording techniques.

What is happening to the disc itself? The recording laser exposes the photochemical layer and burns reflective pits into the surface of the disc.

When the disc is played back, the playback laser illuminates the pits, and then a photosensitive cell interprets the laser light that is bouncing and reflecting off the pits. These very tiny signals are carried to an amplifier and demodulated back into video and audio signals, which we can see and hear (Figure 7–2).

During the recording cycle, concentric lines are written to the disc. It is here that there are some important differences between CAV and CLV discs. A CAV disc, which stores 30 minutes of material per side, rotates once per video frame. This means that each track consists of two fields that make up one video frame. That is, it spins 30 times per second, and this spin speed never changes, whether material is playing from the inner or the outer rings of the disc. The rotational speed of the disc is 1800 revolutions per minute (r.p.m.), and this will not waver. This important characteristic allows a CAV disc to provide steady freeze frames.

CLV discs do not move at a constant revolution. The biggest advantage of these discs is that program time is expanded from 30 minutes to 60 minutes. However, material is spread out farther along the disc, as material is recorded from the inner to the outer edge. As a result, the innermost tracks of the disc move along at 1800 r.p.m. When the outer tracks of the disc are reached, the rotational speed slows down significantly to about 600 r.p.m. So, although it starts out at approximately one rotation per video frame, by the time the end of the disc is reached, it is rotating only once for every three video frames. As a result, the disc is not able to achieve a full video frame for each rotation of the disc. This is why CLV discs cannot offer freeze-frame capability.

Discovision Associates (DVA) Code

Once pictures and sound have been recorded to the laserdisc, special laserdisc codes are added to the glass master. These five-digit codes are encoded onto the glass master by the laser. They represent how the disc is searched. If we want to go to frame 20,000 on the disc, we type that number into the control pad, which commands the disc to perform a search to that frame.

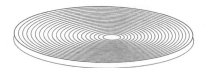

Figure 7–2 Videodiscs have spiral tracks. They play from the inside track to the outside track. Illustration by Jeffrey Krebs.

The time that it takes for the laser beam to leave one spot, get to the next spot, and begin displaying the requested frame is referred to as *access time*. The access time on most laserdisc players is approximately two seconds or less. In the case of more sophisticated machines, such as the Pioneer 8000, access time is less than half a second.

DVA code has a relationship to timecode. Recall that the premaster tape was made in terms of SMPTE timecode. This relationship has to do with a simple equation: Frame N on the laserdisc = timecode N on the premaster tape. Editing with laserdiscs requires that this relationship be established, and either the editor or the editor's assistant must equate the first frame on the disc to the corresponding frame on the premaster tape. If working in timecode is important to the editor, searching for a shot on the laserdisc by using a timecode value will cause the DVA code for that frame to be automatically invoked. The result is that the specific video frame for that timecode will be displayed.

Disc Copies

Disc copies can be made by optical copying or mechanical copying techniques. In optical copying, once the master laserdisc has been made, it acts like a photographic negative that can be contact printed into a series of positive discs. In essence, light is shined through the master and reaches and exposes the other glass discs. In mechanical copying, a silver stamper is created that is the reverse image of the glass master. This allows laserdisc copies to be made in much the same fashion as vinyl audio recordings and at lower costs than optical copying.

OTHER CHARACTERISTICS OF LASERDISCS

Erasable or Non-Erasable

Laserdiscs can be *write once, read many* (WORM), which means that once material is transferred to the disc, the disc can only be played back. No additional material can be recorded to the disc, and it cannot be erased. It is permanently a playback medium. When the discs are no longer needed, they are either stored or discarded.

If a laserdisc is *write many, read many* (WMRM), it is an erasable medium. Material can be recorded to the disc and edited, and new material can be recorded to the disc when the old material is no longer needed. It should be noted that the recording units are smaller and considerably less expensive than laserdisc mastering machines, and the average cost is approximately $40,000.

Dual-Sided Discs

Dual sided discs are created when single-sided discs are glued together. If a player with a single laser is being used, one would simply flip the disc over to play the other side. In the case of a dual-laser machine, the disc would simply be left to rotate, while the two laser beams seek frames on either side of the disc.

Disc Skipping

Occasionally, humidity can affect laserdisc playback. The laser beam must pierce through a 1/8" layer of plastic on the disc, which is placed there for protection and rigidity of the disc. However, when humidity occurs, one unwanted side effect is that the disc may "fog." As a result, the laser's beam may end up being diffused in this fog, compromising the ability to seek to a specific frame reliably.

Bowling

This phenomenon is akin to the warping of a vinyl record. When plastic-based laserdiscs sit on the player's spindle for extended periods, the discs may bowl. Glass-based discs are used in high-service installations and are heavier and provide more rigidity than their plastic counterparts.

Cost

The cost can vary greatly. To make one disc costs about $1,800 and CAV masters (called *one strikes*) cost anywhere from $200 to $300 each. Making plastic-based discs for nonlinear editing systems averages approximately $75 each.

Laserdiscs and Consumers

We are currently experiencing a resurgence in interest in laserdiscs in the consumer marketplace. Because of their superior picture and sound quality compared to traditional consumer VCRs, viewing a feature film on laserdisc has increased in popularity, especially since the public's experience with compact audio discs has also dramatically increased.

THEORY OF OPERATION OF LASERDISC-BASED NONLINEAR SYSTEMS

The second wave of nonlinear editing systems began in 1984 with the introduction of the EditDroid. The CMX 6000, a descendant of the pioneering CMX 600, appeared in 1986. In 1989, the Epix system was introduced. All of these systems are laserdisc based and nonlinear. They also provide random access, with a minimum access time of 900 milliseconds (ms) to a maximum access time of two seconds.

Clearly, the rapid search capability of laserdisc compared to the videotape-based nonlinear systems are an advantage. However, comparisons among the various nonlinear systems is quite a dangerous issue. Because each wave of nonlinear editing systems has capabilities that the others may not have, it is important to understand why the systems were created and what needs they were designed to address. Ultimately, the editor must choose the proper tool.

Laserdisc nonlinear systems share many of the characteristics

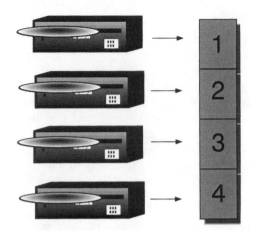

Figure 7–3 The laserdisc-based wave, similar to the videotape-based wave, simulates nonlinear editing by utilizing multiple playback machines loaded with identical source material. Illustration by Jeffrey Krebs.

of the tape-based wave. In the same way that nonlinear editing is simulated through the use of multiple videocassette players, this second wave provides nonlinear random access by having identical material on multiple laserdisc machines, which offer much faster access to the material (Figure 7–3). While systems in this category offer different methods and user interfaces, a general operational description follows.

Transferring Material

Transferring original footage to disc provides 30 minutes of playing time per disc. There can be a number of laserdisc players, and some systems offer up to 12 machines. The amount of footage that can be loaded and be accessible at any one time depends on the number of machines and the method by which the system reads the picture and sound tracks.

How to transfer material is an important issue. When the disc is made, the goal is to fill the disc to its 30-minute capacity. This may not always be possible. If the editor needs ten minutes of footage to edit the sequence, a disc will have to be struck. This is one reason why proper planning, footage designation, and disc creation must be taken into account. The usual method is to group together all the footage necessary for a scene to be edited. Since material is usually shot out of continuity, several premaster tapes usually exist in an in-progress state. When all the selected footage becomes available, the premaster tapes are then transferred to disc. If we didn't group our footage in this way, we wouldn't have all the required elements to cut the scene with the fewest number of discs possible. Instead, we would have footage for the scene spread across many more discs than is necessary.

Storage, Discs, and Disc Players

In the early days of second-wave systems (before 1988), four identical discs were required for every 30 minutes of material. Here's why. Disc machines have one laser, which is used to read the reflective pits. Suppose we want to cut from one sync source to another sync source. Since we want to be able to cut from picture to picture and from sound to sound, we have four sources of information. This will require four discs and four machines. One disc will have the "from" picture. A second disc will have the "to" picture. A third disc will have the "from" audio. A fourth disc will have the "to" audio.

However, there is no technical reason why a laserdisc machine must have only one read laser. The Laser Edit system introduced a dual-headed player that could read both sides of the disc. Other systems such as the CMX 6000 followed this lead. There were two reasons for the design and implementation of dual-headed machines. First, the number of discs needed was reduced to two since the "from" and "to" sources were now on singular discs. Second, having a dual-headed system meant that trafficking problems were less frequently encountered; while one head was playing material, the other head could be seeking different material.

Based on these types of machines, a 12-machine system is capable of three hours of picture and sound since two machines will provide 30 minutes each of picture and sound. The same configuration but with single-headed read lasers would yield one-half the amount of storage; for this reason, the use of single-headed systems has diminished.

Today, laserdisc systems are divided into three categories: those that use WORM discs, which cannot be erased; those that use WMRM laserdiscs, which can be erased; and those that use a hybrid of WORM and WMRM laserdiscs and videotape. Erasable laserdiscs are often rated in terms of thousands of rewrites without quality degradation.

Some laserdisc-based systems combine one player/recorder laserdisc machine using WMRM discs with other player-only machines, using WORM discs. Even though the laserdiscs provide fast access times, there are still complex editing scenarios where the discs will not be able to seek to the material in time. In these situations, like the first wave of systems, the laserdiscs cue to the appropriate sections and a prebuild occurs, resulting in the series of shots laid off to a *slop disc* as it is sometimes called. Once this more complicated series of shots is on the prebuild disc, it can be seamlessly integrated into the normal playback of the sequence, thereby addressing a potential trafficking problem.

A hybrid laserdisc system includes WORM, WMRM, and videotape media. Videotape players and recorders, although they carry the issue of sequential access, do allow the use of material that does not have to be transferred to disc. If an edit is proceeding and some late-arriving material must be used immediately but only exists on videotape, the footage is played directly from the videotape.

TYPICAL SECOND-WAVE EDITING SYSTEM DESIGN

The laserdisc-based and videotape-based waves of editing systems developed simultaneously. By outlining the operation of a currently available laserdisc system, the CMX 6000, many of the traits that systems in this wave share will become evident.

The CMX 6000, introduced in 1986, incorporates the familiar film conventions of film bins, lift bins, and trim bins (Figure 7–4). The basic layout of the system consists of two screens. One screen displays source material and the second screen displays the sequence being created; this is similar to film's feed and take-up reels. After material is transferred to disc, log information is entered by hand. These footage descriptions can vary, but they usually consist of scene and take information.

Once footage has been logged for a specific disc, the footage can be viewed by scrolling to the wanted take via a touch pad. The

Figure 7–4 The CMX 6000 laserdisc-based editing system is shown in the main editing console along with the rack-mounted laserdisc players. Photo by Howard A. Phillips.

material appears in the source monitor and can be played. The film editing model is duplicated in the user interface conventions, which allow the editor to splice or extract material anywhere in the sequence. The concept of the playlist exists, and as a result, multiple versions of a sequence are easily accomplished. By simply adding or removing items from the playlist, a new version is created and can be preserved.

Several nonlinear systems, including the CMX 6000, allow the editor to choose monitor positions. For example, since film is usually edited from the left to right (feed reel on left, take-up reel on right), it may be possible to arrange the source monitor on the left and the record monitor on the right. Depending upon the system, the trimming tools might also emulate the film model, with the tail frame of the outgoing material on the right and the head frame of the incoming material on the left. This may not seem like a very important feature, but it indicates the system's flexibility to be used in different ways according to how the editor wants to work.

Laserdisc editing systems also differ in their methods of displaying optical effects. Some systems utilize an on-board video switcher that performs dissolves; originally, the CMX 6000 offered a common film tool, the Chinagraph marker, to display dissolves. When the system was first introduced, dissolves were displayed as a wipe that moved across the screen. This was basically the same approach that the film editor would have taken by marking the dissolve with a grease pencil. Since film editors were accustomed to seeing transitions displayed in this fashion, the method was familiar and common.

TRAFFICKING

The same trafficking conflicts that arise in the videotape-based systems can surface in the laserdisc-based systems. While additional laserdisc machines can be employed, an additional solution is to prebuild material that cannot be played in real time to the CAV disc through the use of a CAV laserdisc recorder. This CAV disc may be either WORM or WMRM. This prebuilt material is then played back at the appropriate time. These discs often have one-half the capacity of the standard CAV disc, or approximately 15 minutes.

PRE-VISUALIZATION TOOLS

Previsualization is a key element in a mature editing system, and previsualization tools are what separates the basic offline system, the nonlinear system, and the online system. Although we most often think of editing as a process that combines shots in a meaningful manner, editing does not only mean judging and managing how one shot "transitions" to another shot. Editing has evolved into a craft where a variety of different forms of media must be combined into a complete vision. The ability to discover how these disparate elements can be combined before the final assembly phase is reached is enhanced by previsualization tools.

The use of the CAV disc is particularly important because it offers a solution to a perplexing problem. Prior to the incorporation of laserdiscs, nonlinear editing systems had not addressed the ability to play pictures at speeds other than normal play. With CAV discs, editors suddenly had available a variety of motion effects, such as slow motion, fast motion, freeze frames, and reverse motion (called *reverse printing* in film). By adding more of these basic editing features into the nonlinear system, the previsualization capabilities of the system are extended.

Titling abilities are also offered on the laserdisc-based systems. Usually limited to only one type style, the ability to type in characters and automatically center or move them higher or lower in the screen allowed the editor to view where titles would be placed in the program. These titles could then be keyed over the background video and could be moved in and out of the frame at specified durations.

Integrating these, albeit limited, character-generation features provided yet more information to the editor and client about how the final program would be arranged. If the editor knows that titles are going to be required over certain shots, it is most helpful to be able to view these titles as the background pictures are being chosen.

An additional feature that laserdiscs offer is the ability to see film material play back at its native speed. Since laserdiscs have an index that identifies every frame according to its corresponding DVA code, the nonlinear editing system based on laserdiscs may offer a unique playback feature that is not always a standard option: the ability to play back material at 24 fps.

When film is being edited, it is played back at its normal shooting speed of 24 fps. However, in the NTSC television system, when film is transferred to video, as it must be for the videotape- and laserdisc-based waves of editing systems, the videotape plays back at its normal speed of 30 fps. During the film-to-video transfer, film frames are essentially recorded to video more than once.

Since the laserdisc indexing system traces where the additional six frames per second have been placed, these frames can be masked and viewing the frames can be prevented. The result is that if film was shot at 24 fps and transferred to video and then to laserdisc, the material can be viewed at its normal playback speed. This is more acceptable to editors who have spent many years viewing film at a true 24 fps and not at the NTSC video equivalent of 30 fps. Being able to view 24-fps material at 24 fps also means that only video frames with valid and corresponding film frames will be edited. This is particularly important when an EDL from the electronic editing system is used to create a negative cut list.

A

Figure 7–5 The film synchronizer and its software counterpart both allow an editor to view easily the structure of a sequence. Photo by Michael E. Phillips.

GRAPHICAL USER INTERFACE

The use of graphical computer interfaces to create representations of the program being edited began with second-wave systems and, specifically, the EditDroid in 1984. This tradition of incorporating a very visual interface for the user to interact with has been carried on by the digital-based third wave of nonlinear systems.

The essential idea behind the use of computer graphics to represent the program being edited comes from the common film synchronizer. The film synchronizer, acting as a means of keeping the film and magnetic tracks in sync, can also be looked upon as a global view of a sequence being edited. On the nonlinear system, these "software synchronizers" are text representations of the film and magnetic tracks (Figure 7–5).

B

14/1	15/1 B&W		16/1		15/1 B&W	17/3	18/1	20/1		20/1		20,
16/1					16/1							
16/1					16/1							
00:00		01:00:10:00				01:00:20:00			01:00:30:00			

C

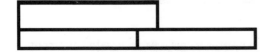

Figure 7–6 By graphically representing a picture overlap, the editor can more quickly determine what was done at a transition without having to interpret the edit decision list.

Recreating the film synchronizer in computer software represents yet another unique tool that the film and videotape editor had not previously been able to enjoy: a graphical representation of the program as it is being created. While the film editor uses the film synchronizer, the ability to see the entire show displayed in this manner is not easily accomplished. Similarly, the videotape editor has no tool that can graphically show the structure of the show from beginning to end. These software synchronizers, or *timelines*, as they have come to be known, show picture and sound tracks.

It is interesting to note that manufacturers of traditional videotape editing systems, both offline and online systems, have recognized that visual tools are important to the editing process. Since 1991, more manufacturers have begun to incorporate visual depictions of the actual structure of a program into the design of their linear editing systems. For example, a picture overlap (split edit) would usually appear as this information to the videotape editor:

TITLE: SEQUENCE 1, SPLIT
FCM: DROP FRAME

| 001 | 051 | V | C | 01:02:49:15 | 01:02:58:06 | 01:00:00:00 | 01:00:08:21 |
| 002 | 051 | AA | C | 01:02:49:15 | 01:02:57:06 | 01:00:00:00 | 01:00:07:21 |

SPLIT: VIDEO DELAY = 00:00:01:00

However, another method would be to combine the usual EDL form with an additional graphical view. When the videotape editor is scrolling through a long lists of edits, rather than trying to decipher the meaning of the edit points themselves to determine which were straight cuts, which were dissolves, and which were split edits, the editor can consult the graphical reference, which is a representation of the type of edit that was done. In Figure 7–6, the bottom section represents a cut being made on the audio track, and the upper section shows the picture extending past the audio cut point. This is a picture overlap.

A dissolve could be represented in a combined text and graphic form (Figure 7–7).

| 003 | 003 | V | C | 00:05:13:07 | 00:05:13:07 | 01:00:42:02 | 01:00:42:02 |
| 003 | 002 | V | D | 03:08:52:00 | 03:09:03:08 | 01:00:42:02 | 01:00:53:10 |

Figure 7–7 A dissolve is represented by a combination of graphics and the more traditional edit list format.

Does representing edit points in this manner make a difference to the editor or help to streamline the editing process? Receiving visual clues from the editing system should allow the visual artist to concentrate better on the creative process and direct less energy toward the management of the technical processes of editing. These clues aid the editor by providing information as to what type of work was done at a particular edit event number. As a result, the editor can devote less time to answering the question, "What was done here?"

GRAPHICAL USER INTERFACE AND AUDIO EDITING

The use of graphics assists the editor who must navigate through the editorial decisions that must be made when creating a program. This can readily be seen when we examine how nonlinear systems that offer graphical user interfaces display audio tracks on-screen for the editor.

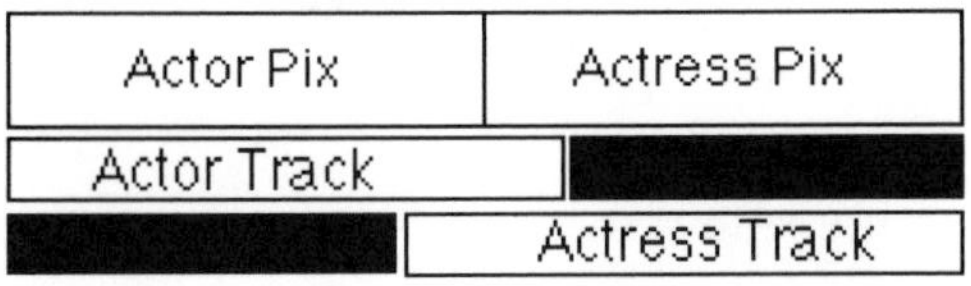

Traditional Film Sound Editing

Let's return to the film editor working with traditional film editing tools. Let's say that a dialogue scene between a man and a woman is being cut. The editor will most likely plan on having the actor's dialogue on one magnetic track and the actress's dialogue on another track. With two tracks of dialogue, the editor can accomplish a sound overlap when she wants the man and woman to be talking at the same time. Structurally, it would appear as in Figure 7–8. The picture cuts from the actor to the actress. While the actor begins to speak, there is corresponding black film leader on the other sound track. While the actor is still talking, the actress begins to talk. This is shown by the overlap where both audio tracks have information drawn in. After the actress has begun to talk, the actor stops talking, and this area is shown with a corresponding amount of black leader.

Let's say that it is important to the rhythm of the scene for the film editor to rehearse some sound effects (sfx). One of the magnetic tracks is removed (the film editor is working on a six-plate editing table: one picture head and two sound heads for a total of six plates), and sound effects, already transferred to magnetic tape, begin to be placed in strategic sections.

For example, when the actor begins to speak, there is a bit of off-screen applause. Later, when the actress is talking, the sound of crashing dishes is heard, and the actress reacts to this sound. If we had the ability to put all these tracks on the system at once, it would appear as in Figure 7–9. Blank leader has been used to allow a sound effects track to be created. Two sound effects have been added: the applause and the crashing dishes. However, since the editor only has two sound tracks, the layout would be as shown in Figure 7–10. The editor has taken down the sound track that has the actress's dialogue. In this way, the editor can see and hear the actor along with the sfx track. Perhaps it may be necessary to adjust the placement of the sound effect slightly. By hearing the actor against the sound effects, these judgments can be made.

By putting the actress's track back up, the editor can judge the placement of the dishes sfx against the actress's dialogue as shown in Figure 7–11. If the dishes sfx must be moved, it can easily be adjusted several frames in either direction until the proper effect is achieved with regard to what the actress is doing

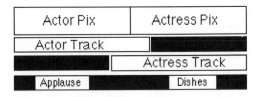

Figure 7–8 A sound overlap occurs where the man and woman talk at the same time.

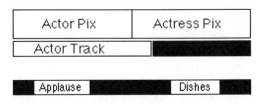

Figure 7–9 Adding a third audio track allows the editor to rehearse sound effects against the actor and actress dialogue tracks.

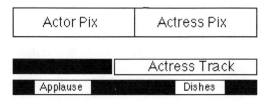

Figure 7–10 By taking down the actress's track, the editor is able to hear both actor and sound effects tracks.

Figure 7–11 By taking down the actor's track and putting up the actress's track, the editor is able to hear both the actress and sound effects tracks.

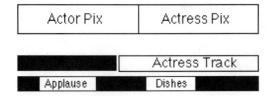

Figure 7–12 After judging that the sound effect of crashing dishes occurs too late in relation to the actress's dialogue, the effect is moved forward.

and what she is saying. If the dishes sound effect occurs a bit too late, it can be moved up. In Figure 7–12, the dishes sfx is moved up and takes place a bit earlier in the whole piece.

This is the method of editing sound on the film flatbed. The editor has complete flexibility in placing audio exactly where it needs to be, and she can cut the audio in increments down to a quarter frame (equivalent in size to one 35mm film perforation). Being able to hear many soundtracks at once usually requires more sound plates. Some editors have loaded additional soundtracks on another flatbed in the same room and started both systems simultaneously to hear more audio tracks!

Second-Wave Audio Editing

There has been a great influx of electronic and digital audio workstation alternatives to the traditional method of sound editing. Note that both tape- and laserdisc-based systems are capable of playing back two audio channels. Although several manufacturers of both waves have also created nonlinear versions of audio editing systems that provide more than two tracks of audio, the systems that combine picture and sound offer only two tracks of audio. Thus, second-wave audio editing is quite similar to film sound editing.

The incorporation of two tracks of sound editing has thus far been adequate. This is changing rapidly as the third wave of nonlinear editing systems has risen in popularity. These systems offer a greater number of tracks. The benefit for the editor is that more audio tracks can be simultaneously available for construction, playback monitoring, and output directly from the editing system. In the previous example, it was necessary to remove certain magnetic tracks since we only had the capability of manipulating two at a time.

In third-wave systems, the editor can work on more audio tracks, and the edit will appear as in Figure 7–9. In this way, the editor can put all three tracks of sound up at one time and, more importantly, can continue to place each element while leaving other elements intact, just as the film editor is capable of doing. This allows the editor to build tracks with the instant benefit of judging how the addition works against the existing tracks.

Audio Playback

One aspect of the laserdisc systems that differs from the tape-based systems is the nature of the audio playback when material is being played at less than normal speed. For example, when the film editor or the videotape editor is trying to locate the beginning of a word, the playback speed is slowed down while the editor searches for the place where the word actually starts. This is called *scrubbing the track*. The film editor physically rocks the mag track back and forth along the sound head and can isolate the point where the word begins. The videotape editor rocks the videotape back and forth over the picture and sound playback heads and can isolate the beginning of the word.

If film is being used or if analog videotape is being used, these scrubbing techniques allow the editor to control the rate at which the audio is heard. As the audio playback is decreased or increased, there is an accompanying change of pitch. If playback is faster than normal, there will be an increase in pitch; if playback is slower than normal, there will be a decrease in pitch. Therefore, pitch can be thought of as the height or depth of a tone (or of sound), depending upon the relative speed of the vibrations that produce the tone or sound.

If the film editor is scrubbing the mag track at less than normal playing speed, the analog vibration that is created produces a pitch that is decreased from the sound's normal pitch profile. For a slow motion image of a person running and talking, one effect often used is to slow down the pitch of the person's speech. The words come out slowly and sluggishly. Similarly, if the scrub process is faster than normal play, there will be an increase in the pitch of the sound. In essence, the audio scrub is accompanied by a pitch change: a benefit film and videotape editors enjoy not as a feature, but simply as a technical characteristic of playing back an analog medium.

Laserdisc playback at slower or faster than normal speed does not recreate the analog vibrations that affect pitch. Instead, the editor receives "sample points" of audio. If the editor is searching rapidly through material on a CAV disc, the laser beam skips from one section to the next. No continuous audio vibration is created from this skipping around. Rather, the editor hears the exact sound at each point at which the laser is currently located. This means that if the editor is trying to find the beginning of a word, the sound he hears is the frame of audio of exactly where the laser is directed. If the laser is moved to a prior frame, that frame of audio will be heard. This frame-by-frame playback of sound does not recreate an audio vibration that will lead to a change in pitch. Thus, the editor hears each frame of audio at its correct pitch.

This is often viewed as being both a feature and a detriment. Hearing only the frame of audio underneath the laser beam removes any doubt as to what frame of audio the editor is "parked on." However, this type of audio referencing takes some time to get used to; the human ear and the craft of film and videotape editing have evolved into hearing sounds in the natural analog representation: with an accompanying change of pitch.

It is worth investigating how a second-wave system handles audio at variable playback. Some systems, when parked on the frame, continue to play that frame until a muting circuit is activated. This *buzzing* as it is sometimes called, can be objectionable. Because pitch change is sometimes important to the editor, some systems offer audio buffering techniques. A portion of audio is read into a memory location, and as the editor scrubs back and forth, not only is the frame of audio at the current position read, but surrounding frames of audio in the memory buffer are also read. These additional samples are used to represent a continuous audio signal. In this way, audio scrub with pitch change is available.

When digital videotape formats such as D1 and D2 first began to appear, they were incapable of buffering some portion of sound. As a result, there were early complaints by editors who were accustomed to being able to control the audio pitch as part of the audio scrubbing process. Forced to listen for audio cueing in a new fashion, some editors found the sampling of audio, rather than the buffering of audio, to be a liability. By 1991, audio buffering techniques for digital tape formats were developed that allowed the editor to return to the more familiar way of locating audio points.

Resolution

It is also important to note that film editing can resolve audio to less than one frame. This means that we can make an audio edit at less than a one-frame resolution. In videotape and on laserdiscs, the smallest amount to which we can edit an audio point is one frame. In the digital nonlinear systems, however, subframe audio editing is possible.

Audio Switching Mechanisms

Additional audio capabilities in the laserdisc-based wave, as well as the videotape-based wave, include using an audio switcher and an audio mixer. The distinguishing factor between them is that the audio switcher is usually on-board, or built into the system and controlled by system software, while the audio mixer is out-board and operated by the editor. The audio switcher is very similar to the video switcher; it has a small series of cross-points between which the system will "transition". These audio switchers are described as N by N. For example, a "two by six" has two buses (from and to) and six cross-points (inputs).

If the editor wants to do a dissolve between two pictures, say between laserdisc players 3 and 5, it may also be desirable to dissolve the audio in the same manner. Audio dissolves are called *crossfades*. In this case, the editing system would program each bus with the appropriate sources. The two on-board switchers, one for video and one for audio, would be set up accordingly, and transitions would be made simultaneously for picture as well as sound.

By having the ability to do audio crossfades during the picture transition, the editor can often get a better idea of how audio crossfades will enhance the picture transition. The flexibility of being able to do a picture dissolve separately from an audio crossfade as well as being able to do an audio crossfade separately from a picture dissolve is also important.

Audio Mixing Capabilities

The first two waves of nonlinear systems provide the ability to do extremely limited and simple mixes between the two tracks of sound that have been used. This mixdown is done manually during the output stage. By controlling the audio mixer by hand, the editor can create a rough mix that represents how a sequence will run when completed.

If the editor has created a scene where two people are talking and wants to run a music track underneath the scene just to give

Actor	Actress	Actor	Actress	Actor
Actor	Actress	Actor	Actress	Actor
Music				

Figure 7–13 A sequence with one track of sync audio and one track of music.

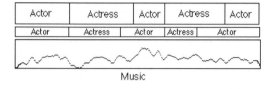

Figure 7–14 Changes made to the volume level of the music track are represented by the waveform. As the sequence progresses, the music level is raised and lowered.

the client an idea of how the scene will work with music, he must be able to do a temporary mixdown (*temp mix*) of the two tracks onto the presentation tape. This temporary version of the mix is intended only to give the listener an idea of how the different sound elements work together.

The sequence shown in Figure 7–13 was edited with monaural sync audio on one audio track and music on another track. Suppose that the editor wants to mix the two sound tracks together while recording the result to a videotape for presentation. Usually, the first- and second-wave editing systems utilize a 3/4" videotape for this purpose. Using this out-board audio mixer, the editor will "ride" the audio levels of each track—left fader being audio channel one, right fader being audio channel two—until the mix is completed. As the sequence is playing back from different videotape machines or laserdisc machines, these faders are manipulated to change the volume of each track dynamically.

If we could represent visually what the editor has done just with the music track and the volume changes achieved by this mixer, we could think of it as shown in Figure 7–14. The editor has started the scene with music down and raises and lowers the volume of the music as the scene progresses. The music is faded down and out at the end of the scene. If the mix is not satisfactory, each system offers different capabilities to redo the mix; some require that it be completely done over, and other systems allow the editor to pick up the mix from just prior to the point of interruption.

Although the audio volume is depicted graphically in Figure 7–14, this feature may not be available in the nonlinear system. Unless the system has the ability to display audio waveforms or unless software has been written to examine the level changes from point to point and then assign a visual representation to these changes, the editor will not have this visual reference. These tools are usually associated with the third wave of systems.

WORK PRODUCTS

When the editor is finished with the session, the work products of the laserdisc- based wave of editing will be a video edit list, perhaps a film matchback list, a printed output of both lists, and a screening copy of the program. When a viewing copy of the program is required, the laserdiscs cue accordingly and play back the sequence, usually to a 3/4" videotape. It is during this stage that the audio can be mixed if so desired. As the various laserdisc machines cue and play back the required source material, the program is assembled onto the 3/4" videotape. With a resolution of approximately 15 kHz, the laserdisc audio is adequate for presentation.

As far as picture quality is concerned, laserdiscs offer better picture quality than VHS tape. Depending upon the type of laserdisc being used and the quality of the pressing, some discs can have defects, in much the same way that videotape has dropouts, or missing pieces of information due to flaked off magnetic particles that are no longer present and therefore cannot record or play back a signal.

With the offline editing process completed on the laserdisc-based system, the next step is the conforming process. The work products from the session are taken to the video online room if a video release is required. If a film release is required, the film will undergo negative cutting. Final audio sweetening will be accomplished either during online or during a dedicated audio sweetening session.

The laserdisc-based wave is being used for a variety of programming types, from commercials to feature films. The majority of programming is long-form work, divided among episodic television shows and feature films.

APPRAISING THE SECOND WAVE

The laserdisc-based wave brought with it new features that were different from the tape-based wave, but more important than the features themselves is the direction that the new functions represented. If a sequence is conceived with many slow-motion effects, these previsualization tools are required on the editing system. Titling capabilities allow the editor to judge better which shots should be chosen as backgrounds. Additional audio features permit a more polished presentation. The laserdiscs afford fast access, and this wave certainly fits into the definition of an electronic, nonlinear, random-access editing system.

The graphical user interface also offers more use of color and shapes to show the structure of the edit in progress. Software representation of the sequence's picture and sound tracks was further developed in this wave. The film and video cutting environments and techniques being recreated and represented by these program timelines also provide a common characteristic between film and video editing.

While the person putting together a presentation may not have a complete idea of how the sequence will eventually turn out when completed, he or she usually has a rough idea of how things may proceed. This layout of the program is done mentally and stored mentally, or it is drawn on paper, showing where cuts may occur and what transitions may be used. With the increased use of graphical interfaces that show the structure of the edit, much of the mental work of keeping track of what shots were used in what sections is handled by the system.

EVOLUTION OF THE SECOND WAVE

What will be the possible evolution of this wave? Certainly, possible strategies include the use of erasable laserdiscs for those systems that still require that discs be pressed. However, one of the common aspects of the evolution of nonlinear editing systems is that what is usually pursued is an improvement of the medium in which the pictures and sounds are stored. The laserdisc technology has not had such improvement; CLV and CAV discs have remained at the same capacities and specifications.

The first and second waves of nonlinear editing systems offer different playback approaches that create a situation where nonlinear editing can proceed. While the first wave uses videotape, the second wave uses laserdiscs. The interfaces and methodologies among the various systems in both waves can be quite different, from the script-based Ediflex to the more evident use of graphical representations of the Epix system.

A major issue that ties the two waves together is that they are intended as offline devices and they serve in the same capacity as offline editing systems: to allow changes to be made during the editing process, to generate an edit list for videotape and film, and to generate a rough cut screening copy for presentation.

The first two waves of nonlinear editing systems also share something quite different from offline videotape systems. It is usually the case that the cost of renting a linear offline videotape room is not as expensive as renting an online room. This is done to encourage using more time to experiment with the cut. Nonlinear offline systems in the first and second wave are usually made available at higher hourly or weekly rates than linear offline systems but usually less than the hourly rate of the online room. The purpose of using a nonlinear system is not to spend more time editing; rather, it is to decrease the amount of time required in the online session without sacrificing the ability to make as many editing changes as necessary to edit the program in the best possible way.

These are important differences, to be sure, but the philosophy of the work products is what influences how the first two waves of systems will evolve. If the intention is to continue to deliver offline work products, improvements to making this a faster process, perhaps with faster playback media, will be sought. However, the normal playback of a videotape is fixed, and the normal playback of a laserdisc is fixed. There is simply not much room for improvement in this critical task. Adding additional player units is an option. New features that increase productivity and allow the editor more options can and should be created.

The decisive factor in choosing to use these systems is that there is no stated intention on the part of the first- or second-wave systems to deliver anything other than offline work products. The final stage of working on these systems is to conform the work in an online environment, whether that is an online video editing room or the negative cut.

Again, choosing the right tool for the program to be edited is the most important decision to make. However, the creation of nonlinear editing systems can be viewed in two ways. The first approach is to conclude that the nonlinear system has some basic set of features but its overall purpose is to serve as an offline system. The system has no further goals than to provide a series of work products to enhance the creative editing process and to streamline the final assembly process.

The second approach, initially viewed as quite a lofty goal, is to gradually build into the system a series of features that rivals the online finishing environment. Depending upon the system and the philosophy of the system's designers, the production and post-production chain can be radically affected by the presence of the digital nonlinear editing system, which has as its ultimate goal a finishing capability. For some types of programs, the offline stage may not require an online stage. This means that the program that is edited on the nonlinear system becomes the final program, and no conform stage is required. The only question as to its acceptability lies in the quality of the finished results from the digital nonlinear system.

This is a possibility, and it is the ultimate goal of the digital-based third wave of nonlinear editing systems. Whether or not this can be achieved based on popular requirements and tastes will remain to be seen. These requirements may stimulate a fourth wave of nonlinear editing systems.

8

Digital-Based Systems

The third and current wave of electronic nonlinear editing comprises digital-based systems. It involves converting analog signals into digital signals and storing these signals on computer disks. The process of analog-to-digital conversion is called *digitization*. The process of reducing the amount of data that represents the original information is called *compression*. Compression is usually, but not always, necessary because the quantity of information in video usually exceeds storage capacities.

Although the compression stage reduces the original quality level of the pictures, a variety of image resolutions are often available on a system; these images are usually more than adequate for making editorial decisions. Once pictures and sounds are in digital form, they can be arranged and rearranged without generation loss and with the same degree of flexibility as film can be reordered, but without the manual labor involved in so doing.

Digital nonlinear editing systems began to appear in 1988 and have caused an unarguable movement away from traditional linear offline techniques. With regard to interest and consistent growth of system use, digital nonlinear has been the hot topic of conversation in the post-production industry as systems came into fruition in 1991 and 1992. Since 1988, over 15 different digital nonlinear systems have been introduced to the marketplace. It has been a remarkably short evolution with regard to the initial acceptance of the digital method of providing nonlinearity in the editing session.

While digital nonlinear editing has proved itself in accomplishing basic editing tasks, a question that often arises is whether digital nonlinear editing will be destined forever to be an offline device. Or, will digital nonlinear editing eventually provide many of the tools heretofore associated with the finishing, or online, suite? This essential point, whether nonlinear editing systems will be able absolutely to provide and replicate the high standards of picture and sound quality associated with the online suite, will continue to be a heated subject of controversy for the 1990's.

Film editing, as we know, is a nonlinear process. We are able to try as many variations on a theme as we need to since we have the ability to physically reorder the strips of film. But film editing is also a destructive process. We are unable to try something without physically cutting the film, splicing it to another section of film, and judging the results. At some point, if we continue to experiment with the same frames of film, we will need to get a duplicate made of that piece of film; it will simply be too marred to be used effectively.

Videotape editing is a nondestructive process. We are able to work with the actual source material without physically damaging it, and since we have the ability to record over material time and again, we are able to preview an edit before actually committing to it. But videotape editing is a linear process. At some point, we must commit to the decision, and we must commit edit by edit. When we do commit, it becomes a time-consuming endeavor to review and make changes.

Digital nonlinear editing methods change the nature of the material being stored and accessed. This is a critical difference in describing the third wave, because the digital editing system is looked upon as being a unique blend of traditional operational methods and advanced digital co-processors, all built upon computer interfaces designed to duplicate the film editor's traditional work place.

"Being digital," it is often said, "is both a blessing and a curse." Traditionally, digital equipment and digital processing techniques were expensive and unproven and meant more work to train and educate staff members. The same concerns apply to digital nonlinear editing systems. However, the relentless move to digital, which is apparent in the professional post-production marketplace, continues to raise the issue of why digital editing systems have caught on so rapidly as people continue to ask: "Why invest in analog directions when the rest of the neighborhood is going digital?"

What does digital afford to the user of the digital nonlinear editing system? Essentially, digital provides greater signal processing capabilities, better search times, and improvements over analog signal loss through successive generations.

HOW DIGITAL NONLINEAR SYSTEMS WORK

General System Objectives

Digital nonlinear systems have often been described as "the word processor for pictures and sounds." Although it has become a cliché to describe this wave in that fashion, it clearly is a deserved analogy. Many word processors have a set of tools loosely termed *cut*, *copy*, and *paste*. They permit the user to easily cut a section of text out of a document and delete it, or to easily copy a section of text from a document and paste it (place it) into a new location. Such moving, refining, and rearranging operations are, after all, at the heart of the editing of images and sounds.

What digital systems sought to do by virtue of their conversion of analog signals into digital data was to offer improvements in random-access. With the picture and sound signals in a digital form, it is possible to have random accessibility to any material in, at the very best, 13 milliseconds (ms) or, at the very worst, 90 ms (1 ms = 1/1000 second).

This obviously represents a considerable time savings over the shuttling of videotapes and is also about ten times faster than the access time of a laserdisc (about 900 ms).

Another objective of the digital wave was to address the issue of late-arriving material that the editor needed to integrate into the program. With the videotape-based systems, integrating such late-arriving material requires that it be duplicated prior to use, and this process can be time consuming. The same dilemma holds true for laserdisc-based systems using WORM drives. With digital systems, material is transferred in real time, and as a result, being able to work with a new piece of footage that is 30 seconds long will take slightly longer than 30 seconds to load into the system.

Also important was to address the issue of erasability. Certainly, videotape is erasable, but with the amount of cueing and shuttling that takes place, it is often the case that tape loads in the videotape-based nonlinear systems cannot be used more than two times. For WORM laserdiscs, little can be accomplished with them after the program has been edited. They can be archived, of course, and if there is ever a reason to go back to the original discs to reedit, the material will be available. For laserdisc-based systems that have incorporated the erasable WMRM discs, multiple uses are available. The magnetic and optical media that are associated with the digital systems are erasable. In the case of magnetic disks, about 40,000 hours of use is a typical rating. Even if a system based on magnetic disks was used for 80 hours a week, 52 weeks a year, it would be approximately 9.6 years before 40,000 hours were reached. Optical disks are rated in excess of 1 million rewrites.

Another issue, misunderstood when digital systems were introduced, was a clear movement away from the traditional use of computers in post-production to control analog devices. Instead, with the digital nonlinear system, the computer is running software that recreates the editing table, the film bin, and the film splicing block. Shown in Figure 8–1 are the general components which will be found in a digital nonlinear editing system.

In addition, the potential for reduced costs in the online stage as a result of a more productive offline stage was another objective.

System Work Flow

Figure 8–2 illustrates the possible work flow paths that can be taken when using a digital nonlinear system. The methods chosen will be based on what end products are required. There are variations, of course, but the five basic paths are as follows:

Figure 8–1 The components of any digital nonlinear editing system include a computer, a monitor that displays the user interface, a set of disk drives for storing material, and a keyboard or similar device for input. A videotape machine is used to load material into the system. Courtesy of Avid Technology, Inc.

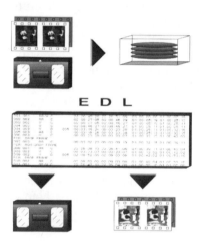

Figure 8–2 This flow of information is possible when using a digital nonlinear editing system. First, the film positive or videotape is transferred to computer disk. Editing proceeds in a nonlinear fashion, resulting in the creation of an edit decision list. The EDL serves as both a video edit decision list and a film cut list. Next, the video EDL is used to conform the original source tapes to create a finished master tape. The film cut list is used to create a finished cut negative. Illustration by Jeffrey Krebs.

Path 1: Originate on Film, Edit Digital Nonlinear, Finish in Linear Online

This method is used when film is the original material, but the end product of the editing process will be videotape. For example, the majority of television commercials are shot on film. They are released and broadcast on videotape. The pathway to and from the digital system is as follows: The film is shot, and the processed film and film sound are transferred to videotape. Next, picture and sound are digitized to computer disk. The editing takes place in the digital system, and the resulting edit list and source videotapes are taken to the linear online suite to conform the finished program. Currently, a major portion of the work being done on digital nonlinear systems follows this path.

Path 2: Originate on Film, Edit Digital Nonlinear, Finish on Film

This method is used when film is shot, and the end product is not a videotape-based release but a film release. For example, for feature films that will be edited in the digital system, there must be a way to return eventually to the original film and to conform it according to the nonlinear edit.

Instead of an edit list that contains the timecodes for the source videotapes, a different list, a negative cut list, is created. It contains the instructions for which frames of the film negative need to be cut and the order in which they should be spliced together. In the case of a theatrical film, when a videotape release is required, the original cut negative is transferred to videotape, and duplicates are then made. This is a developing category of use as more film-originated and film-delivered projects begin to utilize digital nonlinear systems.

Path 3: Originate on Video, Edit Digital Nonlinear, Finish in Linear Online

Here, essentially the same steps are followed as in the first path. The only difference is that the original footage is videotape based, not film based. This method represents a majority of the work being done on digital nonlinear systems.

Path 4: Originate on Film or Video, Edit Digital Nonlinear, Finish in Digital Nonlinear Using Auto-Assemble Option

For situations where the linear online suite can be by-passed, finishing the program can be done on the digital system if it is capable of performing linear auto-assembly. Some systems offer this capability. Here, a traditional three-machine A/B roll assembly is done with a video switcher used to make transitions between the two source machines. All hardware is under the control of the digital system.

This path makes sense if the final product does not require the more extensive capabilities of the online suite. Often, this option is exercised when the user must make a presentation using the picture quality of the original source videotapes and basic switcher effects such as dissolves, wipes, and keys, but not the more sophisticated effects found in the online suite, such as digital effects. It is most likely, however, that the presentation will be reviewed and changes will be made in the digital system. The final step will be to return to linear online to finish the project completely.

For some users, this path can be the final step. It all depends upon the nature of the presentation being made. If a presentation can be completed using the auto-assembly features of the digital system, there is no reason to visit the online suite. However, if the program is more complex than what the auto-assembly process can offer, conforming the program in the online suite is required.

This is a developing category as digital nonlinear users experiment with which programs can be finished within the system and which programs require more sophisticated online.

Path 5: Originate on Film or Video, Edit Digital Nonlinear, Finish in Digital Nonlinear Using Direct Output

This path is the most controversial development surrounding the digital nonlinear systems. After the program has been edited, neither an auto-assembly process nor an online process is used. Instead, picture and sound are taken directly from the computer's disks and recorded to videotape. This tape becomes the show master and is then duplicated for distribution.

The issue, quite simply, is whether the quality of the picture and sound are acceptable enough to justify not returning to the original source videotapes. Again, acceptability depends on a variety of issues: What did the original material look like? What is the ratio of compression? If the ratio is low enough, picture artifacts introduced during the compression stage will be minimized. In addition, what is the life cycle of the program, and how will it be used? If enough of these questions can be answered positively, there is no reason not to exercise the option of taking direct output from the disk drives. The cost savings of this option are clear since, if the direct output can be used as the final product, there is no reason to continue with an online session.

Take, for example, the emerging area of production that is generically referred to as *not for broadcast*. Programs in this broad category range from the two-minute videotape that provides introductory information to new employees to the longer presentations and programs that are done frequently by communications departments in business and educational institutions. Typically, these programs have a rather short life cycle. When the quality is adequate and money can be saved either by avoiding the purchase of additional equipment to be used in an auto-assembly process or by not using an out-of-house facility to finish the program, there will undoubtedly be an increased incidence of these direct-from-disk scenarios.

Using the direct output of a digital system as the finished program splinters the barrier between offline editing and online editing. If a program can be finished in the system that was used to initiate the offline editing process, what label do we assign to the system? Is it offline, or is it online? It is inevitable that improvements to the digitizing and compression stages will ensure that indistinguishable results will be achieved. From the moment that compressed video is no longer apparent, great ramifications will arise for the post-production process. The usual offline to online process will be affected as more users of digital nonlinear editing systems begin to use the direct output option.

PARADIGMS OF THE DIGITAL NONLINEAR EDITING SYSTEM

Four basic concepts can be found in digital nonlinear editing systems. While some of these models evolved from tried and proven concepts in film and videotape editing, others are inventions introduced specifically with the digital-based systems. The four paradigms are the clip, the transition, the sequence, and the timeline. While the actual terms used to describe these models will vary from system to system, the essential meaning of these models remains.

The Clip

When a film editor is making a movie, the shots that are being put together are small pieces of film that are ordered in a specific way. When the film editor has a close-up of an actress that needs to be joined to another piece of film, the close-up will be judged, handled, and spliced into the appropriate location.

As an example, if the editor has a one-second close-up of a character and is working in 16mm film, the piece of film that the editor is holding is about 7" long. To the editor, those 7" represent the one-second shot. As the editor becomes more experienced, initial decisions as to how long a shot should be are often made by virtue of the length of the film strip the editor is holding. Often, when a film editor thinks about how long a shot should be, one image that comes to mind is how much film would be held between outstretched arms. The conversation proceeds as follows:

"How long do you think this shot should be?"

"About that long."

To the film editor, these little pieces of film, whether they measure 7" or 10', represent all the segments of the film that must be joined. Joining these pieces together results in making the movie. A clip is simply a model based on these individual pieces of film, which are joined to make the movie. A clip represents the way that film editors are used to working; a clip starts and ends at a film splice. If a film splice can also be judged as a shot, then a clip represents that shot.

The film editor gets the opportunity to enjoy this simple concept of the clip because film exists in space. The reality of film existing in space rather than in time is a crucial point to grasp. This concept is absolutely essential in understanding why film is a nonlinear editing process while videotape editing is a linear process. Film images, as something that can be touched and judged by their length allow the film editor to make judgments on how long the piece of film should be and, therefore, how long the shot should be held on the screen. These little sections of space can be easily rearranged; they will occupy the same overall space, but the order will be different.

The concept of the clip became the first item in the digital nonlinear tool set and represents terminology that one can find replicated from system to system. When an editor sits down to a digital nonlinear editing system, there will be some reference to the clip, which, to the film editor, simply means that his traditional working methods have been preserved.

The Transition

While the film editor can judge shots based on the space that they occupy, the videotape editor does not have this feedback loop. Our film editor can pick up a piece of film, estimate the appropriate length, and work much closer with, and to, the material. The process of videotape editing proceeds at a greater distance, however, since the editor cannot hold the images in her hands. This is because the process of videotape editing is based not on the clip, but on the transition.

When a videotape editor makes an edit, what is really happening is that a decision is being made about a point in time. While the videotape is being played on the monitors, the editor chooses that point in time, a timecode point, and hits the mark in button on the editing controller. This in mark serves as the location where the previously laid down shot on the videotape master will be joined to this newly chosen point. When the in point has been chosen, the editor usually makes the edit as an open ended edit with no set out point. Next, the recording is made, and when the shot is no longer useful, that is, when the action is over, the editor presses the out button. This action simply means, "Stop. The useful length of this shot has been reached."

When the next edit is made, the editor replays the shot that she has just laid down and judges where the old shot should stop and the new shot should begin. This process, and the thinking that must be done to accomplish this process, is very different from those of the film editor, who picks shots based on the space that they occupy.

The videotape editor is picking shots based on how they "transition" with other shots. She is not picking shots based on the space that each shot occupies. This is because videotape exists in time and not in space. To the editor, videotape images do not exist as images that occupy space. Rather, these pictures exist as images based on the amount of time that they occupy. Therefore, the videotape editor must judge the videotape editing process based on how the shots "transition" to each other. The conversation in the videotape editing room, then, proceeds as follows:

"How long do you think this shot should be?"

"About three seconds."

Videotape editing, based on managing the amount of time that each shot occupies and by virtue of the fact that the shot cannot be held and viewed, must proceed at a greater distance than editing film. Editing videotape means adding or subtracting timecode numbers that realign the transition between two shots. As these numbers are manipulated, the videotape editor achieves the desired transition. The essential point is that these shots are not being judged based on the space that they occupy or the relative length of each shot; they are, instead, being judged based on how the shots will "transition" from one to another. This can be viewed as a distinct or subtle point based on your own experiences, but film and videotape editing are different not because one is nonlinear and another is linear. They are different because of space versus time. Space will always be nonlinear, and time will always be linear.

Consider two shots joined together. The editor wants these shots to "transition" not as a cut, but as a dissolve. The film editor draws a mark across the shots to indicate the amount of space that the dissolve will span over the length of the two shots. The videotape editor types a duration in frames for the dissolve. While the tasks don't sound very different, it is as if the film editor has thought, "This dissolve should happen over this *range of space*," while the videotape editor has thought, "This dissolve should take this *amount of time*." This is a subtle, but quite real, division of thinking.

The concept of the transition is the second item that has found a place in the digital nonlinear tool set. To the editor, transition-based editing tools on the digital nonlinear system consist of a model where shots can be determined by their timecode length and by views that show the transition between the outgoing and incoming material. In this way, the editor's working methods have been preserved.

The Sequence

In film, clips are edited together to create a sequence of shots that forms a scene. This sequence of images becomes the film editor's "cut." In video, transition editing is done, and the videotape editor's "cut" becomes the "master tape." Even though the terms are different, the concept of the sequence represents the editor's completed work.

On the digital nonlinear system, the third item in the tool set is the sequence. When the film editor sits down to the digital system and asks where the cut is, the sequence represents that cut. When the videotape editor sits down to the digital system and asks where the master is, the sequence represents that master tape.

The Timeline

In traditional film and videotape editing, there is no way of viewing the entire span of the project in total. It is not possible or practical to take all the individual clips and the transitions between those clips and spread them out on a table so that they can be viewed.

Before building a house, a blueprint can be viewed and modified. While the house is in construction, the blueprint is consulted at critical points. This blueprint serves as feedback to the builder, but the blueprint cannot be modified; the home must be created based on the rigidity of the blueprint.

In the same way that the builder requires an overall view of the project being constructed, the film and videotape editors can enjoy benefits by having a global view of the program being edited. This is a tool that both film and video have never been able to offer to the editor. To judge the overall pacing of a sequence, the film editor has to run the film over and over, judging the individual length that each shot occupies. The videotape editor must run the tape back and forth to judge the flow from transition to transition.

The timeline seeks to preserve and display, onscreen, the film synchronizer, where the film and magnetic sound tracks are laid out and manipulated. Although the film synchronizer can only be used for one small section of film, the timeline can show an all-encompassing view of the program or a very specific section of the program, in the same fashion that a blueprint can show the overall vision as well as microscopic details. The timeline usually represents pictures in the graphical form of boxes or shows actual images. The audio tracks for the project are drawn on screen. All of this graphical feedback is extremely similar to the blueprint, but while the home cannot veer from the blueprint, the timeline can be easily updated according to what the editor is doing to the sequence. The conversation, then, proceeds as follows:

"How long do you think this shot should be?"

"A little longer than those other shots next to it."

Digital nonlinear systems offer the editor a section on the computer monitor that relates to the timeline. This is a place where the editor can visually locate a clip or a transition and can manipulate the space that the clip occupies or affect the type of transition by changing the profile of the timeline for that section of the program. The essential difference between the editing features that are found in the timeline and editing based on the clip or editing based on the transition is that timeline editing is a direct manipulation of both the components (clips) and the joining points (transitions) in a simultaneous fashion. In this way, the timeline offers something very unique to film and videotape editors: a way to shape a program just as the sculptor shapes the huge mass of clay into something that can be enjoyed.

The timeline is a new tool for the editor. While film editing is deeply entrenched in the concept of the clip, and videotape editing is firmly rooted in the transition, digital nonlinear editing augments these roots with the timeline's simultaneous clip- and transition-based editing tools. Future digital nonlinear systems will make a break from clip- and transition-based editing and offer timeline-based editing in an attempt to create a new method of combining the space versus time elements of film and videotape editing. Timeline-based editing offers that possibility for change.

DIGITIZING AND STORING MATERIAL

What is it like to edit on a digital nonlinear editing system? What considerations come into being when editing moves from the first wave to the second wave and on to the third wave of nonlinear editing? While each digital nonlinear system is different, the available basic work flows, methods of operation, and software tools are fairly consistent among systems.

The informational flow of a digital system begins with the source material. Whether the footage is on film or videotape, the first step in any digital editing system is to transfer these signals to computer disks. Whereas original material was transferred to multiple tape copies in the first wave and to laserdiscs in the second wave, footage must first undergo a transformation for third-wave systems.

We start off with film that has been shot and processed. The film is then transferred to videotape. Either at the time of transfer or later, the sound associated with the film is also transferred to the videotape. The result of this transfer, or *telecine process* as it is called, is a number of videotapes that contain the material to be used during the editing session. The videotape formats that are chosen can be analog or digital.

Clearly, programs do not consist solely of material originated from a film or video camera. Title cards, three-dimensional computer graphics, sound effects from compact discs, graphics from paint systems, and a variety of other sources are used in the creation of what may appear to be even the simplest programs. Incorporating all of these disparate elements into a program requires that they be digitized into the system.

Most digital nonlinear systems require that material that is not based in videotape first be transferred to tape before it is digitized and that the tape has ascending SMPTE timecode. The chief reason for this is that an EDL is one of the work products from the editing session. By first transferring to tape the various items that will be used in the edit, they can be digitized into the system.

The process of converting the analog video and audio signals into digital signals and transferring the material to computer disk before we can actually edit with the material is a real-time procedure. If we have five minutes of material, it will take slightly more than five minutes before we are able to edit with the footage. Once the footage resides on the computer disk, it can be accessed quickly and treated like a word in a document. Think about how rapidly a very powerful computer can spell-check a document or search for a word. This speed allows the editor to very quickly scan through material.

Being in a digital system means that the editor no longer must wait for material to cue or for the system to search to a location. Recall the example of the film editor who could not simply point to a roll of film and go to a specific location within that roll. With a digital system, such a pointing routine is possible. The material instantly cues to that point and is displayed on the screen. The editor can then go about deciding which portion of the footage should be used in the program. While the digital nature of the material now affords distinct access advantages, it is most important to emphasize that the key to the successful use of these systems rests upon the software programming and the tools available to the editor.

When signals are converted from analog to digital, what is lost in the transformation? There are important things to understand about this conversion because just one frame of video takes up an enormous amount of disk space. For a picture displayed on a Macintosh computer at a pixel matrix of 640 horizontal × 480 vertical multiplied by 24 bits of color, one second of NTSC video requires about 30 MB of storage; about 1 MB per frame. How much footage (without audio) will fit on a 600 MB magnetic disk? About 20 seconds.

To provide amounts of storage that are practical, the solution is to compress the video coming into the system. The compromise, of course, is that the original image quality is not available; these compression schemes are termed *lossy* because some amount of the original data for the picture being transformed is lost and irretrievable. Today, most digital systems offer pictures that are more than acceptable for making editorial decisions.

Digitizing the Footage

Loading material into the system is usually a straightforward procedure. In general, a videotape is played back, and the digital nonlinear system is put into record, or digitize, mode. Shown in Figure 8–3 are the three components that are digitized in real time: picture, stereo audio signals, and a timecode reference. Some systems can digitize up to four tracks of audio with picture.

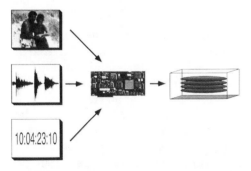

Figure 8–3 Video, stereo audio, and timecode are digitized in real time from the source videotape. The video is compressed, and all signals are stored to either magnetic or optical disc. Illustration by Jeffrey Krebs.

These are processed through processor boards resident in the computer, and the digital data is stored to computer disk. A number of disks can be linked together to provide additional storage capacity.

It is important to know whether the system has built-in tape deck control. This is usually in the form of remote control units that provide the communication link between the computer and the tape machine. Without this link, the tape must be played manually. While this may not seem important at this stage, if any auto-assembly functions are required, in which several tape machines are being directed in concert, this will not be possible without remote deck control capability.

A variety of tape formats can be used with digital systems. Whether the footage exists on U-matic, Betacam, D2, and so on, there should be no difficulty in controlling these machines and processing information from them. Often-asked questions include whether consumer VHS machines can be used and whether machines without SMPTE timecode can be used. Although it varies from system to system, there is no reason why the digitization process should not be successful with VHS material or in the absence of timecode. However, since SMPTE code is a requirement for a meaningful EDL, there will be no usable timecode entry for any edits made using footage without timecode. In addition, if you are working in NTSC, the system should be able to work with both non–drop frame (NDF) as well as drop frame (DF) time code.

Playback Speeds

Although loading material into a digital system should occur at normal playback speed and in real time, the material does not have to reside in digital form at full playback rates. This distinction is important because it has ramifications on the amount of storage available to the user. First, remember that "normal playback" is different in various parts of the world.

Different standards are associated with the normal playback of motion film and motion video. In countries that follow the NTSC (National Television Standards Committee) standards, motion video is normally played back at 30 fps (actually it is 30 fps for non–drop frame timecode and 29.97 fps for drop frame timecode). The scan rate is 525 lines at 60 Hz. The NTSC standard is used in, among other places, the United States, Japan, and Canada.

PAL (phase alternate line) is characterized by a normal playback speed of 25 fps. The scan rate is 625 lines at 50 Hz. PAL is used in Europe and throughout the world. An NTSC tape will not play on a PAL monitor and vice versa.

SECAM (séquential couleur à mémoire), like PAL, has a normal playback of 25 fps with a similar scan rate. There are actually two forms of SECAM; it is primarily used in Eastern Europe and France.

Perhaps the only consistent world standard for playback is 35mm film; the accepted normal playback is at 24 fps. A 35mm motion picture can be made anywhere in the world and can be transported and played on a 35mm film projector anywhere in the world. The original choice for a normal playback speed of 35mm film fluctuated between 16 and 18 fps. The goal was to determine at what speed film could be exposed while preserving the semblance of natural movement. There was also an economic reason to use fewer frames per second: It would save money. Eventually, to ensure that film sound could be adequately reproduced, 24 fps was chosen.

Digital nonlinear editing systems must be specified for use in either NTSC or PAL. While most systems are capable of both NTSC and PAL, some are not, and it is wise to inquire. Although there are three major forms of normal playback, 30, 25, and 24 fps, not all digital editing systems are capable of playing back digitized video files at this speed.

In film, each frame consists of one picture, and a series of these still frames, when projected in rapid succession, provides the appearance of movement. In video, each frame consists of two fields of information. To draw all the pixels of the television signal, the video signal is alternately scanned: First the odd lines of information are drawn, and then the even lines of information are drawn. This sequence of odd, even, odd, even, scan lines are then combined, or interlaced, to present the complete picture. These odd and even lines are called *fields*. When two fields are put together, the result is one video frame. As we know, there can be different playback speeds for video: either 30 frames per second (60 fields) or 25 frames per second (50 fields).

Normally, if we are working on an NTSC system, we expect that the corresponding playback of the digital nonlinear editing system will be 30 frames per second. This may not be the case. If, because of limitations in its architecture, the system cannot play back at normal speed, we may be forced to look at footage at less than normal speed. This is typically some fraction of 30 fps, say 15 fps or 10 fps. The movement is somewhat staccato, and this can be a factor in deciding whether a system can be successfully used.

The image and sound quality chosen affect any digital system's ability to adequately play back the material at normal speed. As the data requirements increase for any frame of picture and sound, it is important to determine if the system will continue to offer playback at normal speed. There is also a cost trade-off in assembling a system that can play back the highest quality pictures and sounds. Some users may opt for a less functional, but more affordable, system.

Systems that can play back at film's native speed began to appear in 1992. A number of advantages are associated with 24 fps playback, including a 20% savings in storage compared to NTSC rates. More important, however, are the related items that must be addressed to make the transition from film to video and back to film again.

Storing to Disk

Loading footage into a system requires a continual process of digitization, compression, and storing the material to disk. How disks are used in the process is indicative of the quality of picture and audio being stored because not all picture and audio qualities will fit equally on different disk types. As a rule, the highest quality images and audio are always associated with magnetic disks rather than optical discs.

Digital systems vary in the number of disks that can be linked together at one time. The usual range is from a minimum of seven disks to a maximum of 28 disks. The greater the number of disks that are simultaneously available, the greater the amount of footage that is available at any one time to the user. The size of each disk drive is also variable, with a range of 500 MB to 2 GB. Obviously, as the disks increase in capacity and as more of them are linked together, an increase in the overall footage capacity of the system is achieved.

Digitizing Parameters

During the process of digitization, the user is presented with a series of choices regarding how the material should be captured. Systems vary in terms of the available choices, but in general, there will be modes for the following (Figure 8–4):

1. Digitizing rate
2. Image quality
3. Audio quality
4. Number of audio channels

Each decision affects not only the quality of the material, to be worked with, but also the available storage capacity of the system. The options offered during or after digitization are necessary because they allow flexibility if the storage capacity begins to become a problem as the disks become full.

Digitizing Rate
Some systems allow the user to specify a capture mode at other than normal play rate. For example, let's say that we have 30 minutes of material to digitize. We only have ten minutes of disk space left. Compromises must be made if we are to load the full 30 minutes. Adjusting the digitizing rate by reducing the number of frames that are captured each second is one solution. In this case, if instead of capturing at 30 fps, we capture at 10 fps, we triple our available storage from ten minutes to 30 minutes and are able to fit the material on the system.

Flexible systems also allow this conversion process to be done after material has been digitized. If there are files on the disk that were originally captured at 30 fps and we need to make space on the disk, we can choose to convert the files and keep every third frame. When files are converted in this fashion, only the video file is affected. If the material has synchronous audio, it is left intact and is not converted; it would not make sense to convert the audio in this fashion. The video file will exhibit a staccato motion as it holds in position for three frames, plays (updates) for one frame, holds for three, plays (updates) for one, etc.

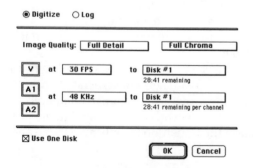

Figure 8–4 The user is presented with a screen that allows frames per second, image quality, audio quality, and number of audio channels to be assigned prior to digitization. Courtesy of Avid Technology, Inc.

There has always been a stigma attached to working at less than normal play speed. Some editors consider it very difficult to work at less than normal speed, while others do not mind at all. It is a very subjective decision. The goal is simply to be able to make the choice if the circumstances require. Certainly, for material where dynamic action is not occurring in the frame and where there is a great deal of footage that must be sorted through, working at less than normal speed could be acceptable.

Image Quality

There are usually a variety of image quality settings from which to choose. These settings are based on the attributes afforded by the compression scheme that the system is employing. Today, image compression in digital nonlinear systems is based on the following methods: software-only methods or hardware-assisted methods, either digital video interactive (DVI) or joint photographic experts group (JPEG).

In general, manufacturers offer a choice of image resolutions designed to offer lower picture quality but maximum storage capacity, a middle resolution with midrange storage capacity, and a high resolution designed to provide the best possible image but with the lowest storage capacity. The calculations of how much information to store and discard for each frame of material are handled by the system's algorithms and are not usually accessible by the user.

Flexible digitizing and compression designs take into account extraneous picture information that can be discarded, thus realizing a benefit in increased storage, and this flexibility extends to choices that the user can make. This availability depends on the compression method used and the digitizing and compression methods that the manufacturer has created in the software.

Some digital nonlinear editing systems offer the ability to choose whether to work in color or black and white. If a picture is compressed and the full amount of color is preserved, the storage requirements will be greater than if the picture is stored as black and white. It is not unusual to have a 15–30% savings if color information is discarded. While it may appear strange to work in black and white in a world of color, it is not at all unusual. Film editors for years have worked with black and white work prints without being hampered.

Audio Quality

A variety of audio samplings are often available. The most common choices are 15 kHz, 22 kHz, 32 kHz, 44.1 kHz, and 48 kHz. When referring to the digital sampling of analog audio signals, it is fair to say that the sample rate should be double the highest frequency that we want to preserve. For example, if the sample is at 44.1 kHz, the highest frequency that will be preserved is 22.05 kHz. The human ear has a range of up to about 20 kHz.

To distinguish them, it is fair to say that a 7.5 kHz audio sample is comparable to AM radio audio, while a 22.05 kHz realized sample (from the 44.1 kHz original sample) is the mastering standard for digital audio compact discs. The audio quality associated with D2 videotape and digital audio tape (DAT) is 48 kHz.

Logically enough, choosing higher quality audio decreases the amount of storage available. The storage requirements for even one frame of full-bandwidth red-green-blue (RGB) video is 921,600 bytes, while one complete second of 48 kHz audio is 96,000 bytes. The disparity in storage requirements provides some idea as to why there is usually not much concern about the amount of additional storage that higher quality audio requires. It also provides some insight as to why so many digital audio workstations are available.

However, a much more interesting observation must be made with regard to why so many audio resolutions are available. Professional-grade digital audio on a digital nonlinear editing system may appear to be logical enough, but early systems introduced in 1988 did not have such capability. To distinguish digital nonlinear systems that serve in a strictly offline capacity, certain manufacturers chose to offer one item usually not found in the traditional linear offline edit suite: high-quality digital audio.

By offering high-quality audio capabilities, it is possible to decrease the amount of duplicative work that is done from the offline to the online process. Instead of giving cursory attention to audio as the program is undergoing its initial creation stages, the ability to work with high-quality audio means that audio joins the EDL, screening copy, and auto-assembly features as yet another work product that is realized from the editing system. Specifically, the high-quality audio files can be brought to the online session, or the audio can be output from the digital nonlinear system to the final master tape. This avoids having to duplicate the audio work during the online session.

There can be confusion regarding the actual nature of the audio entering and exiting the digital nonlinear system. Is the audio digital or analog? We know that the audio is digital once it is resident in the system. However, whether the system can input and output digital audio depends upon the audio interface between the nonlinear system and the video or audio device serving as the playback source.

For example, if we have a D2 videotape and we want to preserve the digital integrity of the audio, we first choose to digitize the audio at 48 kHz, which is its usual sample rate. However, while the D2 machine can provide digital outputs for the two tracks of audio, the nonlinear system may not have audio hardware that can process audio digitally. That is, it may only have an analog-to-digital (A/D) convertor. If so, there is only one way to bring audio into the system: in analog form.

Once the audio is in the system, it exists in digital form. However, the audio output capability is limited by the type of audio convertor. If only a digital-to-analog (D/A) convertor is available, this digital audio will become analog as it exits the system. Thus, we started with digital audio on the D2, it exited the D2 not as digital but as analog, it was converted to digital as it entered the nonlinear system, it left the system as analog, and the audio was recorded back to the D2 as digital. The goal is to avoid such a convoluted pathway because a loss of signal quality results.

If preserving the integrity of the digital audio signal is of concern, audio interfaces to the digital nonlinear system should provide AES/EBU (American Engineering Society/European Broadcasting Union) inputs and outputs. Two connectors are usually found: one for input and one for output. Each connector carries both the left and right channels. Direct digital interfaces are commonly available.

It is also important to realize that devices that are capable of playing back digital audio may not be capable of outputting a digital audio signal. Take, for example, the audio compact disc. The signals on this medium are digital, but most compact disc players only provide analog audio outputs. It makes no difference if the medium is digital if the conduit to other devices is an analog output. A professional-grade player must be used that provides digital outputs. Only then will the digital integrity of the signal be preserved when transferring the audio from compact disc to the nonlinear system. Ideally, the digital nonlinear system should be capable of preserving the digital integrity of the audio signal by providing digital inputs and outputs.

Number of Audio Channels

Most systems are capable of simultaneously digitizing two channels of sound. However, some are capable of digitizing as many as four channels of sound at one time. As most productions involve just two tracks of sound on the original material to be digitized, four channels of sound are rarely required during digitization. One developing aspect is the production use of more than two tracks of sound during the shooting stage. With videotape formats that offer four tracks of sound, we have begun to see additional microphone setups and recording being done on audio channels 3 and 4. Obviously, as more audio channels are digitized, overall storage capacity will decrease.

Storage

At the conclusion of the digitization and compression stage, material has been transferred and changed from an analog state to a digital state. The pictures and sounds to be used in the editing process now reside on computer disks. How much footage can be stored at any one time? The answer is not as straightforward as one would expect. Storage capability varies from system to system and is dependent upon the choices that are made regarding seven variables. Depending upon the system, and in special circumstances, there will be an eighth variable.

1. Image quality
2. Digitizing rate (e.g., 30, 15, 10, 5, 1 fps)
3. Audio quality
4. Number of audio channels
5. Capacity of each disk

6. Total number of disks available
7. Media portability
8. Fixed or variable frame size

In tape-based nonlinear systems, the amount of tape available on each cassette is fixed. In general, 4.5 hours are available. With laserdisc systems, each disc holds 30 minutes and a number of machines may be available. In general, a three-hour system is common. There are no choices to be made with regard to changing picture quality or audio quality. Therefore, the amount of overall time available cannot be extended in the first wave since tapes with more capacity per cassette are not used. (Although six-hour tapes are available, they are typically not used due to the thinness of the tape and the rigorous shuttling that occurs and that could lead to tape damage.) For disc systems, the only possible method of increasing storage capacity is to add additional machines since the disc itself cannot be extended to offer more capacity.

With digital systems, however, there are many variables, and most of these variables, in the form of choices the user can make, will have significant effects on the total amount of storage available. The computer disks can be filled to capacity slowly or quickly depending on the nature of these choices. However, the ability to influence the capacity of a system is a feature that should be placed in the user's hands.

The ability to take control over the amount of storage available from moment to moment also represents a clear and decisive change from the two previous nonlinear waves. With digital systems, capacity can be changed as we decide that one section requires high image and audio quality and another section can be digitized at lower image and audio quality. Admittedly, reducing the quality of picture and sound is not the ultimate solution; we always want the best possible quality for both. Improvement in compression technology and disk capacity will work in conjunction to achieve this goal. However, being able to increase the amount of material stored is important.

When investigating a digital system, it is important to keep these eight variables in mind as most specification sheets have a range of storage numbers based on some of these different conditions. If we read that a system offers a seven-hour capacity, we must ask about those eight variables. Otherwise, we may find that the seven hours refers to files that are video only at five frames per second!

Media portability refers to whether the computer disks used to store the material are fixed or can be removed from the disk drive mechanism. In the case of optical discs, portability is easily achieved. We can edit a project based on footage stored on optical discs, and when another project needs to be edited, the optical discs for the first project are removed, and the second project's discs are inserted into the drives.

Fixed magnetic disks do not offer this same portability. Faced with the same scenario, the material resident on the disks would have to be erased, freeing up storage, and the new material would have to be recorded to disk. With little time between projects, this may be difficult to schedule. If magnetic disks must be used, the solution is to employ removable magnetic disks, which are easily portable like optical discs.

It should be noted that the most critical variable of all is the image quality. The image quality choices that are offered to the user are based on calculations that the manufacturer has made, and they can be fixed calculations or variable calculations. It is important to note that if we are about to digitize footage and we have one minute available on a disk drive, we may or may not achieve that one minute of capacity. We may find, in fact, that we have stored less than one minute or that we were able to store more than one minute. This will depend upon the complexity or simplicity of the series of moving pictures. A simple series of images may allow us to store more than the one minute estimate, while a complex series of images may require more storage and provide us with less than one minute of material.

There may or may not be that eighth variable in the digitizing process. Can the system being used distinguish between complex and simple material and adjust the compression process accordingly (fixed or variable frame sizes), thereby making the best use of storage space? If so, storage statements from the manufacturer will be estimates and should best be taken with the advice that is often given to a new car buyer: "Your actual mileage may vary!"

Equally important is the ability to erase material from the system and to free up storage space. It should be possible to erase any video or audio file easily and to gain back that storage space. If we have digitized take 3 which is 35 seconds long, and we find that take 4 is a better take and is roughly the same length, we can erase take 3 from the system, momentarily gain back the 35 seconds of storage, and then digitize take 4. This process of erasing, freeing up storage, and digitizing new or alternative material should not be easily dismissed as it may be done quite often during an involved edit session for which a great deal of footage has been shot.

Storage in Early Digital Nonlinear Systems

During the first two years of digital nonlinear, 1988 and 1989, a serious stigma was attached to these systems. At that time, digital nonlinear was described as being most appropriate for editing television commercials. Much of this has to do with three aspects of television commercials: Most tend to include a fast pace of cutting, there is usually a small amount of selected footage from which the commercial will be edited, and the images tend to concentrate on single items in the frame rather than on a great deal of action taking place in the frame. There is also usually a shorter post-production schedule for commercials than, say, a one-hour dramatic television program.

These characteristics of commercials usually were dealt with quite successfully by digital nonlinear editing systems. Because the environment is digital, fast cutting scenarios are facilitated, and there are fewer trafficking problems than with the first two waves. Because large amounts of footage are not required, the number of computer disks can be kept to a minimum. Last, simple images tend to require less storage per frame as the images are digitized and compressed.

For these reasons, during those first two years, digital nonlinear made its reputation with the editing of commercials. This changed considerably during 1991 and 1992, as digital nonlinear began to be used for a wide variety of programming: commercials, documentaries, television shows, and feature films. Improvements made to compression technology and disk capacity during 1991 were the leading reason for the change.

How Image Complexity Affects Storage Requirements

We have identified image quality as one of the eight variables that affects storage capacity. The eighth variable has to do with whether the system is capable of dynamically adjusting how much information is being allocated to the picture frames as they are digitized and compressed.

If the system is capable of using a variable amount of storage as the action in the frame decreases or increases, it is very important to understand how what your footage looks like will affect how much storage you will have. For these purposes, let's identify three categories of programming: commercials, episodics (one-hour television shows and theatrical films), and news gathering (also includes sports and documentary footage).

Tests have indicated that the type of programming has a great deal of influence on the storage numbers that can be achieved. This factor will make more sense after a more in-depth discussion on how digital video compression works. For now, consider the test results discussed below.

Commercial Footage

In television commercials, detailed foregrounds are usual, and little if any action occurs in the background since the intention is to direct the viewer's attention to one simple visual, usually placed in the center of the screen (Figure 8–5). In addition, there is usually very little simultaneous action from more than one figure on the screen. Camera movement varies, but it is usually static, and when movement is present, camera moves are often very dynamic and short in duration. Art direction obviously plays an important role in commercials, and vibrant color schemes are common.

For the test, a 500 MB disk, was filled to its formatted capacity (approximately 470 MB) with original footage. In Table 8–1, the kB/frame column refers to the number of kilobytes required for each frame. Video and two channels of 22 kHz sound were captured. Initially, the disk default for the material was 30 minutes, 57 seconds. The amount captured was 41 minutes, 53 seconds. *Disk default* refers to the average amount of information that can be stored based on the digitizing parameters chosen and the capacity of the computer disk.

In this case, the nature of the action occurring within the frame

Figure 8–5 Commercial footage typically consists of one principal item of interest and a nondescript background, which serves to center attention on the item being advertised.

Table 8–1

Test 1: Commercial Footage

Digitize Segment	Duration	kB/Frame
Segment 1	6:06:10	4937
Segment 2	4:08:01	4308
Segment 3	9:14:13	4872
Segment 4	7:14:13	4880
Segment 5	5:00:00	5030
Segment 6	3:00:00	4783
Segment 7	2:59:17	5120
Segment 8	3:14:13	4847
Segment 9	56:10	N/A
Total	41:53:17	

Figure 8–6 Scenes that are longer in duration and that involve dialogue passages are representative of episodic footage.

Table 8–2

Test 2: Episodic Footage

Digitize Segment	Duration	kB/Frame
Segment 1	20:00:00	6993
Segment 2	5:00:00	5899
Segment 3	4:59:24	6259
Segment 4	2:30:00	7136
Segment 5	5:00	7876
Total	32:34:24	

Figure 8–7 News footage is often associated with dynamic action in both the foreground and the background with unplanned and unpredictable camera movement.

was not as complex as the video compression algorithms assumed by the system. When less complex pictures presented themselves, the system needed less space to store the frames; as a result, more material could be loaded than the disk default indicated. If we examine the kB/frame column, we can see how the action occurring within each segment changed the amount of kilobytes required to store that segment. These examples ranged from 4,308 bytes to 5,120 bytes for each frame stored. Lower kB/frame values indicate more storage; higher kB/frame values indicate less storage.

Episodic Footage

Medium and wide shots tend to be used more often in episodics than in commercials (Figure 8–6). There is a modicum of camera movement, and when movement is present, it is usually slow. Instead of a lot of camera action, frames are usually static, and actors enter, read dialogue, and exit. These longer sections are usually balanced with shorter, more action-driven scenes.

For the test, a 500 MB disk was filled to its formatted capacity (approximately 470 MB) with original footage. Video and two channels of 22 kHz sound were captured. Initially, the disk default for the material was 30 minutes, 57 seconds. The amount captured was 32 minutes, 34 seconds (Table 8–2).

Although more material could be stored than the disk default indicated, the episodic footage required much more storage than did the commercial footage. In the first test, the total storage capacity was 41:53. The episodic footage, being more complex in terms of the action occurring within the frame, required a greater amount of data to be stored per frame. The kB/frame column indicates this, and we can see that a larger number of kilobytes were required to store this type of footage than were required for the commercial footage. As a result, we achieved less overall storage even though no other parameters changed in the experiment. These examples ranged from 5,899 bytes to 7,876 bytes for each frame stored.

News Gathering Footage

In news gathering, sports, and documentary footage, the action occurring in the frame is often of an unpredictable nature. Action may take place in the foreground and also in the background. There may be dynamic camera movement (Figure 8–7). There may also be movement on several planes of action: a building on fire in the background, water shooting out of a fire hose in the midground, and fire fighters in the foreground of the frame. Such complexity will certainly affect the amount of storage required to preserve the detail in these frames.

For the test, a 500 MB disk was filled to its formatted capacity (approximately 470 MB) with original footage. Video and two channels of 22 kHz sound were captured. Initially, the disk default for the material was 30 minutes, 57 seconds. The amount captured was 29 minutes, 47 seconds (Table 8–3).

In this example, less material was stored than the disk default indicated. When we examine the kB/frame column, we see the largest amount of data required per frame compared to the other

Table 8–3

Test 3: News Gathering Footage

Digitize Segment	Duration	kB/Frame
Segment 1	8:27:06	7428
Segment 2	4:21:05	7616
Segment 3	4:18:05	7620
Segment 4	8:19:06	7426
Segment 5	4:00:00	7502
Segment 6	22:00	6696
Total	29:47:22	

two types of footage. These examples ranged from 6,696 bytes to 7,620 bytes for each frame stored. This results in a decreased capacity of about 12 minutes compared to the commercial footage. As a result, the last segment could not be stored in its entirety.

Not all digital nonlinear editing systems take into account this dynamic quality of footage. If the system cannot adjust instantly to a picture's simplicity or complexity, storage will be used unnecessarily, or portions of an image will not be drawn intact. The type of footage for which a digital system will be used is yet another factor in determining how to judge that system and how you should approach the configuration and operation of the system.

While it may appear to the user as an uncomplicated procedure, the digitization stage itself and the hardware and software that direct the effort are extremely crucial. Successfully digitizing video, audio, and timecode necessitates that all elements work together, from the digitizing interfaces (which convert the analog picture and sound to digital) to the timecode reader (either internal to the system or in the videotape machine) to the ability of the computer disks to sustain the storage of video and audio second after second.

THE USER INTERFACE

The manner in which footage is displayed and can be searched so that the right piece is found and the tools that are provided to mold some larger amount of footage into a concise presentation, whether it be a 30-second commercial or a two-hour feature film, all fall into the domain of the user interface (UI). What elements of film and videotape editing do manufacturers of digital nonlinear editing systems hope to preserve? What are the tools that both film and video editors use that should be present in the digital nonlinear environment? These are some of the questions that point directly to a well-designed system that should offer an easy solution to the common necessities of editing.

A machine can be more powerful than another, a computer can be faster than another, and a computer software program can offer better features than another, but if the door into that power, speed, and depth of features requires too much effort on the part of the user to open, the door will remain closed (and the system will remain in a corner, unused). There are more than a few owners of post-production facilities who will attest to this!

Digital nonlinear systems must take into account the operability of the system. Whether the user is a film editor, a videotape editor, a producer, a director, or a writer makes little difference. If the interface and design are truly intuitive, it matters little if one has actual training in the craft of editing. The system should be capable of easy learning and operation for individuals who are not classically trained as editors. At the same time, the system must retain the familiarity with classic film and videotape editing paradigms.

To some, this may appear to be a blasphemous assertion. However, the rapidly evolving digital nonlinear culture, as opposed to the film culture or the videotape culture, will surely mean that a greater number of individuals will begin to use these digital machines. There will be good and indifferent results. Putting technology into the hands of individuals who were never inclined to learn or master the technical details of film or videotape editing will allow some of them to exercise their creativity. For others, the ability to put together a presentation will be satisfying enough. The results may not be professional, but they will be adequate for that user's requirement.

For the professional, artistry is enhanced as housekeeping activities are eliminated. For the online editor who must be concerned with the technical details of signal management, color frame indications, and generation loss if a change is required, the ability to put aside these details and concentrate only on the cut can be felt as a burden being lifted from the shoulders.

Will the professional film and editing community survive? Yes, of course these disciplines will survive. Nonlinear editing systems will only remove technological impediments. Professional skill and creativity will still be required to produce professional work. It is inevitable that the user interface will become the most important aspect in assessing any system. When the competitive war of image quality is over and systems provide similar picture quality and similar storage capacities, the one overriding factor will be the tools and the method of operation that the system affords. How well manufacturers take this into account and plan for the invisibility of hardware and the visibility of software will signify whether they will remain in business.

The big words in computers since the late 198's, regardless of what they are used for, have been *graphical* and *intuitive*. Judging a digital nonlinear system's overall interface and how it accomplishes its operations must include an evaluation of the system's intuitiveness. For example, faced with an editing operation, is the usual method that the editor would use something that the system designers took into consideration? Or, rather, is the manner in which the system achieves the operation not as logical in its approach, causing the user to wonder, "Why does it work that way?"

A system's interface must be viewed as much more than the manner in which buttons are displayed on a computer screen. When someone is heard to say, "I like that system; it's got a great interface," what the person often means is that the system is not only presented in an easy-to-comprehend manner visually on screen, but also that it operates in a sensible fashion. It is important to consider and evaluate any system's interface from a complete perspective. Interfaces, whether they are text-based or graphical in nature, refer to the complete working methods of the system and how intuitive their operations are. The true test will always be if the system can be operated without referring to any documentation.

If the system is truly well designed, the editor won't even have cause to remark that "it makes sense that it works that way."

Rather, that operation will just go unnoticed and accepted. These questions have nothing to do with the features of one system over those of another; instead, they are concerned with the intuitiveness of the system in achieving and offering its entire range of functions to the user. This is the true interface that must be judged.

GENERAL DESIGN OF DIGITAL NONLINEAR EDITING SYSTEMS

Although each digital nonlinear editing system is unique, three areas are usually represented in some fashion on the computer monitor of almost every system:

1. An area where footage is displayed
2. An area where the main editing tasks are accomplished
3. An area where a graphical representation of the edited sequence is shown

As discussed above, the analog to the film editor's canvas bin is an electronic "bin" where footage is usually displayed. The images shown are only pictorial representations of the actual video and audio files that are stored on the computer disks. These shots can be thought of as "pointers" to the original footage. In Figure 8–8, the representations of the original footage are shown in their bin display form. These shots point to the digitized video and audio files that now exist as digital data on the disk. When the user indicates that a shot should be played back, the system associates that shot with one or more digital files on the disk (combination of video and audio tracks), and the user sees and hears the shot playing back from the disk.

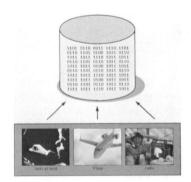

Figure 8–8 Digitized representations of the original images point to data files on the computer disk. Illustration by Rob Gonsalves.

There are usually two designs to represent the actual area where editing takes place. The first approach is to create one main editing screen, which serves as the destination monitor or where the program being edited is displayed (Figure 8–9). This convention is in keeping with the idea of a film flatbed, where one screen is available, and the film feed reel and take-up reel take their position on either end of the screen.

The second approach is to display two monitors: one that serves as the source display and one that serves as the edited program display (Figure 8–10). This convention is in keeping with the idea of a source machine and a record machine in videotape editing.

Which approach is better? It is impossible to answer this question fairly; neither approach is better or worse than the other. The way to approach the question is to judge which display makes sense based on one's background and work experience. One important thing to remember is that these display portions of the interface are created by software. Some systems allow the user to access portions of the software to, in effect, create a particular editing environment. For example, depending upon how extensively the manufacturer allows the user to create this customized "editing room," a one-display editing room can become a two-display editing room.

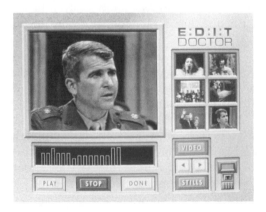

Figure 8–9 A single-display editing interface. The screen image is an example of DVI-based compression. Courtesy of Intel Corporation.

Figure 8–10 A double-display editing interface: the EMC2 digital nonlinear system. The screen image is an example of JPEG compression. Courtesy of Editing Machines Corporation.

This is the real power of software-intensive systems: More and more items can be left up to the discretion of the user. A key point of the third wave of nonlinear systems is that more of the traditional offline and online tools will become resident in the software environment of the digital system. The software-intensive systems will be less reliant on dedicated hardware and a fixed method of user operations.

Digital nonlinear, like its counterparts in the first two waves, must take into consideration the environment in which film and videotape editors currently operate. Whether the project being edited is a commercial, a music video, or a feature film, basic tools are needed.

Representing the Footage

Once all the selected footage (or "selects") has been loaded into the computer disks, editing can proceed. Since more material is always shot than can be used in the final program, there is usually no reason to load all the original footage into the system. Instead, selects are chosen. The material to digitize and the material to leave out of the system is selected by the editor, and these decisions are usually influenced by notes made about the take during shooting and comments made by the director.

When all of the footage shot for the production must be available in digitized form, the editor must decide about the types of storage devices used. Since magnetic disks are more expensive than optical discs, the decision will ultimately be influenced by budgetary constraints.

The display of footage can be done in a variety of ways. Typically, a "window" serves to emulate the film editor's canvas bin (Figure 8–11). A window can be thought of as a graphic box within the computer screen. This "electronic bin" is really no different from the canvas bin; it serves to store material and allows the editor to order the material in the bin. Usually, multiple bins can be open, and this is just as if several canvas bins were brought into the film cutting room and were being used to

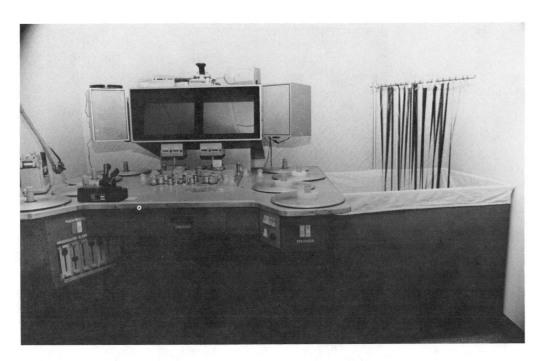

Figure 8–11 The film editor's bin is made of canvas, and individual strips of film are hung into it until they are required. The concept of the film bin is preserved by many digital nonlinear systems. Photo by Michael E. Phillips.

hang strips of film. The film editor has random access to any strip in any bin; digital systems offer the same flexibility.

Since this window becomes the editor's view into the footage, the manner in which the organization and presentation of the material is achieved is extremely important. Editing is a craft dependent upon not what footage to include, but on what footage to exclude! If the footage is easy to identify and locate, less time will be spent looking for the right shot for a section of the program.

Digital nonlinear systems usually have extensive logging and database features. If the editor does not want to look at all of the shots in the electronic bin, but instead wants the system to display only shots for a specific scene, these conditions can be entered, and the database can be searched. Only the specific shots relating to that scene will be displayed.

As befits an icon-based user interface, footage that resides in the system usually can be represented (symbolically labeled) by titles or by an individual representative frame. If the editor is accustomed to working from log sheets and is in the habit of referring to shots by scene and take numbers, this method of titling the shots can be used. Another method is to give the shots more descriptive names, such as "Good Take," "Don't Use," or "Close Up, Reporter Speaks."

If, on the other hand, a visual way of working is desired, the shots can be viewed in a visual layout. Once these individual frames are displayed, their order can be rearranged. This is known as *storyboarding*. Storyboarding has become a basic method of preediting in which individual shots are arranged and rearranged. The purpose of storyboarding is to obtain a visual impression of how shots will work together in sequence. The benefit of

NAME	Description	Script Lines
Music	Stereo	
regular slate	-	
Speedometer	Good Take	
2 guys rear angle	Not Great	"I didn't find the money..."
2 guys talking	Nice	"I didn't find the money..."
13/2	Beach	
13/2 B&W	Beach	
14/1	Boom in Shot	
15/1 B&W	Statue	
16/1	Best Take	"Went on vacation for a week..."
17/3	-	
18/1	Eyes / Keeper	
19/1	Logo Draw	
19/2	Logo Expands	
20/1	Eagle	
20/2		

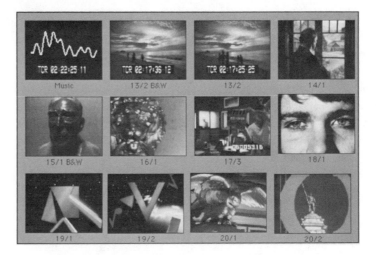

Figure 8–12 Individual shots can be displayed with text descriptions or in a visual format. The drawn storyboard, which served as a guide during the shooting stage, can be recreated on the computer screen by rearranging the shots.

storyboarding is that, often, a clear path through the footage can be achieved without making the first edit (Figure 8–12).

Take, for example, the sequence or commercial that is being edited using a storyboard as a guide. In much the same manner as an advertising agency presents a concept for a campaign to a client through the use of storyboards, digitized shots can be arranged to pose "what if" questions. This can be beneficial to ensure, even before actually editing anything, that the flow of action will be correct and will be similar to what the director or advertising agency intended. In this way, the editor can look at the arrangement of shots and try variations: "What if we put this shot here and that one there? It wasn't shot with that order in mind, but it's a better approach."

The Editing Interface

Once footage is displayed or arranged in the manner in which the editor is accustomed to working, the next stage is to begin editing. How best to recreate the film editing table or the computerized videotape editing system in software is a task with which manufacturers of digital nonlinear editing systems must constantly struggle (Figure 8–13). There are a limitless number of

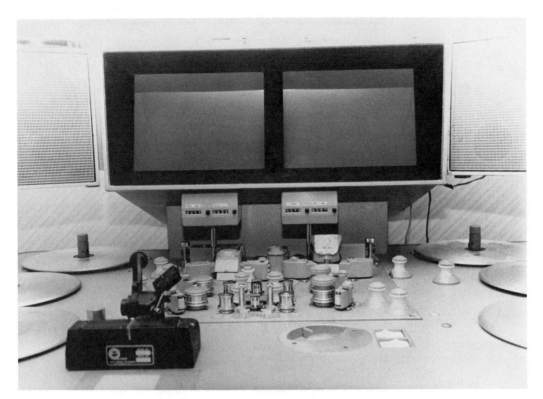

Figure 8–13 A standard editing environment for the film editor features a Steenbeck editing table. Photo by Michael E. Phillips.

alternatives. Should the editor look at one screen, or do two screens make more sense? Does the system take more of a film approach or a videotape approach to editing? Is it a combination of both crafts? What improvements over these methods can the system offer? What tools does the editor use constantly? How should they work? And so on.

Digital nonlinear systems are characterized by their interface (Figure 8–14), and a computer's interface is typically divided among two types. The first is text-based: the directions by which the user commands the computer to do something are all written.

Figure 8–14 An editing interface display of DVision, a digital nonlinear editing system. Copyright © 1991 by TouchVision. Photo courtesy of Intel Corporation.

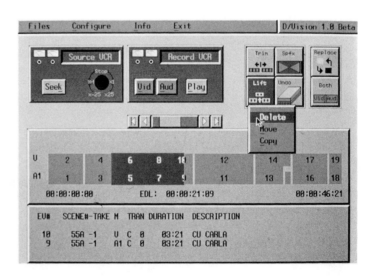

Something is typed, and something happens as a result. The second type utilizes graphical user interfaces, which are characterized by pictures and graphical icons that signify operations that do not require text-based commands to be accomplished.

An example of a very questionable user interface would be an electronic painting system that requires that the user type in the word ERASE to delete a picture element. A more successful graphical user interface implementation would simply be to provide the user with an eraser symbol that can be manipulated on the screen and that would be used to erase the unwanted sections of the picture.

Often, digital systems include graphical icons that emulate a common tool (Figure 8–15). For example, the film editor's take-up reel may be shown on screen as a spool that rotates with a strip of film hanging from it, or the videotape editor's show master may be drawn on the screen as a videocassette. Tools that represent editing operations are represented on screen by symbols. An example would be using an audio speaker symbol to represent a tool that would be used to monitor sound.

SAMPLE PROCEDURE FOR DIGITAL NONLINEAR EDITING

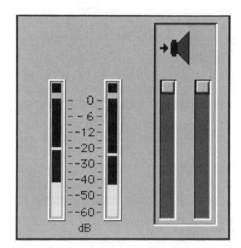

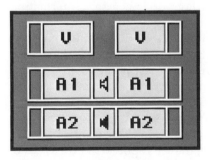

Figure 8–15 Familiar editing tools can be represented in software. These graphical icons represent some of the common icons available on digital nonlinear editing systems. Courtesy of Avid Technology, Inc.

Editing on a digital nonlinear system is often described as a combination of the nonlinear aspects of film editing, the repeatability of computer-controlled timecode videotape editing, and the ease of changing the order of a sentence when using a word processor. In keeping with this description, digital nonlinear editing systems usually include a combination of the work methods associated with both film and videotape editing and borrow heavily from common word processing routines, such as the ability to move large sections of material from one location to another. As a result, both film and videotape editors usually are able to find familiar work methods and conventions available in the same system.

What is the editing process like on a digital nonlinear editing system? To get a better understanding of the common tools that are available, let's consider how a digital system would be used for a complex editing scenario. For this representative procedure, let's assume that we will start with film footage, edit on a digital nonlinear system, and finish on videotape and film. This method is used when film is shot and the goal of the digital nonlinear edit will be several work products, leading to the two desired end results: a finished program ready for videotape distribution and a finished program ready for film distribution.

The pathway to and from the digital nonlinear editing system is as follows: First, film is shot, and the processed film and film sound are transferred to videotape. Next, picture and sound are digitized to computer disk. The editing takes place in the digital system, and the resulting edit list and source videotapes are taken to the linear online suite to conform the finished program. During the digitizing stage, specific information regarding the relationship of the film frame to the video frame was captured.

This information is used to relate the original film to the work that was done in the nonlinear system. The result is a negative cut list, which is used to cut the original film negative to facilitate a simultaneous release on film.

Step 1: Transferring Material from Film to Videotape

Since we are interested eventually in being able to create a finished video and film program, the first step is to transfer the selected takes from the film negative to videotape. This transfer is a one-lite or best-lite transfer. This is the most inexpensive type of film transfer; minimal set-up time is required to provide a viewable picture.

By transferring directly from negative to tape, the step of printing a film work print of all selected takes can be avoided. This can result in a considerable economic savings. If a lower budget feature film shoots at a 9:1 ratio (18 hours of shot film to two hours of finished film), approximately 100,000 feet of film will be exposed. Depending upon the telecine facility, perhaps as much as 25¢ per foot of film can be saved by not making a film work print. Multiplied by 100,000 feet, a savings of $25,000 can be realized.

However, not cutting a work print is not entirely realistic. Images typically play differently from the smaller screens of the digital editing systems to the large screens of the movie theatres. As a result, edits may need to be revised based on how they play on the big screen. Therefore, the decision is made to cut a work print. However, the work print will not be ordered until the digital nonlinear session is completed. By waiting for the majority of editing to be completed on the digital system, the footage of the work print ordered will be considerably less than if all selected takes were printed.

During the telecine transfer, the following items are addressed. First, original audio recordings, either from 1/4" audio tape or DAT, are synchronized to the appropriate picture. Second, if automatic film logging hardware and software are resident on the telecine machine, relationships among the film frame, video timecode, and audio timecode will be captured. These relationships are important and will provide necessary information when the time approaches for the creation of a negative cut list.

The lab report from the telecine session will include some type of print-out of these relationships. If the lab is not automated, there will be a sheet of hand-written notes. While these are notes acceptable, error can occur. The preferred method is to use some type of automatic film logging software. A variety of manufacturers provide both hardware and software for these purposes; popular systems include Evertz, Pogle, and Time Logic Control.

If the lab uses these systems, the lab report will be a series of computer-printed sheets that have all the necessary information to create the negative cut list. Additional developments include the creation of files that can be stored on computer floppy disks. These files contain the same information as the computer print-outs, but are more convenient because the data can be transferred from one system to another without involving any additional manual data entry.

The videotapes from the one-lite telecine session will not be used for the online session because the color quality is marginal. The sole purpose of these tapes is for direct input into the digital nonlinear system. One note should be made about the film-to-tape process. There are several reasons why it would be desirable to bypass the tape stage completely and transfer directly from the telecine to the digital nonlinear system. The major reason would be to achieve further cost savings and to avoid the step of digitizing from videotape to digital disk. Instead, during telecine, footage would be digitized to the computer disks, and editing could proceed directly after the telecine session.

Step 2: Digitizing the Material

The next step is to transfer material from videotape to the digital system. This is a real-time process, and there are several goals. First, there should be compatibility between the floppy disk created at telecine and the nonlinear system. This is a great advantage because it means that the telecine files can be read by the nonlinear system. By transferring files in this manner, the stage of creating the footage database is well underway (Figure 8–16).

Automatically creating this database also provides the digital nonlinear system with all the information required to digitize the material from the videotapes automatically. Failing the automatic file transfer, the material would be logged manually from either the telecine report or from scanning the videotapes for timecode start and end points. Clearly, the time savings can be significant under such an automatic procedure.

Next, the digitizing parameters must be chosen. The type of image quality and audio resolution that will be used depends on the various items that affect the digitizing process. How much material has been shot, and how much must be accessible at any one time? If 100,000 feet, or 18 hours, of circled film takes have been shot, it may be possible to have all the material digitized and available at one time, depending upon the image quality chosen and audio resolution and number of audio channels.

Alternatively, if it is decided that editing will only proceed

Figure 8–16 Information about each film segment can be captured automatically during the telecine session. This data is then transferred to the database of the digital nonlinear system.

	NAME	Tracks	Start	End	Edge	Tape1
	16/1	VA1A2	00:12:38:25	00:13:16:16	KJ289027-5073+00	001
	17/3	V	00:12:43:01	00:13:15:10	KJ289027-5254+00	001
	18/2	V	00:15:40:23	00:15:49:27	KJ289027-5296+00	001
	18/3	V	00:18:08:16	00:18:11:16	KJ289027-5340+00	001
	20/2	V	00:19:57:11	00:20:03:23	KJ289027-5414+00	001
	20/3	V	00:20:23:08	00:20:30:08	KJ289027-5456+00	001
	13/2	V	02:17:24:25	02:17:29:15	KJ289027-5612+00	002
	13/3	V	02:17:35:01	02:17:41:05	KB390897-5265+00	002
	14/2	V	02:22:25:11	02:22:29:25	KB390897-5386+00	002
	19/1	V	15:24:35:08	15:24:40:16	KJ220591-5077+00	015
	19/2	V	17:00:40:00	17:00:48:11	KJ220591-5219+00	017
	21/1	V	18:08:49:16	18:08:57:06	KJ220591-5297+00	018
	21/2	V	18:09:09:05	18:09:11:24	KJ220591-5391+00	018
	21A/1	V	18:34:21:26	18:34:24:09	KJ220591-5447+00	018
	21A/2	V	18:34:27:01	18:34:31:02	KJ220591-5540+00	018

with the best possible image and sound quality, simultaneous access to all the footage may not be possible. This is only due to the fact that the cost of magnetic disk storage may be cost prohibitive and that there will be some limitation on the overall amount of disk drives that can be linked together. Several scenarios can be employed regarding footage.

First, if we are able to choose digitizing parameters that offer sufficient picture and sound quality for our needs, we may be able to fit all the selected material onto the system's computer disks. If so, we can proceed directly with the editing process.

However, if we are unable to fit all the selected footage on the system because we need to work at better image and audio qualities or because of a limitation on the number of disk drives that can be online simultaneously, there are additional options. Most likely, we will choose to work in sections or acts. We may be able to fit on disk all the selected material for a series of scenes that make up one act or one film reel, representing approximately ten minutes of finished material. If so, we would digitize the selected material for these scenes and proceed on an act-by-act basis.

Another alternative would be to work at lower than normal play rates, such as 15 frames per second, thereby doubling storage if working in the NTSC domain. While this may not be appropriate for all projects, it is an option.

Either removable magnetic disks or optical discs will enable footage to be easily removed in the same manner that videocassette loads from the tape-based wave of nonlinear systems are removed and replaced with different loads.

In the digital nonlinear environment, there has been a great deal of concern that all the circled footage will not be available to the editor due to the expense of maintaining digitized material on computer disks. This is alleviated somewhat by the less-expensive optical discs, although they are clearly more expensive than videocassettes. Again, each nonlinear wave has different methods of storing pictures and sounds. Any editing situation that has a substantial amount of footage would be approached on a reel-by-reel basis anyway, especially if editing occurs during the shooting stage.

An additional tool that may be available as a feature of the nonlinear system is the selective deletion of digitized material. If an act has been edited, it may be possible to delete the material that was digitized for the act but not actually used. This option usually includes an ancillary choice of including some additional material before and after a cut point. These "handles" allow the editor to make slight trimming adjustments. The ongoing process of deletion and digitization of new material is easily achieved.

The general rule during the digitizing process is that all material that is digitized, whether it is from the telecined videotapes or is later-arriving material such as temporary music, sound effects, and wild tracks, should have some form of identification. This will almost always be in the form of SMPTE timecode. If the goal is to generate either a video EDL or a film cut list from the materials used during the editing process, these materials must have some reference that can be used for identification at a later time. The user can decide if material should have this reference. If the editor must try some temp music against a scene and does not care if the EDL will provide accurate events for the music section, no timecode is necessary.

Depending upon the system used, there may be a separate digitizing station that does not offer any editing features but serves as a mechanism to create the disks that will be utilized by the main editing system. This helps to streamline the process of footage preparation.

For those cases where the digitized picture and sound quality are acceptable enough for finished results, edit lists will not matter, and there can be less reliance on valid timecode or references to the original source materials. As image quality continues to improve with better video compression methods, there will be many applications where finished results from the nonlinear editing system are acceptable as a final product. When the goal is to leave the system with the finished program, it will make little difference whether input material has valid timecode.

Step 3: Editing

At some point, the editor will have reviewed the footage about to be edited. Most often, the editor will have watched the video-cassette dailies and or made notes while reviewing the footage as it was digitized. Once the footage is resident on the system, the editor has a variety of tools available for editing, depending upon the feature set of the system chosen.

The interfaces of the digital nonlinear editing systems vary considerably. It would not be practical to discuss each and every feature that can be found in the digital editing systems. Instead, a general overview is provided. There will be one or two computer screens. Where two screens are available, each screen serves a specific purpose; one screen displays footage in text or pictorial representations, while the second screen displays the main editing window, where the actual editing is done. Thus, one screen is a footage display screen, and the other represents the editing work space. There may be an optional NTSC or PAL monitor, which is used to refer to the digitized picture as a full-screen display.

Editing Tools

One common method of working is to treat the main editing window as a virtual cutting room that comprises two sections: a source window and a record window, or a film feed reel and a film take-up reel. The basic concept remains the same: Transferring selected material from the source side to the record side creates an edited sequence.

Many concepts of film and video editing techniques are preserved. Tools are provided for the tasks of *splicing*, in which material can be added anywhere within the sequence, and *extracting*, in which material can be removed from anywhere in the sequence. Input devices include a keyboard, mouse, trackball, and dedicated key pad or controller knob.

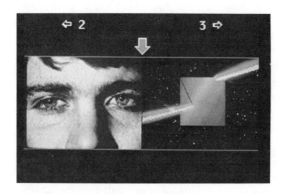

Figure 8–17 Depending upon the trim mode capability of the nonlinear edit system, it may be possible to see both the tail of the outgoing shot (left) and the head of the incoming shot (right). Courtesy of Avid Technology, Inc.

Trimming Edits

The manner in which shots are trimmed actually represents something of an improvement over the film and videotape models. When a cut is made on film or videotape, the editor must run the material back and forth over the cut to get an idea of how the edit point should be trimmed. Being able to see both the outgoing and incoming material displayed adjacent to one another on a digital nonlinear system is a valuable feature. In this way, the editor is able to see both components of the transition simultaneously.

In Figure 8–17, the editor is able to see the exact frames of the edit point. The left side shows the tail frame of the outgoing shot, and the right side shows the head frame of the incoming shot. Being able to see both frames of the transition is especially useful in match action cutting, where it is important for action in the outgoing frame to match action in the incoming frame. By adding or removing frames to either side of the trim mode, the edit is changed.

Additional tools may include script integration, in which the script is displayed on the computer monitor and allowing the editor to point to a certain section of the script so that the computer can display all shots relating to that section.

Timeline Views

As editing proceeds, a graphical representation of the edited sequence may be available. These views, usually called *timelines*, display picture and audio tracks (Figure 8–18). Shot descriptions can be viewed, and there may also be audio waveform displays as well as sophisticated waveform-based manipulation and editing. The timeline also serves as a means of identifying which particular frame or portion of a segment is being played back. A moving indicator is usually drawn onscreen.

Timeline views can be extremely helpful in judging the overall pace of a sequence and seeing, in a representative form, how much time is allotted for a series of shots. Timeline displays are unique to digital nonlinear systems, although manufacturers have begun to include timelines on many different types of editing systems, both linear and nonlinear, of the three waves.

A further benefit of a timeline display is the ability to take macroscopic or microscopic views of the program. Shown in Figure 8–19 is a 30-second sequence. Note the very dense section indicated by the dark band on the top row toward the end of the sequence. Looking at the sequence in this view provides little information regarding what shots are included in that dense section.

By expanding the timeline, it is possible to look at the very

Figure 8–18 The timeline is a unique contribution of the nonlinear editing system. Here, a sequence is depicted in graphical and pictorial form. Courtesy of Avid Technology, Inc.

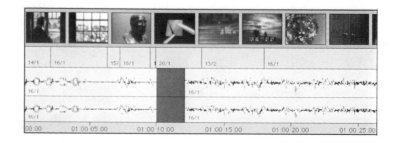

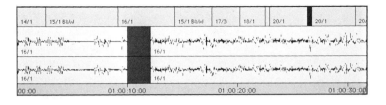

Figure 8–19 In this 30-second sequence, a dense section is evidenced by the dark band on the top row toward the end of the sequence.

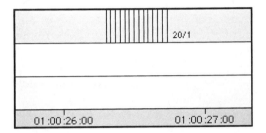

Figure 8–20 When the timeline is expanded, it is apparent that the dense section is composed of many edits.

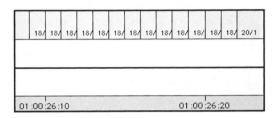

Figure 8–21 Further expanding the timeline reveals that the dense section is actually a series of edits that are one frame in duration.

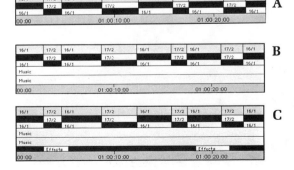

Figure 8–22 A timeline display of virtual audio tracks as well as the picture track.

dense section and determine exactly what shots were used (Figure 8–20). By doing this, it is easy to see that the dense section is actually comprised of many edits that are one frame in duration (Figure 8–21). This ability to magnify a sequence can be very helpful, especially when fast cutting scenarios are used or animations are being done.

Timeline views of the editing structure do not really have counterparts in the film or videotape worlds. This is a true contribution of the computing environment. The only way to see the structure of the edited sequence in film is to run the film back and forth in order to judge the relative length of the different film segments. Many film editors actually listen for the film splices as the film moves through the projector. The cadence of the splices can be useful in determining the cutting rhythm. Determining the structure of an edit in videotape means either running the tape back and forth or reviewing the edit list to judge the relative durations of the individual shots.

Virtual Audio Tracks

Digital nonlinear editing systems may offer more than the standard two channels of audio and the standard two tracks of audio. For our purposes, *channels* is used to indicate the number of audio voices that can be simultaneously monitored. *Tracks* refers to the number of audio tracks that can be edited onto as the program is being edited.

As editing continues, it may be necessary to edit additional tracks of sound. If we are using two tracks of audio to edit overlapping sync dialogue and we want to place stereo music and one track of sound effects within the scene, we will need three additional audio tracks to edit onto. Virtual audio tracks can be created that then become part of the sequence being created. *Virtual* means that the tracks are not channels, that is, there will almost always be more tracks than there are channels on a system.

Shown in Figure 8–22a is the sequence with audio tracks 1 and 2. We add stereo music to the sequence, and tracks 3 and 4 are displayed on the timeline shown in Figure 8–22b. Next, we add a fifth track for sound effects, and the resulting timeline display is shown in Figure 8–22c. Systems vary in the amount of virtual tracks offered, usually between eight and 24 tracks. Simultaneous channels of sound, the number of tracks that can be monitored at one time, will range from combinations of two, four, or eight.

Optical Effects

In film, effects such as dissolves are not seen during the editing stage until the optical effect is created by the lab and returned to the editing room. In video, dissolves are easily viewed through the use of a video switcher. If, while editing, an optical effect is needed, a variety of effects are available. In general, the user should have access to cuts, dissolves, fades, and keys. Keys are in the form of graphic files or title cards that can be superimposed over background images.

Depending upon the digital nonlinear system being used, optical effects can be shown in real time in much the same way that a video switcher operates, or they will involve precompute time, where the A side material is combined with the B side material to create the optical effect. Precompute time depends on the computer processing power, the number of pixels that must be combined, and the length of the optical effect. These precomputed effects then reside on the computer disks in the same fashion as the original footage (Figure 8–23).

When a film editor specifies how long a dissolve should be, his interface is the film itself and a Chinagraph marker. The film editor makes a dissolve mark over the frames that should be dissolved together. The videotape editor must choose a number of frames for the dissolve duration. The digital nonlinear system offers a variety of interfaces regarding the creation of optical effects.

In the example shown in Figure 8–24, the editor chooses a type of transition between two shots. She is given a choice of duration (videotape concept) and a choice of how the transition should be positioned, for example, centered around the cut point (film concept). This interface design combines both the film and videotape concepts for how optical effects are created.

Multiple Versions

As with all nonlinear editing systems, multiple versions can be easily made and stored. Changes can be made at any time and to any version, independent of other versions.

Step 4: Output

The fourth step is to create several work products. The first work product is the video EDL for the program. A variety of EDL

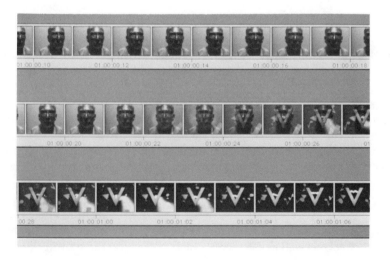

Figure 8–23 Optical effects such as dissolves can be created directly on the digital nonlinear editing system without the use of an external video switcher. Here, frames of the statue are dissolving into the graphic animation.

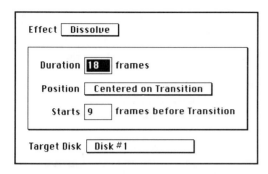

Figure 8–24 An example of the interface for creating optical effects on the digital nonlinear editing system.

formats should be available. The most common formats are Ampex, CMX, Grass Valley Group, and Sony. Because all the editing done in the nonlinear system is virtual, the edit list will usually be a "clean" EDL with no overrecords or extraneous information. Depending upon the system, EDLs will be available on DOS, Macintosh, or RT-11 floppy disks. The RT-11 format is used for CMX and Grass Valley Group editing systems.

The second work product is a screening copy. The screening copy is a direct-from-disk version of the program. Here, picture and sound are played directly from the nonlinear system and recorded to videotape. The ability to accept the picture resolution of a screening copy will influence whether sections of the program can be judged based on these screening copies or whether an auto-assembly of the program is required.

The third work product is an auto-assembly. Here, the nonlinear editing system acts as an edit controller. One or two source videotape machines, one record machine, and an optional external switcher can be controlled as the program is built from the original source tapes to a master tape. The purpose of the auto-assembly feature is to provide the best possible presentation or finished videotape.

Step 5: Transferring to Film

Recall that there are two objectives of our editing session: to create both a video release and a film release. The first work product, the video EDL, is used to determine what material was actually used in the program. A work print is made from only the sections of film that will be used in the final program. This transfer is done because the goal is to conform and project the work print. The film cut list for the work print is yet another work product of the nonlinear system. Using the correlations between the film frames and the videotape frames, a film cut list is created that provides the necessary instructions for how the work print should be conformed to agree with the video EDL. Usually, the audio track that was generated for the video program is used during this screening stage.

Once the film work print is conformed, it will be projected. Any changes that have to be made will now be made, either by keeping the project in a film edit or by recalling the sections to be changed in the digital nonlinear system, reediting them, and creating a new work print cut list. This cut list is now called a *change list*, since it only lists the changes made to the original work print cut list. The work print is conformed again and projected. This process continues until the picture editing process is completed.

Step 6: Cutting the Negative

As part of the negative conform process, the digital nonlinear editing system should provide comprehensive work products as they relate to the film finishing process. Optical effects lists and additional lists that the negative cutter requires should be available from the digital editing system.

Negative cutters most often work with two work products: the negative cut list and a videotape copy of the auto-assembled show. The videotape copy includes burn-in titles showing video timecode and the edge numbers of the film frames. If the negative cutter specifically requires this videotape, an auto-assembly must be done at the time of editing.

Of course, the eventual goal is to use the direct output of the digitized images with system-created burn-ins of video timecode and film edge numbers. In this way, the entire time and expense devoted to the auto-assembly stage will be removed. When direct-from-disk pictures are acceptable to the production team, they will usually be acceptable for passing on to the negative cutter. Until that time, the auto-assembled videotape copy will continue to be the requested item.

Step 7: Producing the Finished Products

Once the answer print stage is complete and the color grading and sound mix have been approved, the release print stage occurs. It results in the film duplications needed for distribution. At this time, the film can also be transferred to videotape for foreign distribution.

Conclusion

While the steps outlined in this scenario can vary greatly based on the system used and the editor's preferred methods, these steps represent accepted working methods. The goal in using digital nonlinear editing systems—indeed, in using any nonlinear editing system—should be to enjoy greater creative freedom. If money can be saved as a result of better working routines, this is an added benefit. In moving from any editing system to another, it is often necessary to discuss four reasons for making the switch: increased creativity, a time savings, an even trade in terms of costs incurred, or a cost savings.

This chapter has been concerned with a description of the third and current wave of electronic nonlinear editing systems: the digital-based wave. Digital nonlinear is at the beginning stages of its evolution and has already rapidly captured the attention of the post-production marketplace. The most critical argument of debate regarding the third wave of nonlinear editing systems is that the pictures are not the quality of the original.

The answer to this problem relies on more efficient compression schemes and hardware. Manufacturers of digital editing systems are promising that full-resolution digitizing and compression will eventually be achievable.

TALKING WITH EDITORS

How Did Editors React to the First Appearance of Digital Nonlinear Editing Systems, and What Predictions Do They Make for the Future of Nonlinear Editing?

Tony Black, ACE, Washington, DC:

While I saw and was impressed with the early nonlinear systems, they were never practical enough, due to their expense, for me to use until the personal computer–based systems came onto the scene. Systems like these are moving toward full video and audio transparency. In other words, whatever quality footage you play into the system you get an exact, no-loss copy of it. We already have that audio capability and are moving closer to the video equivalent. Storage capacities keep increasing and digital systems will dominate.

Basil Pappas, editor, New York:

I saw my first digital nonlinear system in 1989. Having been through videotape and laserdisc approaches, I felt that digital nonlinear systems offered a way of providing nonlinear editing in a compact system which offered a small up-front cost to the producer. Erasable media was a key, and this, along with the speed of being digital, made me feel that digital was the way to go. These systems have moved along at an incredible rate, and the totally flexible way of working and the graphical and visual interface were something that I had never been able to enjoy before.

Paul Dougherty, editor, New York:

It was love at first sight. Offline editing was always a step down in using good tools. Now, with digital nonlinear editing, even when doing a less complex project, I'm getting a kind of craft satisfaction that is unprecedented. There's less drudge work and more fun to be had with this complex and capable tool. With an eye toward the future, the only alternative to digital nonlinear is to go straight to online. Nonlinear is the way editing was meant to be. Once you've done it on a good system (good interface), there's no going back. It's the only way to deal with revisions.

Peter Cohen, editor, Los Angeles:

I experienced the original EditDroid and the CMX 6000. With the introduction of all digital nonlinear offline, I felt that the cutting edge of technology had found new horizons. Some systems go to too much of an extreme in being electronic flatbeds. There are many valid concepts in video editing, why not use them? The digital wave will most definitely dominate. As video compression continues to get better, so does the acceptability of the output of the digital system. The digital systems have the best chance of becoming the "one box," a nonlinear, online graphics, effects, paintbox, and audio sweetening machine.

Alan Miller, editor, New York:

The initial reaction to digital nonlinear editing was acceptance by both groups with a slightly greater acceptance from traditional film editors, mostly those doing commercials. System limitations restricted certain programming forms, such as long form work, and it's only since 1990 that the ability to cut all forms of programming has become practical. In the future, edit rooms will essentially be workstations. Editors will have the capability to do as much as they want to do.

9

Editing on a Digital Nonlinear System

Editing on a digital nonlinear system is characterized by a certain work flow as you go through the process of creating a presentation. In the previous chapter, the major components of this work pattern were outlined. To review, it begins with the input stage, in which the picture and sound elements are transferred to computer disks. Next, the editing stage allows complete flexibility in terms of easily placing picture and sound elements in different order and duration. Last, the output stage provides several editorial work products.

Although we have described the processes involved in each of these three stages of the digital nonlinear system, it is very useful to outline a representative edit session. Understanding what it is like to edit on a digital nonlinear editing system requires an examination of how we can approach building a sequence. In so doing, we can obtain a better notion of the decision-making process and the creative flexibility that digital nonlinear editing provides.

THE INPUT STAGE

A very common computer term is *garbage in, garbage out*, which refers to the inability of a computer to make sense of information that is nonsensical. If we feed the computer nonsense, we may get a highly stratified, highly structured return on our input, but nevertheless, the data will still be nonsense. The same can be said for any presentation being created; if the shots we have available to us are not appropriate to the spirit of the piece, we cannot come away with the best possible results for that program.

Making sense of the digital nonlinear universe into which you will be throwing yourself means being conscious of the material that you will transfer into the computer disks. Although the following thoughts have been stated earlier, they bear repeating now. It is very important that you begin with all the potentially useful footage that you can possibly load into the system. Remember, selected footage can be deleted at any time and replaced by alternative footage. This process of "going for tonnage" with regard to footage allows you to enter the more creative process of winnowing down the footage, differentiating the secondary footage from the primary footage. Once through these stages, the process of honing can begin.

THE EDITING STAGE

Once picture and sound are resident on the computer disks, the editing stage begins. It may be helpful to arrange the various shots that will be used in the sequence. Recall that storyboarding these shots is easily accomplished by displaying footage in a representative frame mode (see Figure 8–12) and rearranging the shots until they are in their intended order. At the same time, the database capabilities of the nonlinear system can be accessed to display footage relating to a certain scene. Now is the time to take advantage of these database and storyboard tools.

What quickly becomes apparent as we begin to edit in a digital nonlinear system is the reliance on the clip, the splice, the transition, and the timeline. We use these models as we create our sequence.

Shown in Figure 9–1 is a representative digital nonlinear editing interface. The upper portion of the screen is composed of a source window (left) and a record window (right). This upper section is where the majority of the editing process occurs. In the center of the screen is a track panel, where video and audio selections are made on an edit-by-edit basis. Below is the timeline. Nothing has yet been drawn in the timeline, but a graphical view of the structure of our sequence will be shown here when we begin editing. Also shown is a clip, which has been loaded into the source window. This shot of a horse has been selected from a bin. The record window is blank, signifying that we have not yet begun to edit our sequence.

When we edit our sequence, we do so by moving material from the source window to the record window in much the same way

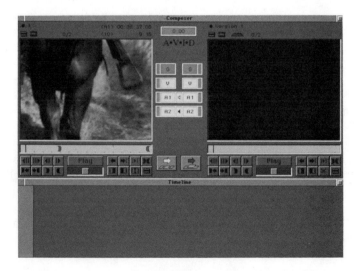

Figure 9–1 A representative editing interface of a digital nonlinear system. Courtesy of Avid Technology, Inc.

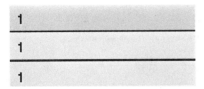

Figure 9–2 After making the first splice, video is shown as the top band with two audio tracks drawn below.

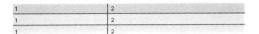

Figure 9–3 After making another splice, the sequence now consists of two shots, with clip 2 longer than clip 1.

that the film editor moves material from the feed reel to the take-up reel. As we edit, the structure of our sequence is reflected in the timeline. Let's edit a sequence that illustrates the basic concepts of a digital nonlinear edit session.

Splicing Shots Together

We begin by choosing the desired in and out points of the horse clip, and we make a splice of video and two audio channels (VA1A2). Shown in Figure 9–2 is the view that is drawn in the timeline after we make the splice. Video is shown as the top band, and two bands, one each for A1 and A2, are shown below. For purposes of this exercise, the horse clip has been labelled 1. Note that the timeline preserves the name given to this clip.

Next, we choose in and out points for another clip, and we splice this clip after the previous clip of the horse. Shown in Figure 9–3 are the two splices that we have made. Note that the clip 2 is longer than clip 1. Without having to refer to any film footage or timecode numbers, we can see the relative length of these clips.

We choose in and out points for a third clip and splice this after clip 2. Shown in the Figure 9–4 are the three clips in our sequence. Clip 3 is approximately the same length as clip 2. Note also that each clip was spliced into our sequence with video and two channels of audio.

Each of the shots that we used in the sequence was accessed from a bin, and we were able to edit quickly with the material. We did not need to change source tapes or film rolls, and the shuttling time was dramatically reduced. As a result of this decreased time in "getting to the shot," we had more time to figure out how to use the shot and what portions of the shot should be used.

Even though we have three shots in our sequence, we have yet to do any nonlinear editing! We have instead proceeded in a very traditional, linear method of editing, in which we chose points and appended them to the last frame of the previous shot. However, suppose that we now want to shorten the duration of a particular shot in our sequence. Let's say that we want clip 2 to be shorter. If we were editing linearly, we would have to either rerecord clip 2 with a shorter duration and then rerecord clip 3, or we would copy the material, thus losing a generation.

Trimming a Shot

Because we are editing nonlinearly, changing the duration of clip 2 is easily accomplished. We utilize the paradigm of the transition to easily add to or delete from any clip in our sequence (see

Figure 9–4 The sequence now consists of three clips.

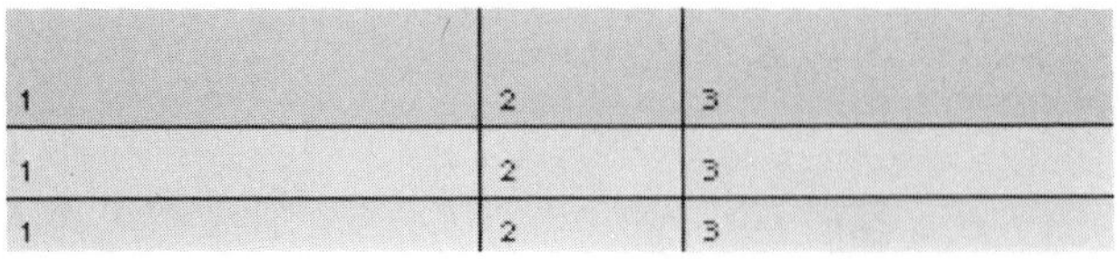

Figure 9–5 Trimming clip 2 is accomplished independent of clips 1 and 3. The length of the sequence is shorter as a result.

1	2	3
1	2	3
1	2	3

Figure 8–17). Several tools are used as part of the transition editing device. There may be numeric trim keys, which are used to type in the number of frames that should be added or deleted, or there may be a graphical icon, such as a pair of scissors, which is used to "cut out" the unwanted frames of picture or sound.

As shown in Figure 9–5, clip 2 is now shorter in length than when we originally edited it into the sequence. All we had to do was delete frames, and clip 2 became shorter in duration without affecting clips 1 and 3. This is nonlinear editing: making a change anywhere in a sequence without having to redo work downstream of the change. This operation took mere seconds to accomplish as opposed to the many minutes it would have taken to rerecord the material in proper placement had we been working in a linear fashion.

Customizing the Timeline

It may be possible to customize the layout of the timeline to apportion more screen space for specific tracks. For example, if you compare Figure 9–4 with Figure 9–5, you will note that the timeline display between the two examples has been changed. In Figure 9–5, the timeline has been elongated vertically. In so doing, we have given the video track more screen space than the audio tracks. Having this flexibility is a very useful tool, especially when the number of cuts increases and we want to expand a section of the timeline to better judge what clips have been used in the sequence.

This ability to take both a macroscopic and a microscopic view of a sequence is yet another tool that can give the editor an idea of how much screen time has been afforded to a particular shot or series of shots. Judging the pacing of a sequence becomes easier not only by seeing the images playing back, but also by viewing the graphical structure of the sequence. It is a tool that has heretofore been lacking in both film and videotape editing.

Rearranging the Order of Clips

Trimming shots anywhere in our sequence is not the only nonlinear editing operation available to us. Consider the order of our sequence: clips 1, 2, and 3. We originally spliced our shots in this order, and we then trimmed clip 2 to shorten its duration. What if we suddenly decided that we did not like the order of the clips? Do we have to reedit the material into a new order? No. Instead, we can use the power of digital nonlinear editing to rearrange the shots in our sequence. Before we proceed, we may

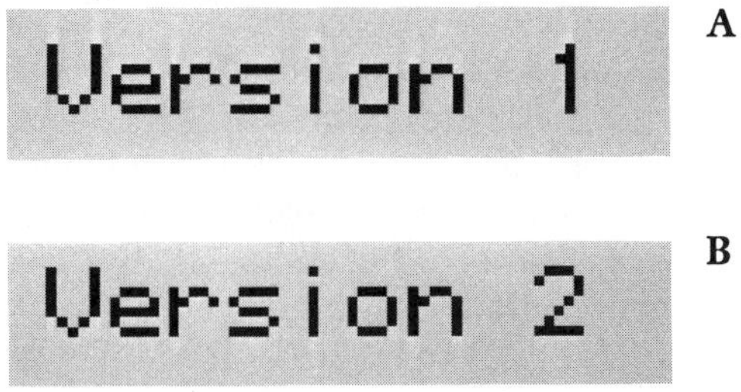

A

Version 1

B

Version 2

Figure 9–6 Making use of the multiple version capability of the nonlinear system, the sequence is duplicated and is labeled Version 1.

decide that we want to take advantage of another feature of the nonlinear system: multiple versions.

We label our original sequence *version 1* (Figure 9–6a). We use the *duplicate* feature of the nonlinear editing system to create an exact copy of version 1. The duplication procedure is almost instantaneous since no actual video and audio files (now stored as digital files) are being copied. Instead, the duplicate function merely gives us another playlist that we can manipulate. This playlist represents a sequence, and we label this sequence version 2 (Figure 9–6b).

Version 1 is our original sequence: clips 1, 2, and 3. Version 2 is an exact duplicate. Since we are now working in a system that provides us with multiple version capability, we can make changes to version 2 while leaving version 1 untouched. To do this in film requires that the required sections of film be duped; with videotape, we would create a second master tape. In either case, a significant amount of time and, therefore, expense must be realized.

Shown in Figure 9–7 is yet another view of the timeline, which displays clips 1, 2, and 3 in pictorial rather than in graphic form. These different choices for viewing footage within the timeline allow the editor to look at text or pictures as the editing process continues. Note that in version 1, clip 1 is the shot of the horse, clip 2 shows clappersticks in the frame, and clip 3 shows a different angle of the horse.

Rearranging the order of shots is a difficult procedure in linear videotape editing. Rearranging the shots in film is easier, but it is still time consuming. However, on the nonlinear system, rearranging shots simply means a quick restructuring of the playlist.

In Figure 9–8, the order of the clips has been changed from 1, 2, 3 to 1, 3, 2. In version 2, we see both shots of the horse first and the shot of the clappersticks becomes the last shot. Depending upon the editing system, rearranging shots in this manner may be as easy as moving (usually with a computer mouse or trackball) a representative frame of one clip from one location to another. The result is that the playlist and the sequence is reordered. The

Figure 9–7 This timeline view displays clips 1, 2, and 3 in pictorial rather than in graphic form.

Figure 9–8 Rearranging the order of shots is easily accomplished. Here, the order of the clips has been changed from 1, 2, 3 to 1, 3, 2.

power of such rearrangement cannot be overstated. Just the ability to easily rearrange clips in this manner can save a significant amount of time while allowing more creative flexibility in investigating the best way to arrange shots in a sequence.

If we begin editing a sequence by first choosing all the shots that we want to use, we can splice shot after shot as we build up the sequence. Next, before any fine trimming is done, we can rearrange the order of these shots. Once we are satisfied that the shots are in the correct order, we can then proceed with transition editing to hone the edit points.

The reordered sequence is shown in the timeline in Figure 9–9 in a graphical view instead of a pictorial view. This is still version 2, but the sequence is now displayed in a different fashion. Again, the important aspect is flexibility, not only in terms of editing functions, but also in terms of allowing the user a choice of how material will be displayed.

We have now experienced two methods of nonlinear editing: the transition-based trimming tools and the ability to rearrange any component in the sequence. It should be noted that this flexibility pertains equally to the audio and video elements in a sequence. Audio from one take can be used to replace audio from another take, and these replacement and rearrangement operations are easily accomplished.

Splicing New Material into an Existing Sequence

Thus far, the nonlinear editing operations we have done have been after the initial, linear edits were made. We made trims to an existing sequence, and we also rearranged the order of a sequence. However, when we are editing, it is a normal routine simply to splice new material anywhere within the sequence being built in the same fashion that a film editor makes splices anywhere in a sequence.

In Figure 9–10, we have our sequence, which has been rearranged to show a clip order of 1, 3, and 2. Let's say that we want to splice a new shot somewhere within clip 3, at the point where the line divides clip 3. Doing this will insert the new clip, and the result will be that our sequence will show clips 1, 3, and the new clip, then resume with clip 3, and finally finish with clip 2. The result is displayed in the timeline shown in Figure 9–11.

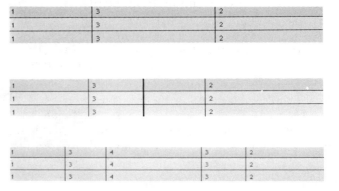

Figure 9–9 The reordered sequence shown in a graphical view instead of a pictorial view.

Figure 9–10 The line within clip 3 signifies where a new shot is to be spliced.

Figure 9–11 After the splice, clip 4 divides clip 3.

With nonlinear editing, we can easily splice new material anywhere within a sequence. In this case, we in effect split clip 3 into two sections and inserted clip 4. This splicing operation results in a longer sequence. Conversely, if we wanted to remove clip 3 and leave clips 1 and 2, we could easily do the reverse operation of the splice function, resulting in the extraction of clip 3. These splicing and extracting operations are the basic tools that the film editor constantly employs, and they are methods that are preserved in the nonlinear system's toolbox of features.

This type of additive splicing can be made anywhere in our sequence. We have complete flexibility to go to any section of the sequence we are working on and add material. The sequence simply expands accordingly to make room for the new material. The exact opposite operation, the process of extracting material, results in the sequence contracting accordingly. Clearly, this is not possible in videotape, which cannot expand and contract in structure.

Adding and Deleting Material without Affecting the Length of the Sequence

It may be necessary to selectively remove material from our sequence without changing the length of the sequence. For example, if we want to replace a section of clip 4, we can remove just that section. If we were editing film, we would remove the necessary number of frames and replace them with film leader (blank sprocketed film) to keep all material downstream of the change in proper synchronization. In the nonlinear system, we have the same flexibility.

In Figure 9–12, a portion of clip 4 has been removed. We want to replace this section with different material. However, if we aren't exactly sure what should be placed in this hole, we do not have to fill it immediately. Instead, we can simply leave this "electronic leader" within clip 4 until we find a shot that is appropriate. This procedure is very similar to what the film editor would do in the same circumstance. Illustrating the hole as a graphical "gap" is also useful as it instantly shows the editor what holes in either picture or sound need to be filled.

When we do find an appropriate shot to fill this hole, we simply insert the shot into the correct space, as shown in Figure 9–13. Clip 5 now resides where the blank leader used to be. If we

Figure 9–12 Selectively removing a section of clip 4 leaves a space that will be filled when an appropriate shot is found.

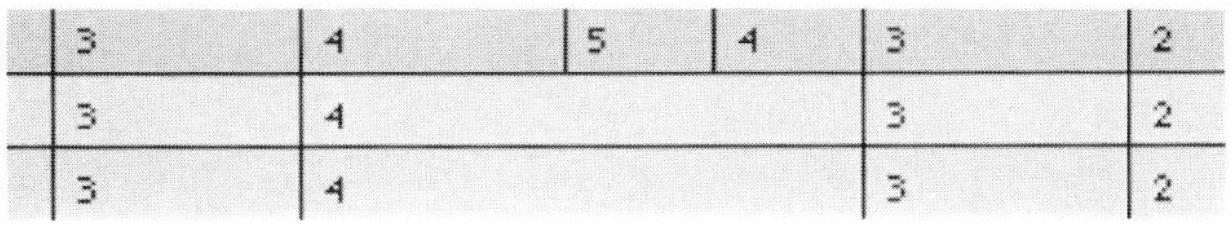

Figure 9–13 Clip 5 is positioned in the sequence.

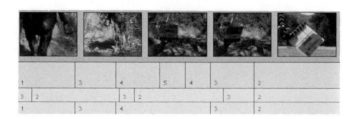

Figure 9–14 A combination of the graphic layout that shows picture and soundtracks and the pictorial representations of the actual footage.

need to shorten or extend clip 5 by a few frames at either the head or the tail, we can easily do so in the transition editing mode.

This type of operation, lifting out a section of a shot, filling it with another, and honing the transition points, is commonly done when we are inserting cutaways from one shot to another.

Combining Pictorial and Graphical Views of the Sequence

As we continue to edit, we may find that in addition to viewing our sequence in a graphical form, we want to view the sequence in a pictorial form. Shown in Figure 9–14 is a combination of the graphic layout that shows our picture and sound tracks and the pictorial representations of the actual footage. Each horizontal bar shown represents an edit point for either picture or sound. We can choose to look at every frame of our sequence in this manner, or we can, as in this figure, see representative frames of the sequence. A view of this type is helpful because it gives the editor feedback on two levels. What the editor is used to seeing, pictures, continue to be shown, along with the additional graphic feedback about the structure of the sequence. The question, "What did I do there?" becomes easier to answer when the layout of the sequence's structure is accessible in this fashion.

Audio Editing

Audio editing can be made easier by giving the ear an additional tool: the eye. The display of the audio waveform for a passage of dialogue, for example, may make it easier to identify where an audio cut should be made. This is especially true if we are cutting between syllables when doing a particularly precise audio edit. By displaying the digital audio signal as a visual waveform, it may be easier to hone in on a particular sound that we want to extract from our sequence.

In Figure 9–15, the audio waveforms are displayed as yet another component of the timeline. It is usually possible to

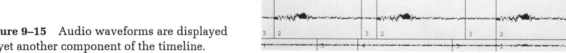

Figure 9–15 Audio waveforms are displayed as yet another component of the timeline.

expand the scale of these waveforms to better judge a section of audio and to edit out samples of audio at very small increments of time. In film, audio edits can usually be made down to 1/4 of a frame resolution. On videotape, audio editing is limited to one frame resolution. With digital nonlinear systems, audio editing can be done at the sample level, usually on the order of 1/100 of a frame resolution.

Audio volume settings from shot to shot can also be changed, and the result is that it becomes possible to create either temporary or finished audio mixes on the digital system. In addition, the straight bars that separate one audio cut from another often can be manipulated to create audio crossfades. By changing the angle of the straight bar, a crossfade is usually created. This allows one audio component to dissolve, or blend, into another audio component.

Basic Digital Nonlinear Editing Operations

These basic methods of splicing, extracting, rearranging, and viewing the structure of a sequence form the tools and operations available on the digital nonlinear editing system. A variety of additional features is available, many of which are described in Chapter 8. By providing the ability to view optical effects and create finished audio mixes, the digital nonlinear system offers increased previsualization of the sequence being created.

When more ideas can be entertained and more types of previsualization can occur in an edit system, there will be a much greater chance of achieving a better program as a result of trying things in one position, rearranging them, and honing them to perfection. This is what the digital nonlinear editing stage provides to the user.

THE OUTPUT STAGE

When we are satisfied with our sequence, we determine which editorial work products are required. If we are going to the online suite, we may only need a video EDL. If the goal is to leave the digital system with a program for presentation, we record onto videotape the direct output of the disk drives as our sequence is played back. If we must deliver both on videotape and film, we require the video EDL and a film negative cut list.

WILL THE PROJECT BE FINISHED IN LESS TIME?

One common question is whether the digital nonlinear editing system will save time over conventional linear methods. Another question is whether the saved time will be used to create extra versions that would not normally have been attempted due to lack of time. The answers to these questions are very subjective, but there seems to be several ways of addressing these issues.

For some users of digital nonlinear editing systems, there is a time savings on the order of 30–40%. For other users, the time factor is equal to linear editing, but in that span of time, more versions are attempted, and a better finished piece is created due to the additional time available to experiment with the footage. One point that is not open to interpretation from users of the digital nonlinear system is that the increased work products, such as the video EDL and the rough cut of the sequence, decrease the experimentation done during the online session.

SYSTEMIC ISSUES

The Operating System versus the Software Program

The editor, over time will be aware of and responsible for some fundamental issues. Although, in theory, it would be nice to think that the person putting together a program would only have to think about the creative decisions involved in that process, the fact is that all nonlinear editing systems use computers. Computers, in turn, utilize two important, and different, routines.

The first routine is the computer's operating system, often referred to as the *OS*. This operating system is the set of instructions that forms the basic communications link among the various components of the computer. The second routine is described under the generic banner of *software programs*. Software operates within the restrictions of the OS.

What's important to keep in mind is that the methodology of the OS is usually different from the methodology of any of the various software programs that operate within the OS. When the two routines work together well, the manner in which the nonlinear editing system accomplishes its tasks can be invisible to the user. However, when there are conflicts, it is often necessary to have developed some expertise in both routines.

File Organization

As you begin the nonlinear session, be aware of whether the system automatically organizes where material is placed (the files that make up your program) or requires you to organize this material. Good software programs make a first attempt at organizing your files in an efficient and logical manner. If this is not satisfactory to the user or if the user prefers a different approach, the user can take charge and dictate how and where files should be organized.

Documentation

Computers are machines, and they require maintenance. The hardware, the OS, and the nonlinear software program can experience difficulties. In all cases, there is usually a body of experience available to you from other users or from the manufacturer. When a software program is released, it is accompanied by release notes and "bug" notes. Release notes outline the specific features of that software program, and these notes are usually a vastly abbreviated form of the documentation for that software program. "Bug" notes outline any known problems in that software release. These could be simple problems or more complex problems. In any case, these various documents represent a body of knowledge that many users simply dismiss with the statement, "I don't have time to read the material. I have to get right to work." As with any machine, taking some time to read about the known methods of operation can pay great benefits at a later time.

THINKING NONLINEARLY

Often, first-time editors, writers, producers, and directors will bring to the nonlinear editing session a linear philosophy. They start with an orthodoxy of presumptions that are based in the linear mode of the editorial process and, consequently, the storytelling process. That is *wrong.* An open mind is required to play "what if" scenarios with footage, especially if you are very close to the footage. Often, the first-time user of the nonlinear system begins in a linear fashion. Only after significant minutes or hours have passed do the first-timers magically discover that editing nonlinearly is easy; it's thinking nonlinearly that can be difficult at first.

Once you believe in your heart of hearts that you're really done with the nonlinear session, try something else. In going for the single solution or direction, you can fall into an orthodoxy. Therefore, it is extremely important to remind yourself constantly to remain as flexible as the nonlinear system should be. If you do this, you'll get the best possible results from your footage.

This discussion of a representative nonlinear editing session has been brief. Many more operational features are usually available to the editor. However, the various operations described here serve to illustrate the nonlinear tools and editing concepts that form the basis of a typical nonlinear edit session. In addition, the more philosophical concerns must not be overlooked.

10

Digitization, Coding, and Compression Fundamentals

Digital nonlinear editing system operation is based on a fundamental principle: Pictures and sounds are first converted to digital data and are then manipulated in the same way that a writer manipulates words with a word processor. The process by which moving pictures and sounds are transferred from their originating form into digital data is known as *digitization*.

BASIC TERMS

It is helpful to outline some basic terms relevant to the digitization process:

Analog   In analog recordings, the changes to the recording medium are continuous and analogous to the changes in the waveform of the originating sound or to the reflectance of the original scene.

Compression   To reduce in volume and to force into less space.

Digital   In digital recordings, numbers are used to represent quantities, and numbers in rapid sequence represent varying quantities.

Digitize   To convert continuous analog information to digital form for computer processing.

Sample Points   The amount of data used to represent an original form.

AN EARLY SAMPLING EXPERIMENT

In the early 1970's, Bell Laboratories in Murray Hill, New Jersey, performed several studies as part of their research for the picturephone. The goal was to transmit a caller's likeness over standard

155

phone lines. These studies sought to ascertain how much information would have to be sent over the phone line for the person to be recognized. Additionally, what methods would have to be developed whereby light information could be turned into energy that could then be transmitted?

The examples shown in Figures 10–1 to 10–5 recreate these experiments. One can see varying degrees of recognizability based on how much information is present in each of the photos.

Bell researchers, seeking to preserve the intent of the image (the recognizability of the face) with as little information as possible (thereby limiting the amount of data that would have to be transmitted), came to the conclusion that the samples of a

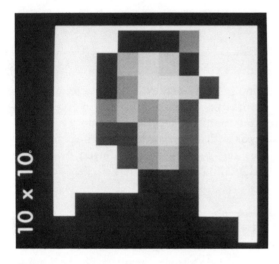

Figure 10–1 A 10 × 10 sampling of a portrait. How much information must be present to recognize a picture? This basic question is at the heart of this chapter.

Figure 10–2 By increasing the number of samples to 20 × 20 (400 sample points), more of the image becomes apparent. By holding the image a distance from your eyes and squinting, the subject of the photo may become clear.

Figure 10–3 By increasing the sampling to 40 × 40 (1600 sample points), the picture becomes clearer.

Figure 10–4 In this 80 × 80 sampling, the subject of the photo should be clear.

Figure 10–5 The normal, unsampled view (307,200 sample points) shows President Abraham Lincoln.

picture could be limited and that recognition would be aided by the tendency of the eye to blend contiguous areas. They concluded that, while it is difficult to identify minimally sampled pictures, blurring and blending the steps of such samples will improve recognizability.

Pixels and Sample Points

A *pixel* is the smallest element that makes up a picture. In essence, pixels are the patterns of dots that make up an image on a viewing screen. *Pixel matrices* are calculated by multiplying the number of pixels that are contained both vertically and horizontally over the span of the viewing screen.

The Bell Labs experiment, one of the first to involve the digitization of a picture, first reduced a photo to a series of 100 blocks (10 vertical × 10 horizontal = 100), hence the process of selective sampling. Each block had its own shade of gray, and when viewed normally, the photo looked like 100 blocks of information with varying shades of gray. But when the interconnecting blocks where blurred, a portrait became more obvious.

This early Bell Labs experiment forms the basis for representing visuals in a digital form. The number of samples that are assigned to a signal and the storage of this data are two fundamental issues that the digital nonlinear editing system must address. A great number of samples for a signal have to be stored; too few samples may not adequately represent the original signal so that it can be recognized.

How are units of information defined within a digital video medium? Each pixel that makes up a portion of the image that will be converted from its original form to digital data is categorized based on its brightness value. Once these digital data have been stored to computer disk, pictures and sounds exist as numbers to the computer operating system. These digital data are a result of the sampling process. A "sampling" includes the following characteristics: how often a signal is measured, the quantity of its amplitude, and the definition of its frequency over the span of the signal.

THE FLASH CONVERTOR

How do you get pictures and sounds into a computer, and how does the computer figure out what to preserve in the picture and what to discard? The photo of Lincoln can be sampled into the computer with all the information necessary to reproduce the photo without a noticeable change. Conversely, the original photo can be sampled into the computer with fewer references to the original. In that case, the integrity of the photo is compromised.

A method of converting video signals into data that computers could process arrived in the late 1970's. *Flash Convertors* were expensive and able to process only black and white pictures, but they represented a very fast way of converting analog signals to

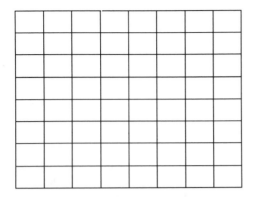

Figure 10–6 The flash convertor divides the computer screen into a grid pattern to define each pixel in the grid's array. Illustration by Rob Gonsalves.

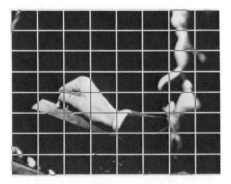

Figure 10–7 The flash convertor is then used to divide the picture into its precise elements: the pixels of the image. Illustration by Rob Gonsalves.

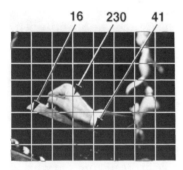

Figure 10–8 The flash convertor next evaluates the brightness of each pixel and assigns to it a number. Illustration by Rob Gonsalves.

digital signals. With a flash convertor, it was now possible to convert one frame of video at a time into data that could then be interpreted by computers. Although they could quickly convert analog signals to digital signals, these flash convertors were not able to keep up with the stream of 30 NTSC video frames each second. They were only able to handle about one frame each second.

How Does a Flash Convertor Work?

The flash convertor first divides the computer screen into a grid pattern so that each pixel making up the grid array can be located (Figure 10–6). As a picture is processed by the flash convertor, what becomes apparent is that a picture is composed of precise elements, a number of pixels (Figure 10–7). At its simplest level, the flash convertor takes each pixel and converts its brightness value into a number (Figure 10–8). The establishment of these numbers for brightness values is based on a gray scale chart consisting of 256 steps, in which 0 = full black, and 255 = full white.

Finally, the flash convertor processes the image and assigns values for each pixel of the image. These numbers are put into memory, and at that point, the image is regarded as a set of values (numbers) that the computer then can manipulate. Early flash convertors did not preserve full detail of the picture. With only 256 steps to represent an image, which may originally have had thousands or millions of shades of gray, the reduced sampling level compromised the full detail of the original image.

When any signal is flash-converted, or digitized, the signal is converted into binary data via an analog-to-digital convertor (ADC). Like all ADCs, the flash convertor's function is simply to take the instantaneous value (pixel duration sample) of an analog signal voltage and convert it into binary data form. One byte can express a range of values between 0 and 255, but the video scale is expressed between 0 and 100 IRE units. If one section of a picture is measured at 47 IRE units, how is that portion of the picture and the pixels that make up that portion converted to digital data?

Let's mathematically convert one video pixel with an IRE value of 47 units into a byte.

Pixel, or P = 47.
Byte range, or B = 100 units.

We begin by dividing B (100) by 2 to find the midscale (50). Is P now larger than this new value of B? The first bit will be set to a "1" if the answer is yes. Since a byte consists of eight bits, we will repeat this test eight times at eight decreasing test values for each bit. Before each successive test, we will again divide B in half. Whenever P is larger than B, we will set the corresponding bit's binary value to a "1," and we will subtract B from P. Otherwise, the bit remains unchanged at "0," and P also remains unchanged (Table 10–1). Thus, 47 IRE units of video is expressed as the byte 01111000.

Table 10–1   Pixel-to-Byte Conversion

B = 100	P = 47			
B = B ÷ 2 = 50	Is P larger than B?	NO	BIT 1 = 0	no changes
B = B ÷ 2 = 25	Is P larger than B?	YES	BIT 2 = 1	P = P – B = 22
B = B ÷ 2 = 12.5	Is P larger than B?	YES	BIT 3 = 1	P = P – B = 9.5
B = B ÷ 2 = 6.25	Is P larger than B?	YES	BIT 4 = 1	P = P – B = 3.25
B = B ÷ 2 = 3.12	Is P larger than B?	YES	BIT 5 = 1	P = P – B = .13
B = B ÷ 2 = 1.56	Is P larger than B?	NO	BIT 6 = 0	no changes
B = B ÷ 2 = .78	Is P larger than B?	NO	BIT 7 = 0	no changes
B = B ÷ 2 = .39	Is P larger than B?	NO	BIT 8 = 0	no changes

The conversion task, as expressed in Table 10–1, is quite close to the way in which even a simple computer language such as BASIC would handle it as a software routine. Instead of using software, the flash convertor is a dedicated piece of high-speed hardware that rapidly samples the video signal as pixels and outputs binary values as bytes that are stored in a section of memory reserved for use as a frame buffer.

Color video consists of three distinct analog signals of red, green, and blue (RGB). Thus, digital conversion is often conducted by three flash convertors operating in parallel.

To get our original analog signal back, the computer performs the tasks in reverse via a digital-to-analog convertor (DAC).

Bit No.	Bit Values
1 = 0	50.00
2 = 1	25.00
3 = 1	12.50
4 = 1	6.25
5 = 1	3.12
6 = 0	1.56
7 = 0	0.78
8 = 0	0.39

When we take the combined bit number for bits 1 through 8 we have the following string: 01111000. However, since there are no changes to be made for bits 1, 6, 7, and 8 (see Table 10–1), we only combine the bit values for bits 2, 3, 4, and 5. We therefore add 25.0 + 12.5 +6.25 + 3.12 for a total of 46.87 IRE units.

Frequency Definition and Amplitude Definition

The number of samples assigned to represent a signal can vary. One very simple form of a digital sampling system is the facsimile machine used to transmit documents. Fax machines that operate in black and white mode have only two samples that can be assigned to a pixel: 0 = black, and 1 = white.

The goal of incorporating such a scale is simple: to transfer high-contrast information reliably. The more information preserved with regard to samples of a pixel, the greater the perceived result. In general, here are some sampling guidelines with regard to pictures and sounds. For pictures, if the sampling frequency is very high, the perceived picture will have a lot of spatial definition. If the sampling frequency is very low, there will be fewer stages of brightness to recreate the original picture; therefore, it will be compromised. For sounds, if the amplitude definition is very high, the perceived fidelity of sound will be high. If the amplitude definition is very low, the fidelity of sound will be decreased; therefore, it will be compromised.

We can begin to see how pictures are affected by the number of samples used to represent the picture. There are a number of image plane schemes. With RGB signals, a typical color picture is composed of 24 bits. Each chrominance component has eight bits each. With eight bits × three colors, we arrive at our 24 bit picture. Next, to calculate the number of potential colors that the 24 bit picture could represent, we have the following:

8 bits = 256 colors. Each color can represent 256 shades of gray. Therefore, 256 for red × 256 for green × 256 for blue = 16.7 million potential color combinations.

Many products store varying numbers of bits per sample. Examples of videotape formats and post-production hardware that are based on a specific number of bits per sample and total color array follow:

D2 videotape = 8 bits/sample = 256 colors at 4 FSC

(4 FSC refers to 4x subcarrier frequency; 14.3 MHz NTSC and 17.7 MHz PAL. This is the sampling frequency that is most often used for digitizing composite video signals. Note that D2 is not limited to 256 colors because color information is represented by modulation of the subcarrier).

Quantel Paintbox = 24 bits/sample = 16.7 million colors

(The Quantel Paintbox stores three channels for picture information; one for luminance and two separate channels for the color difference channels. These two separate color memory planes are in full resolution).

DF/X Composium = 32 bits/sample = 16.7 million colors + 8 bit alpha channel

Devices used as compositing systems, such as the DF/X Composium and Quantel Harry store four separate channels in a 4:4:4:4 ratio. Each channel has an equal number of samples. In a 32 bit system, eight bits per sample are allocated for luminance, the color difference signals, and a new, fourth, component: the key channel information.

With the flash convertor, we achieved the ability to transfer the pixel array of an image into data that illustrates how to represent analog-based information as digital data. However, the governing process, the ability of the computer to process this data, had yet to be addressed.

COMPUTERS AND VIDEO

In the late 1970's, video shapes could not be drawn without distortion on computer screens. The main reason was that there was no agreement with regard to the pixel matrix of the computer and the pixel matrix of video systems. It is important to remember that computers and video are quite far apart with regard to how they display pictures; they simply do not have much in common. If we look at a few of the differences, some of the problems of exchanging information between the two become apparent:

Computer Screen	Television Screen
Noninterlaced display	Interlaced display: 2 fields
Scan rate varies; Apple Macintosh = 66.7 Hz	Scan Rate: 59.94 Hertz (Hz)

Just in terms of the manner in which images are displayed, there is an incompatibility between computer and television screens. With the video system of displaying images, the entire screen is covered in a methodical pattern, and information for field 1 is scanned and combined with information for field 2. This is known as an *interlaced display*; the two fields are combined to form one view. Images on a screen are comprised of pixels. A pixel is one (single) point of an image's makeup.

However, computers are normally used in a *noninterlaced display*, where there is only one "field" of information. For a computer to be able to display an entire screen's worth of information as it relates to the video world (two fields), it must remember the total array of pixels over the screen. (This is why it is not possible to record the output of a computer directly to videotape without a conversion process, since we are attempting to record a noninterlaced display on a medium that requires an interlaced signal.) Doing this was not always possible, and it took large (and expensive) amounts of computer memory to display full screen pictures.

In the late 1970's, computer memory became more powerful, and the cost of computer memory began to decrease. (This trend has remained consistent, and we can generally plan on computer memory becoming about four times cheaper every two years.) Once memory became affordable, large screens became possible, and displaying a picture over the span of the computer screen's display became easier to achieve. *Raster* refers to a pattern of scanning lines covering the area upon which an image is displayed. Being able to display just one picture on the computer raster was a very time-consuming feat, on the order of two to five minutes, depending upon how complex the picture was. In addition, the computer could not change from picture to picture with any degree of speed since the new picture had to be "painted" onto the raster in the same time-consuming fashion.

In the early 1980's, a development that would move computers closer to video occurred. Companies such as Xerox and Apple

Computers began to develop computer screens that had a pixel matrix of 640 horizontal samples × 480 vertical samples.

This represented a very big breakthrough. In the computer world, there was now a rectangular grid, and all the pixels on this grid would directly relate to how a video image could be replicated on the computer screen. Another way of thinking about this compatibility is to consider how television images are displayed. Since the traditional format is four units wide by three units high (aspect ratio = 4 × 3), when one divides 640 × 480 by 16, an aspect ratio of 4 × 3 is achieved.

It is also important to remember that we are still not talking about displaying video on the computer. Computers were still limited to displaying computer-generated and computer-scanned images. More importantly, in the late 70's, there was still no means of transforming video signals into data that a computer could process. In fact, there wasn't much interest in reversing the process: recording computer signals to videotape.

Eventually, with the emerging use of video in business and training applications, interest developed for both tasks: getting video into the computer and recording computer signals back to videotape. For example, computers could be used to create presentations that would then be distributed on videotape. Conversely, as it became desirable to include moving video in presentations, bringing video into the computer environment had to be accomplished. This could be done with the flash convertor, but there were still limitations in displaying the data fast enough to accomplish moving pictures.

Computer Limitations

By the mid-1980's, being able to generate and process computer images and display them on screen was a task that required dedicated workstations, such as Apollos and the Cray 1. It was not unusual to spend hours to generate just one frame. When many frames needed to be combined, as in the case of computer-generated animations, days could be spent in the process.

The main limiting factors included the computer's inability to retrieve a frame from disk and display it on the screen while searching for the next frame in the series. There was no way to get pictures onto the screen rapidly. Computer disks and their data read times (how fast the disk can provide its information) were another factor since, regardless of how quickly the computer requests the data from the disk, the disk has limitations on how quickly it can provide the data.

Still, the results of such experiments were impressive. Computers could render animations one frame at a time to film cameras to show them in public demonstrations. Some computer animations of landscapes and ocean settings were so extraordinarily lifelike that observers asked where the film had been shot!

As the late 1980's approached, two problems had yet to be solved: how to digitize pictures into the computer in real time, at 30, 25, or 24 frames per second, and how to display these pictures to the screen in real time. Improvements in hardware were necessary to address these shortcomings.

Evolution of the Flash Convertor

By the mid-1980's, we were no longer looking at black and white computer terminals; we were experiencing the explosion of the entry of color into computer displays. Manufacturers of flash convertors began to add color capabilities in the same fashion as color was added to the computer. Prior to this, digitized images were limited to black and white. Although they cost thousands of dollars, flash convertors that handled color began to appear in the mid-1980's.

The pathway for moving from video to the convertor is as follows: A composite video signal is made up of four elements: red, green, blue, and a sync (stabilizing) signal. It is necessary to break this four-item signal into separate components. We began to see flash convertors that had, instead of just one frame-grabbing unit, three units, one each for the red, green, and blue signals. However, with the complexity of color images, decoding and encoding units became necessary.

Through the use of a decoder, the composite signal is separated into its components, and the signals are prepared for the flash convertors. Cost was a factor. These early three-color flash convertors ranged in price from $5,000 to $8,000.

In the late 1980's, the combined functions of a digitizer, an encoder, and a decoder appeared in an affordable unit. Examples include the VID I/O™ from Truevision, Inc., which decoded composite video into red, green, blue, and sync signals and then encoded the red, green, blue, and sync signals so that it was possible to record material from the computer to videotape (Figure 10–9). Digitizing was accomplished through the use of a separate computer card. More remarkable was the price: under $1000.

In 1991, computer boards began to replace the external encoder and decoder units. These computer boards house chip sets that

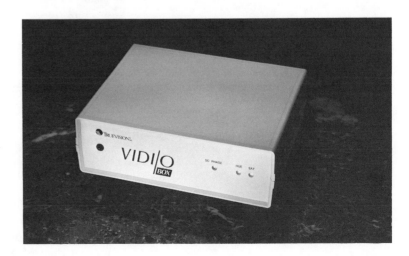

Figure 10–9 The functions of digitizer, encoder, and decoder combined into a single unit. Courtesy of Truevision, Inc.

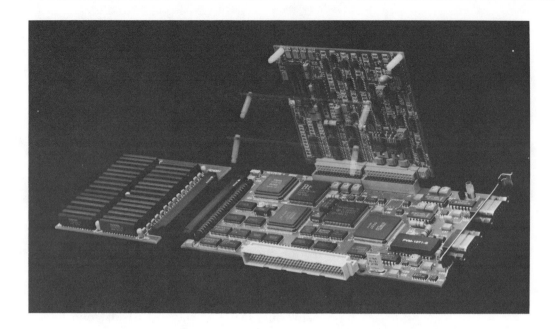

Figure 10–10 Digitizing, encoding, and decoding functions are resident on the computer board. Courtesy of Truevision, Inc.

perform the digitizing, encoding, and decoding functions. Examples include the NuVista+ card from Truevision, Inc. (Figure 10–10).

Approaching Real-time Digitization

By 1985, digital timebase correctors were in wide use. In addition, digital video effects devices had become very popular. Both digital TBCs and DVEs helped set the stage for the real-time digitizing of analog video and audio signals. In the late 1980's, improvements in several areas allowed the digitization process to proceed much more rapidly. Central processing units (CPUs), computer buses (which transfer information from one portion of the computer to a different, internal, part of the computer as well as between internal and external devices), and computer disks all began to get much faster.

Heralded by these developments, it became possible to display pixels on the computer screen much more rapidly. For the first time, it was possible to do picture subsampling from videotape rather than from computer-generated images. This subsampling was still being done in greater than real time, but much more rapidly than ever before. For example, although digitization may not have been at a full 30 frames per second, it was possible to digitize at about nine to ten frames per second. This was a significant breakthrough; one frame per second had been the norm.

The important aspect is that real motion could be seen and detected, even if it was somewhat staccato in nature. It was a revolutionary development: Live action, moving pictures stored on computer disks!

These early attempts at digitizing video were accompanied by a great deal of quantizing artifacts, including spatial aliasing, temporal aliasing, and color aliasing (inaccurate representation of color, absence of color, or blending of colors). In addition, no form of compression was being applied to the frames as they were digitized, and therefore, the files were extremely large. In 1989, a 380 MB magnetic disk drive was considered large, but it was filled very quickly with uncompressed digitized video frames! Reducing the amount of information (samples) became a necessary objective.

COMPRESSING AND CODING

The Bit

ASCII (American Standard Code for Information Interchange) represents the manner in which binary definitions are assigned to numbers and letters. It provides a standard for exchanging different types of files: ASCII code. Two main benefits arise from representing information in ASCII form. First, as a common labeling scheme, it ensures that the letter F in ASCII will be given the same ASCII symbol, 70, regardless of who is doing the labeling. Second, ASCII ensures that files saved and transported in their ASCII form can be accessed by software programs that may not be compatible in and of themselves.

ASCII assigns binary definitions to numbers and letters. *Binary* means numbers that use the power of 2. Computers can only use two numbers, either 0 or 1. The smallest piece of information in the digital world is called a *bit*, which is short for "binary digit." Creating words from bits involves first creating letters; you need a series of bits to represent a letter and, obviously, a series of letters to represent a word.

For the word, *FILM*, for example, the ASCII codes that make up the letters, *F*, *I*, *L*, *M*, are as follows:

Letter	ASCII	Binary Code (Bits)
F	70	01000110
I	73	01001001
L	76	01001100
M	77	01001101

It takes eight bits to represent any letter. If we were to save our ASCII file of the word *FILM* and bring that file to another computer, anywhere in the World, the word *FILM* would be displayed when the ASCII file was read.

This is the point: Any message can be reduced to and represented by a series of bits. It makes no difference if the message is the word *FILM*, a paragraph, a Wordsworth poem, a graphic off a printed page, a color photograph, or a full-resolution video frame. All information, at its heart, is a series of data in the form of bits.

Once information is represented in bit form, many things can be done with the data. One common example is transmitting the information via a modem (modulator/demodulator), which can be used to communicate from computer to computer over telephone lines. When a modem is used, letters are represented by codes, and these codes are sent over transmission lines.

There are two major concerns. First, will we be able to represent a message accurately by reducing it to a series of bits? In the case of our word *FILM*, it is straightforward: We have four letters, and each letter is represented by eight bits. Second, if we store this data or send this data, will the integrity and order of that data be preserved? If we are successful, we send a colleague 32 bits of information, and she receives 32 bits of information, in the correct order, and she sees the word *FILM*.

So far so good, but what happens when we are faced with situations where we may not be able to represent a message accurately by reducing it to a series of bits? How will a color photograph be represented by a series of bits? We now know that the flash convertor is used to represent a picture as digital data, and that one byte's worth of information relates to each pixel in the picture, for example, to provide numeric values for shades of gray. However, what considerations are made, and what compromises occur? Is there a standard such as ASCII code for representing a frame of video that needs to be digitized and sent somewhere?

Compressing a File

Being able to reduce a message into a series of numbers is a critical step in converting an analog signal into a digital signal. The quantity of numbers assigned to the message can be a fixed standard. In the case of the word *FILM*, representing the word in ASCII code leaves no arbitrary decisions to be made. If we want to represent that word and we want to adhere to the standards of ASCII, we must use the correct binary information. As mentioned, when we make this transformation, we send a series of bits, and the same series of bits is received, and the word *FILM* is decoded; we sent 32 bits, and 32 bits were received.

However, there are times when we want to reduce the size of a message being sent, but at the same time, we do not want to change the meaning of that message. The first goal is size reduction, and the second goal is content preservation.

Let's say that we have a 50-page, single spaced document that we want to send to someone over a modem. We finish typing the document on our word processor and save the file in ASCII format. We do this because the person who will be receiving our document is not using the same word processing software as we are using. However, the word processing program she is using will open up ASCII files.

After saving the file, we determine its size, and it turns out to be approximately 100 kB (100 kilobytes or 100,000 bytes), a relatively small file. We then run a communications software package that, before engaging the external modem, gives an estimate of how long it will take to transmit the file based on the size of the file and the speed of transmission. The estimate is ten minutes to send the file.

While ten minutes isn't an extraordinary amount of time, if we have many documents to send or if we simply want to work in the most economical fashion and save telephone costs in so doing, it would be advantageous if we could reduce the transmission time. What if we could reduce the time to five minutes? Two methods quickly come to mind: either find a way to transmit the file twice as fast or find a way to reduce the file from 100 kB to 50 kB.

Let's say that due to the capabilities of the modem we are using, we cannot transmit the message twice as fast. Our alternative is to figure out a way to reduce the size of the file from 100 kB to 50 kB, or, if a 50% savings isn't achievable, to something between 100 kB and 50 kB.

Coding Techniques

Coding, in the context of reducing the size of a file, refers to how information is represented. If we had to repeat a very long word, such as *brontosaurus*, many times, we could choose to code a shorter version. Instead, we could say *bronto*. We have just created a method of labeling an original name (file) while reducing the number of syllables that we must vocalize.

There are software and hardware routines that we can use to reduce the file size. For now, we will concern ourselves with software methods, which are typically marketed under the general category of file compacting programs. They use certain algorithms (an *algorithm* is a procedure or rule for solving problems, as for finding the lowest common denominator) to determine the degree of frequency with which items appear (Figure 10–11).

Huffman coding is one such coding technique used by these compacting programs. In the case of our 100 kB document, we have to be very careful in how we reduce its size. We cannot arbitrarily chop out letters and characters because by doing so we will change the message that is received. We must have a way to reduce the 100 kB file temporarily, and then the receiver must be able to expand it back to its original size, losing no information.

If we use a computer program that employs Huffman coding techniques, we will confirm the following presumption, and we will realize a benefit from it:

> Letters and words in our document are not equally likely; there will tend to be much repetition of a certain class of letters (such as vowels) as well as words.

The previous sentence contains the following letters (for now, upper- and lower-case letters are not differentiated):

a   appears 8 times
b   appears 1 time
c   appears 3 times
d   appears 5 times
e   appears 16 times
f   appears 1 time

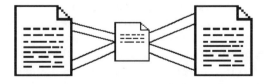

Figure 10–11 With software compacting programs, the file size can be temporarily reduced and then decompacted after the transmission of the file is complete.

g	appears 0 times
h	appears 1 time
i	appears 7 times
j	appears 0 times
k	appears 0 times
l	appears 11 times
m	appears 2 times
n	appears 7 times
o	appears 10 times
p	appears 1 time
q	appears 1 time
r	appears 10 times
s	appears 12 times
t	appears 12 times
u	appears 5 times
v	appears 1 time
w	appears 5 times
x	appears 0 times
y	appears 2 times
z	appears 0 times

When we apply Huffman coding techniques to our 100 kB document, we have only one goal: make the overall message shorter. The fact that certain letters are more likely to appear than others allows us temporarily to reduce the overall document size. Huffman coding assigns codes to the letters such that (1) the most probable letters (those that appear most frequently) will be assigned short codes, and (2) the least probable letters (those that appear least frequently) will be assigned long codes.

In this way, when we need to transmit that the letter *s* appears 12 times, we transmit this information through a short code for that letter. Why transmit a long code when you have to transmit it 12 separate times? Conversely, when we have to report that the letter *z* appears zero times, we use a long code because, although it will take more time to transmit this information, we only have to transmit the long code one time.

For example, if we wanted to assign some codes to the message we want to transmit, we could proceed as follows: The letter *t* appears 12 times, so we assign it a transmission value of 1. We are therefore sending the smallest code for a letter that appears frequently. The letter *p* appears once, so we assign it a larger transmission value of, say, 7. Continuing in this fashion, we will ensure that the most frequently required characters have lower transmission values than the least frequently required characters.

Software programs that work in this manner and make use of Huffman coding techniques analyze the statistics of the message and determine the frequency of a message's components. Then, after ascertaining how many different codes there are, the greater number of messages receives short codes and the less frequent number of messages receives long codes.

It is not at all unusual for a document to be reduced by 40–50% after the redundancies have been analyzed and short and long codes have been assigned. Instead of using the word *reduced*, we need to begin to use the word *compressed*. Huffman coding techniques typically save about 50%, while a similar technique, LZW (Lempel-Ziv-Welch) encoding, typically saves about 75%.

We have achieved our aim, and the 100 kB file has been temporarily reduced in size by 50%; instead of having to send 100 kB over a modem, we are now sending 50 kB. The transfer takes place twice as fast. The person who receives the 50 kB file, now in its compressed form, cannot read it. To read the complete and unaltered file, it must be restored to its original form, and its original form consists of 100,000 bytes, not 50,000 bytes.

After the compressed file is decompressed, the original frequency of messages is restored, and the file again becomes 100 kB in size. The person who has received the file can proceed to read a document that has all the letters, sentences, and paragraphs in the correct order with no loss of information.

SAMPLING

Sampling is one of the first steps that needs to be taken when an analog signal is converted to a digital signal. The process of sampling involves measuring an analog signal at regular intervals and then coding the measurements. A sample is merely defined as a smaller part of a whole that represents the nature or quality of that whole. There are a variety of sampling techniques, each yielding benefits, detriments, and artifacts; some are specific to visual signals, others to audio signals.

When we are presented with video and audio signals that must be converted from analog to digital, we must represent the analog waveforms of these signals by numbers. Where to assign the numbers and how many numbers to assign are very similar to the assigning of codes to the document that we compressed (Figures 10–12 and 10–13). Numbers represent the essential characteristics of the audio signal, but the number of sample points, the number of digits assigned to represent the signal, can vary. Sampling not only refers to how frequently a signal is measured, but also to the degree of measurement that is provided for the amplitude.

Consider, however, a representation where far fewer sample points have been assigned to the same waveform (Figure 10–14). More interpretation time must be given to this second example since there are fewer sample points to indicate the original path of the audio waveform. The process of sampling—indeed, any time an analog signal is turned into a number (when a digitizer turns a signal into a number)—yields a loss, and sampling creates a phenomenon known as *quantization*. This term refers to the limitation to which the analog signal can be sampled by the analog-to-digital convertor in a samples/time relationship. Quantization can also be viewed as an acceptable loss in the original quality of the signal.

It stands to reason that if one A/D convertor can only plot three points whereas another A/D convertor can plot 300 points for the same analog signal, there will be more numeric samples in the latter sampling. Accurately representing and recreating an original signal with just a few samples as opposed to many samples is an

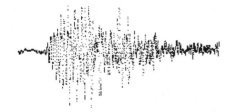

Figure 10–12 The process of sampling begins with an original signal. Here, the waveform of an audio signal is displayed.

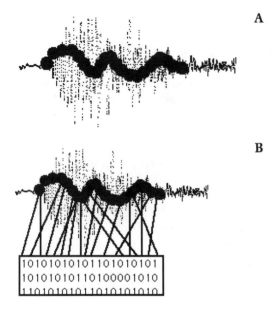

A

B

Figure 10–13 (a) Sample points are now taken along the waveform of the signal. These numbers facilitate the conversion of analog data to binary information. (b) Once sample points have been taken, the binary information is then stored to computer disk.

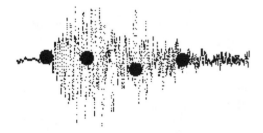

Figure 10–14 Here, the audio signal is represented by only four sample points versus the original sampling of more than twenty points.

impossible task. We always attempt to have as many samples as possible to represent original information. As we now know, to reduce the overall size of a digital file, fewer samples are often taken than would be optimal.

Quantization can yield several noticeable *artifacts*, which are discrepancies between the original signal and the representation of that signal. Perhaps one of the most recognizable artifacts is called *aliasing*, in which images can be distorted or appear disjointed.

Sampling, Subsampling, and the Sampling Theorem

Sampling is the technique by which an analog signal is measured at intervals. These measurements are then assigned codes to help us store and transmit the data. The sampling theorem is a rule for obtaining acceptable representation of a signal with regard to the regularity with which sampling should be done.

The sampling theorem states that a signal must be sampled at least twice as fast as it can change (2 × the cycle of change) in order to process that signal adequately. Sampling the signal less frequently will lead to aliasing artifacts. For example, if we want to preserve an audio signal to the range of human hearing (approximately 20 kHz), we would sample the signal at least two times the desired result. For audio compact discs, the sampling rate is 44.1 kHz, which quite adequately preserves signals that the human ear can process.

Subsampling refers to a technique where the overall amount of data that will represent the digitized signal has been reduced. More generally stated, when more samples are thrown away than meet the sampling theorem, we are subsampling. When the sampling theorem is violated, many types of aliasing may be noticeable.

Spatial Aliasing
In spatial aliasing, the perception of where items are positioned in two- and three-dimensional space becomes distorted (Figure 10–15). When enough samples are provided for the signal shown, it will be represented as a long and solid object. However, when too few samples are provided for the signal, that is, when the sampling theorem has been violated, the solid object is perceived as a series of objects.

This spatial aliasing phenomenon is also very noticeable when looking at computer type after it has been recorded to videotape. On the computer display, the letters look fine, with smooth edges even on fonts with fine serifs. After being recorded on videotape, these edges take on a stairstepped appearance, often called *jaggies*.

Temporal Aliasing
In temporal aliasing, the perception of movement over time is distorted (Figure 10–16). When enough samples are provided for

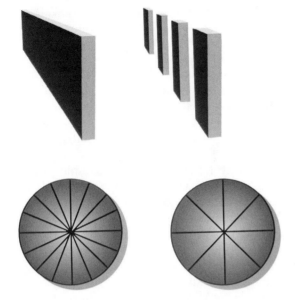

Figure 10–15 Spatial aliasing affects the perception of the three-dimensional space that an object occupies. When we remove sample points, a solid object appears as a series of individual objects. Illustration by Jeffrey Krebs.

Figure 10–16 Temporal aliasing alters the perception of movement over time. As the number of samples (wheel spokes) decreases, there is ambiguity regarding the direction of the movement of the wheel. Illustration by Jeffrey Krebs.

the signal, the perceived direction of the wheel is not ambiguous; it appears to be moving forward. However, when the sampling theorem is violated and there are not enough samples, the direction in which the wheel is traveling becomes ambiguous. It is not clear if the wheel is moving forward or backward. If the sampling theorem were not violated, we would have enough information to be able to judge the correct movement of the wheel; but, if the wheel has traveled more than halfway (180 degrees), and we do not have at least two times the cycle of change represented as information to us, we will not be able to judge the proper direction. Alternatively, as long as we can view the wheel before it reaches the halfway mark, we will be able to perceive the wheel as going forward.

Nyquist Limit

When sampling techniques are used, the sampling theorem asserts that we must sample a signal at least twice as fast as the signal can change. When we are subsampling, we are in violation of this theorem. Note, though, that the theorem states, "at least" twice as fast. There are reasons why we would want to sample a signal more frequently than exactly twice the rate at which the signal can change.

The Nyquist limit refers to one-half of the highest frequency at which the input material can be sampled. For example, if the optimal number of samples for a signal is 44, the Nyquist limit would be 22 (44 samples ÷ 2 = 22 samples).

Figure 10–17 Sampling a signal at less than and greater than the Nyquist limit leaves no ambiguity as to how the signal should be interpreted. Illustration by Jeffrey Krebs.

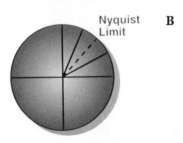

Figure 10–18 The Nyquist limit is shown as the midpoint between one of the four cycles of change. Illustration by Jeffrey Krebs.

Figure 10–19 Sampling at exactly two times the rate of change yields ambiguity. It is impossible to judge the direction of the movement of the wheel. Illustration by Jeffrey Krebs.

The problem with only sampling at two times the rate of change is that conditions can occur in which the resulting information is of an ambiguous nature. When that happens, it is often difficult to interpret and understand the message. Returning to our example of the wheel moving forward, we can sample points between 12 o'clock and 3 o'clock. In Figure 10–17, we have sampled the wheel two times, at less than and greater than two times the frequency of its cycle (at 12 o'clock and 3 o'clock). Because these sample points exist, there is no ambiguity in judging the direction of the wheel. It is moving clockwise. If we draw a dotted line halfway between our two sample points (within 12 o'clock and 3 o'clock), we reach the midpoint of one of the four cycles that the wheel requires to rotate completely (Figure 10–18). This line represents one-half of the highest frequency that can be sampled. It is the Nyquist limit.

However, if we only sample at two times the rate of change, it is not at all clear as to how the wheel is moving (Figure 10–19). We only have a sample point midway through the cycle. Is the wheel moving forward or backward? We do not have enough information to make a judgment. The message presented to us is ambiguous, and as a result, we cannot begin to make an interpretation. One of the possible ramifications of sampling at *exactly* two times the rate of change is that ambiguities could occur.

Affect of Sampling and Selective Removal of Samples on the Message

The act of sampling analog signals to convert them into digital files is a process that brings with it benefits as well as possible detrimental side effects. The degree to which the integrity and intent of the information is preserved and can be adequately deciphered by the viewer/listener is dependent upon the number of samples (represented by bits) in the digitally converted file.

Once an analog video or audio signal has been adequately sampled such that aliasing has been kept to a minimum, the overall number of samples can then be further reduced, and certain samples can be discarded. This process is called *subsampling*. This reduction is not arrived at by randomly removing samples. Rather, such reduction techniques are based on how images are perceived and how the human eye is attracted to certain areas of an image with regard to visual perception methods.

For example, if we remove pixels in violation of the sampling theorem when we sample a solid color background that has been generated on an electronic paint system, instead of a solid background, we would see a quantized background that would show aliasing. Instead of our solid background, we would see stripes of colors, most likely evidenced in a stair-stepped pattern.

Consider what would happen if samples assigned to an image were removed without ascertaining whether the samples were expendable or not. For the example shown in Figures 10–20 to

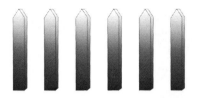

Figure 10–20 Six pickets in a fence. Illustration by Jeffrey Krebs.

Figure 10–21 Sample points between pickets. Illustration by Jeffrey Krebs.

Figure 10–22 Arrows show which points will be removed. Illustration by Jeffrey Krebs.

Figure 10–23 Three pickets are lost in the reduction process. Illustration by Jeffrey Krebs.

10–23, consider the drastic effects of an instruction that states, "remove every other sample." Shown in Figure 10–20 is a row of six pickets in a fence. We attempt to represent the fence by issuing samples. We assign sample points between the pickets (Figure 10–21). We assign three samples per picket, well below the Nyquist limit, for a total of 16 sample points. However, we decide that the resulting sampling is too large, and we attempt to reduce the size by 50%: eight fewer samples. We issue a command to remove every other sample point.

The arrows in Figure 10–22 indicate the sample points that will be removed. However, note that three sample points that will be removed fall on pickets. Our original instruction was to remove every other sample point. As a result, the original image of the picket fence will be altered.

In the resulting image (Figure 10–23), we have the eight sample points that we wanted, but we have only three of the original six pickets! We have lost three pickets during the sample reduction stage. Deciphering the resulting message becomes difficult; it is not at all clear where actual pickets should be placed since we are now in violation of the sampling theorem. This phenomenon of subsampling is actually known as *picket fence effect* and is often seen when computer images are taken from their original environment and recorded onto a medium that may compromise the sampling theorem.

Sampling pictures and sounds is an integral step in transferring analog signals into data that computers can process. The act of quantizing such signals yields artifacts, some of which can be ingeniously masked and filtered. When an analog signal is digitized, samples must be assigned to recreate accurately and adequately the original signal. Lossy compression techniques can further affect the integrity of the signal. Finally, when subsampling or discarding samples to reduce the size of a file further, as evidenced by the picket fence example, entire portions of an image can be lost forever. For these reasons, it is vitally important to understand why we cannot indiscriminately throw away portions of a picture and expect to have it always look acceptable or decipherable.

Decimation

Decimation means removing a great proportion of the elements that make up a whole. This form of subsampling occurs at the pixel level. If we have a 24 bit picture (8 bit samples for RGB) and we subsample the picture to 15 bits (5 bit samples for RGB), then to 12 bits (4 bit samples for RGB), and so on, the original picture will be drastically altered simply because much of the information necessary to represent the picture accurately will have been removed (Figure 10–24). In this example, the reduction of the number of bits per pixel shows significant quantizing artifacts. Although the original quality of the 24 bit picture is affected, these pixel subsampling techniques create visual effects that can be used for artistic purposes. Many of these techniques are employed by digital video effects systems to achieve creative visual effects.

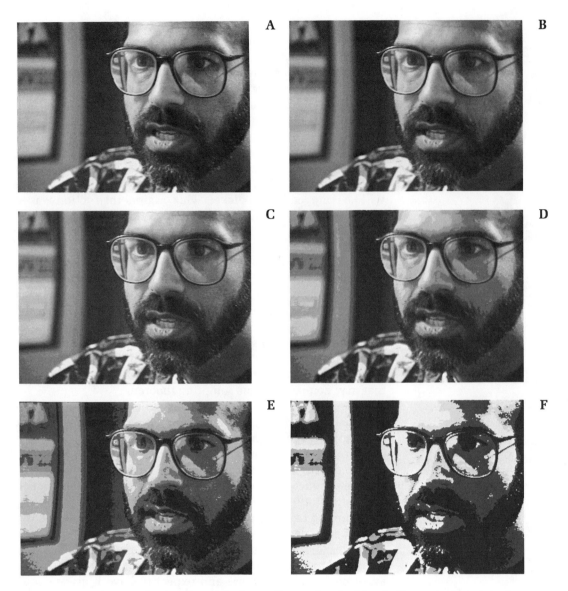

Figure 10–24 Decimation is a form of subsampling at the pixel level. Shown is the original 24 bit, 8 bits per pixel image and subsampled images at 15, 12, 9, 6, and 3 bits per pixel.

Error Masking Techniques

When a number of samples is missing from a signal, there may be ways to mask, or "repair," the affected areas. Digital recordings, unlike analog recordings, do not recreate video and audio signals by decoding waveforms for those signals. Digital recordings assign numbers to represent picture and sound. There is a specific benefit to this process: Numbers don't degrade. Digital videotape systems and compact audio disc systems still rely to some degree on *data error masking techniques* to compensate for media microfailures (dropouts), which are any brief failures of the medium to provide the data that the system is requesting.

Although video and audio signals are infinitely more complex to store as digital data than text, they enjoy an interesting advantage

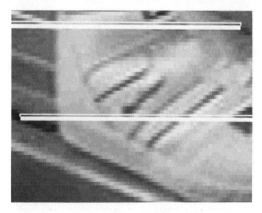

Figure 10–25 Error masking techniques can be used when an original image experiences failures in the medium to preserve recorded information. The white areas in the second image are "repaired" by filling in the defective portions with surrounding lines of video. Illustration by Rob Gonsalves.

A over characters, words, sentences, and paragraphs. Video and audio signals are much more tolerant than text or ASCII files or word processor documents because there is a degree of predictability in video and audio. If errors occur, such as videotape dropouts, and information cannot be retrieved, it may be possible for the problem areas to be repaired. These areas can be filled in (masked) by using information from surrounding areas.

In Figure 10–25, the first image of the sneakers shows the original picture, which has no flaws. The second image shows defective areas that can result from microfailures in the medium. In the third image, the defective portions of the video have been "repaired" by using surrounding lines of video to fill in the defective areas. This technique is a form of error masking.

B Errors in playing back samples of audio can readily be heard if they fall within the normal range of human hearing, about 20 kHz. Error masking techniques are prevalent in digital audio systems, such as audio compact disc players. Compact audio discs are manufactured at a standard sampling rate of 44.1 kHz. Rather than count the exact number of samples—44,100—compact disc players often "oversample" the amount. It is normal to enter a retail store and see machines that offer "2 × oversampling" or "8 × oversampling."

In the case of a compact disc player that offers "2 × oversampling," more samples are taken in (counted) than will be offered as signals in the output data. The disc player reads the CD and processes 88.2 kHz, but it only plays back at a rate of 44.1

C kHz. If an error occurs while the machine is playing back a section of the CD, there will be redundant samples to fall back on. The benefit to the listener is that the error is internal; there is no apparent break in the sound. The "extra" samples don't actually exist on the disc; rather, they are created in the player!

The concept of generation loss or no loss in a digital system also provides insight with regard to lossy and lossless digital compression techniques. The same repetitive and predictive nature of video that makes it fault-tolerant and the beneficiary of error masking algorithms (techniques) also makes video, as well as audio, a highly compressible signal requiring minimal data to represent the level of information perceived.

When a video signal is processed by the flash converter and the analog signal is converted to digital data, all the pixels take the same amount of space. Absolutely nothing is done to take into account any redundancy in the makeup of the video frame. This is because the digitizing process has no way of recognizing that there can be areas of a video frame that are similar and, therefore, that methods may exist by which this redundancy could lead to ways of reducing the data to be stored.

At this stage, we have done nothing to reduce the amount of data it takes to represent a frame of video at its original and full resolution. We have only taken one critical step, the digitization of analog signals and the storage of those resulting data points as digital information. However, creating any type of system that permits the digital editing of pictures and sounds requires that something be done to address the problem of storing the enormous amount of data that results from the digitizing stage.

DIGITIZING THE AUDIO SIGNAL

While frames are being digitized and stored to disk, the accompanying audio signals are also being digitized. A simple pathway that two audio channels from a videotape recorder would take is as follows: The two audio outputs from the VTR enter an A/D convertor for audio. In much the same way that our analog RGB video signals are flash-converted and turned into digital data, the audio signals are sampled and converted. The resulting digital signals are then passed from the A/D convertor to a complement card in the computer that processes the data by storing it to computer disk (Figure 10–26).

Sampling an audio signal is very similar to sampling a video signal. We can assign more or fewer samples, depending upon the audio fidelity that we wish to preserve. Fortunately, the bandwidth requirements of even the most professional quality of audio pales in comparison with the requirements of full-resolution images. Whereas one second of our 640×480 video requires 27.64 MB, one second of two channels of compact disc–quality audio (44.1 kHz) at 16 bits per sample (2 bytes/sample) requires:

$44.1 \text{ kB/sec} \times 2 \text{ bytes/sample} \times 2 \text{ (audio channels)} = 88.2 \text{ kB/sec} \times 2 = 176.4 \text{ kB/sec}$

Clearly, maintaining high-quality audio does not require nearly the amount of data that full-resolution video requires. The ability to store sufficient amounts of audio and to offer random-access audio editing is the leading reason why the development of digital audio workstations has progressed so rapidly. While many digital audio workstations utilize hard disks, the relatively low bandwidth of 176.4 kB/sec (or about 10.5 MB for each minute of stereo audio) allows a variety of disk drives to be used to digitally edit audio.

The limit of human hearing is approximately 20 kHz. Recall that the sampling theory holds that we must sample a signal at least twice as much as the highest frequency that we want to preserve. Professional audio recording systems sample slightly above two times, to over 40 kHz.

Tables 10–2 and 10–3 show the audio frequencies and sampling rates of various formats. The digital samples of 44.1 kHz to 48.0 kHz are designed to provide a realized resolution in excess of 20 kHz.

The number of audio channels that can be digitized simultaneously varies, depending upon the hardware interface and the bandwidth. As audio requires a certain number of kilobytes per second, this additional requirement to the overall bandwidth of signals to be processed must be factored into the digitization process.

Hardware interfaces permit the simultaneous digitization of multiple tracks of sound. Some digital nonlinear systems digitize two tracks of sound at a time, while others are capable of digitizing four tracks. Dedicated digital audio workstations (DAWs), on the other hand, can simultaneously digitize eight or more tracks since images are not involved.

represented digitally by
```
10101010
00101010
10101011
01010101
00110101
```

Figure 10–26 The sampling of an audio signal is similar to the sampling of a video signal. The analog waveform is represented digitally by zeroes and ones, which are stored to computer disk.

Table 10–2   Audio Frequency Rates

Format	Audio Frequency Rate
Film sound optical, 16mm	7 kHz
Film sound optical, 35mm	12–15 kHz
VHS linear tracks	10 kHz
3/4"	10–15 kHz
Betacam	12–15 kHz
Betacam SP	15 kHz
Laserdisc	15 kHz
1" type C	15 kHz
1/4" audio tape (at 15 ips)	20 kHz
Hi 8	20 kHz

Table 10–3   Audio Sampling Rates

Format	Audio Sampling Rate
Compact disc	44.1 kHz
D1	48.0 kHz
D2	48.0 kHz
D3	48.0 kHz
Digital audio tape	32, 44.1, 48.0 kHz

Most often, the A/D convertor operates in concert with an additional sound card that performs operations on the incoming audio data. The capabilities of the sound board vary from card to card; some are only capable of passing the digital data on to the disk drive, while others are capable of further filtering and compressing the data.

It is important to note that the audio signals have, thus far, only been digitized. Digital audio data do not have to be compressed since digital audio data files are much smaller than digital video data files. If we elect not to compress the audio data, the information is passed from the sound card to the disk drive, and the digitization process is completed. However, compressing the audio signal can be accomplished with negligible quantization. Digital compression of audio is a common procedure, and as there is an accompanying savings of storage, it is an option to consider.

BANDWIDTH AND STORAGE

Bandwidth, in the context of the digitization process, refers to the number of bits per second of material that can be processed or transmitted within or between computers. The CPU is tasked with processing a number of bits per second when digitizing. The computer can process only a certain number of frames and a certain amount of information for each frame every second. These factors, in turn, affect how close to the original the pictures will look once stored on disk.

Basic Storage Terms

Bit   The smallest amount of information for computer.

Byte   Eight bits.

Kilobyte(kB)   1,000 bytes.

Megabyte(MB)   1,000,000 bytes.

Gigabyte(GB)   1,000,000,000 bytes.

Note that these figures are approximate! The exact numbers are as follows:

1 KB = 1024 bytes
1 MB = 1024 × 1024 bytes
1 GB = 1024 × 1024 × 1024 bytes

Let's say that we have a computer display that provides a pixel matrix of 640 × 480. This equals 307,200 sample points. If we have red, green, and blue versions, we multiply 307,200 by 3 to get 921,600 bytes of information for each frame of video, if we assume eight bits per color per pixel. If we have 30 frames of video each second, we end up with a total of 27.648 MB of information; this is an enormous amount of data, and we haven't even begun to consider the requirements for sound!

The constrictive factors of bandwidth limitations, economies of processing power, and disk storage capacities must be addressed to solve the essential problem: how to digitize full-motion video with sound and offer enough stored material so that an editor could work with a viable amount of footage? To accomplish the editing of digitized pictures, what is needed is a way to reduce drastically the amount of data stored for each video frame.

INTRODUCTION TO COMPRESSION TECHNIQUES

Lossless Compression

Returning to the example of the document that was reduced in file size and then expanded back to its original file size, we can classify the file compacting program as a *lossless compression technique*. An original message was shortened, compressed, transmitted, and decompressed, and the original message was not changed. We ended up with the exact message that we started with. We did not receive a document with letters in the wrong order or with missing sentences.

Lossless compression techniques are clearly favored over losing information in a message, but lossless compression has boundaries. First, to be lossless, a lot of analyzing must be done. The message must be looked at, the statistics must be gathered, and the codes must be assigned. Depending upon the information to be analyzed, this will require computer power and time. In certain circumstances, such as when video is traveling at full frame rate, there won't exist the luxury of being able to analyze each video frame at less than normal play speed; there will only be a very small amount of time to analyze the message and determine how it should be compressed.

Second, a stage called the *entropy* of the message will eventually be reached. Entropy is the measure of the frequency with which an event occurs in a system. Originally, we had a 100 kB file, and we compressed it to 50 kB. Now, we try to compress the 50 kB file again. We may find that we are able to further compress it slightly, perhaps by several kilobytes, but we will not receive dramatic results and achieve, for example, a 25 kB file. What has happened is that we have reached the entropy point of the message that we are trying to compress. The resulting message is at its optimal point of compression.

It is important to note that the entropy of a message is different from the compression limits of the system performing the compression task. If a software compacting program has, as its limitation, a maximum file size reduction routine of 50%, this is the inherent capability of the system performing the compression, and it is a factor independent of the entropy point of the message being compacted.

Lossy Compression

When we make the decision to compress a file, whether that file is a document, a photograph, or a frame of video, we may have to make decisions that take us out of the realm of lossless compression and into the domain of lossy compression. Lossy compression simply means that we have lost information; the amount of information with which we started is not equal to the amount of information with which we ended.

Why would we want to lose information? Why not always work with lossless compression techniques? Let's return to the 100 kB document. What options are available to us when we still require a file that is less than 50 kB in size? We try to compress it further, but we reach the entropy point of the message. We still want to reduce its size. The remaining option is to choose to lose some information but to leave enough clues so that the message can adequately and accurately be deciphered.

For example, consider our original sentence:

> Letters and words in our document are not equally likely; there will tend to be much repetition of a certain class of letters (such as vowels) as well as words.

What if we were to compress the sentence and make just a few changes that are lossy? We would lose information, but not so much information that we would change the readability and the decipherability of the message. For example, we could try the following:

> Lettrs and wrds in our docment are not equaly likely; there will tend to be much reptition of a crtain class of lettrs (such as vowels) as wel as wrds.

Clearly, we have lost information, but the message can probably be read and understood even with the missing letters. This compressed file will never be the same as the original: we cannot replace information that has been deleted. Regardless, we have achieved our aim: The file is now represented by a smaller amount of data.

The readability and decipherability of this sentence depended on some very careful choices during the deletion stage. If too many letters were removed or if letters were removed that would leave the meaning of the word ambiguous, we would have rendered sections of the message illegible. Assumptions regarding which letters to remove from which words were based on how we process language.

Assumptions regarding how much information is enough to receive the message successfully have to be made regardless of the item being compressed. Information is going to be lost, and we must attempt to ensure that it will be information that is expendable!

These assumptions about how information is processed by the human eye and intellect, theories concerning how much information is enough, and compromises therein, are decisions that will have to be made when the material to be compressed in size is no longer textual, but visual: film and video frames.

Once this lossy compression stage is complete, the losses are confined to one instance. As long as the digital data continues to be processed in a digital environment, there will be no further degradation of that data. Regardless of what we do to the digital samples, no further loss will be realized. This is particularly necessary when the information we are processing is visual material. If we compressed a picture of flowers and then electronically painted glows over the flowers, we would not want to see any further signal loss occur in the picture. By remaining in the digital domain, loss is confined to the initial compression stage. However, faced with the enormous amount of information inherent in each frame of video, digital video compression methods are required to store large amounts of moving pictures and sounds to computer disk.

PRODUCTS AND CAPABILITIES BASED ON DIGITAL MANIPULATION

While the basic digitizing properties of the flash convertor permitted single-frame grabs, sophisticated machines began to appear that could digitize video in real time. This brought forth a series of products that allowed for the management and modification of the resulting digital signal. These devices began to appear in the early 1970's.

Products such as digital video effects units (which rapidly decode, digitally manipulate frames, and then encode), digital still stores (which store full-resolution video frames), digital paint systems (which are sophisticated electronic painting systems), and digital compositing devices (which allow digital images to be layered on top of one another without generation loss), are all examples of products based on digitally manipulating video frames.

Digital video tools are so ingrained in current methods of production and post-production that the effects that they create may be familiar. These tools are used to create many of the special effects we regularly see, for example, in television commercials.

Being able to control components of a picture based on digitally manipulating the numbers that make up the picture provides the user with much more choice and control. Contrast this to the manner in which manipulating analog signals occurs: mostly by brute force approaches that do not permit for easy repetition of the task.

However, with digital techniques, modifications to a frame or a series of frames can be handled more intelligently, precisely, conditionally, and repetitively. When we have the ability to manipulate a frame of video digitally the digital signal sifting becomes much easier to accomplish than trying to accomplish the same sifting in an analog world.

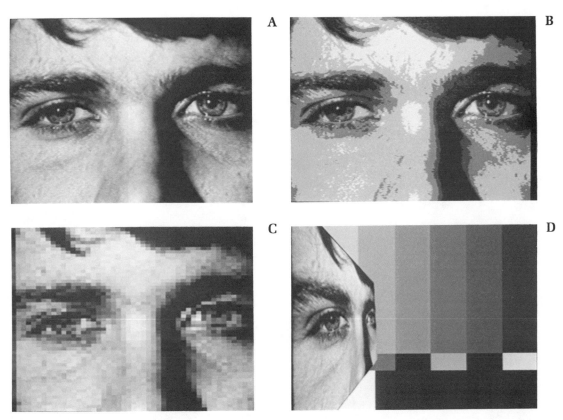

Figure 10–27 By manipulating the digital data that represents a picture, several creative visual effects can be achieved.

Shown in Figure 10–27 are several examples in which manipulating the digital data that represent a picture can lead to interesting visual effects. The first picture is the original 24 bit image. The second picture shows the results of manipulating the degree of chrominance, contrast, and luminance in the image to create a visual effect. In fact, the subsampling technique of decimation is utilized. The third picture is a mosaic version of the original. This effect was created by decreasing the number of horizontal and vertical samples used to display the picture. The fourth picture is an example of adding perspective to the original image by changing its normal aspect ratio of four units horizontal by three units vertical.

Making changes to a frame of video by rearranging the digits that make up the frame has led to the creation of digital effects units that have become standard in the post-production world. In the ever-changing world of new machines that allow the user to paint, bend, enlarge, move, reposition, and combine frames, there are undeniable advantages of being in a digital environment to manipulate frames of video.

11

Digital Video Compression

Digital nonlinear editing systems offer different levels of picture quality. While one system may offer pictures that have been stored in black and white, another system may provide very high quality pictures that are indistinguishable from the original images. The amount of data that is preserved for an image affects the quality of the picture to the human eye. The extent to which the original signal is compressed has an impact on the quality level of the pictures.

Video signals, on a frame-by-frame basis, represent a large number of pixels. Large amounts of computer memory are required to store several seconds of video (Figure 11–1). Depending upon the pixel matrix used—RGB, NTSC, PAL, and so on—the amount of information per frame can vary. For example, if we are storing video signals that have been flash-converted to RGB, we can easily determine how many bytes are required to store the data.

NTSC
It is important to understand that an RGB picture has no inherent resolution with regard to pixel dimensions. For example, if we show an NTSC picture on a Macintosh computer that has dimensions of 640 × 480, we can determine the storage requirement by performing the following calculation:

640 × 480 = 307,200 pixels × 3 (RGB) = 921,600 bytes (921.6 kB)

Thus, it takes slightly under 1 MB to store one video frame at this pixel matrix. Each second of video requires 27.648 MB of storage (921.6 kB per frame × 30 frames per second).

PAL
A PAL RGB picture may be calculated at a pixel matrix of 768 × 576.

768 × 576 = 442,368 pixels × 3 (RGB) = 1,327,104 bytes

With PAL signals, 1,327,104 bytes are required to store one video frame. Each second of video requires 33.178 MB of storage (1,327,104 bytes per frame × 25 frames per second).

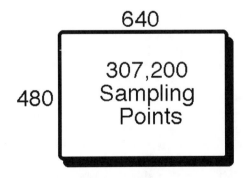

640

480

307,200 Sampling Points

Figure 11–1 The data required to represent one frame of video is enormous. The 100 sampled points of the Lincoln photograph would fit into this same space over 3,000 times!

CCIR 601, D2, Component Digital NTSC

For CCIR 601 signals, we use the following calculation:

720 × 486 = 349,920 pixels × 2 (RGB) = 699,840 bytes

With CCIR 601 signals, 699,840 bytes are required to store one video frame. Each second of video requires 20.995 MB of storage (699,840 bytes per frame × 30 frames per second). We multiply the number of bytes by 2 since samples of R-Y and B-Y are alternated in the 4:2:2 environment.

High-Definition Television

In HDTV, the number of total sampling points with regard to horizontal and vertical lines has yet to become standardized. There are various proposals at hand for an HDTV standard, including these:

1. 1125 lines at 60 Hz in a 16:9 aspect ratio (30 MHz bandwidth for RGB and luminance)

2. 1250 lines at 50 Hz in a 16:9 aspect ratio

For the purposes of this calculation, let's take the realized resolution of proposal 1 and calculate the amount of storage required for one frame and for one second.

We begin with 1,920 horizontal sample points and multiply by 1,035 vertical sample points to get 1,987,200 pixels. We multiply this by 2 because it takes 1 word (2 bytes) to store each pixel in a digital composite format (4:2:2, where 4 equals luminance samples, 2 equals red minus luminance [R-Y] samples, and 2 equals blue minus luminance [B-Y] samples).

1,987,200 × 2 = 3,974,400 bytes

Therefore, to define one entire HDTV frame, we require 3.974 MB. For 30 frames of video, just one second, we need 119.232 MB (3,974,400 bytes/frame × 30 frames) of storage.

EDITING FULL-RESOLUTION FULL-BANDWIDTH DIGITAL VIDEO

Not all applications are based on editing digital video that has been subsampled and compressed. Rather, an entire segment of the post-production industry is concerned only with the editing of full-resolution digital video. When editing short pieces that rely heavily on graphic compositing in which layers of video and graphics are intertwined—indeed, in any circumstance where the program will involve many generations—the only solution to preserving the full bandwidth of the video is to edit in a lossless environment.

A variety of systems digitize analog signals and do not subsample or compress these signals. A variety of systems operate only in a full-resolution, fully digital mode. These are digital systems such as the Quantel Harry, a digital layering device, and the Abekas A62, a composite digital recording and playback device, which in essence comprises two magnetic computer disks that store digitized information.

There are many such products that, as a first step, have something in common: They let the user digitize and store video at full resolution, that is, at full bandwidth. These full-bandwidth signals require a large amount of storage. The Abekas A62, for example, holds only 100 seconds of video on its two internal magnetic disk drives.

When it is necessary to have more storage available, these types of devices have only one recourse: to offer more storage capacity by adding more or larger disk drives. With some systems additional disk drives cannot be added, and capacity is limited to the original offering. Under such circumstances, the user needs to get a bit more creative. If ten minutes of full-resolution digital storage is necessary and only five minutes of disk capacity is available, work must proceed in smaller sections.

It is important to realize that these full-resolution, full-bandwidth systems do not remove any information from the original picture. Recalling our statistics, these systems must be capable of storing about 1 MB of information for each video frame. Because of the limited amount of storage, these devices are typically used for shorter programming, such as ten- to 30-second graphics openings for television shows and for segments that involve many layers of composited images.

Digital nonlinear editing systems must provide a reasonable amount of footage with which the editor can try options for the program being edited. Is it possible to edit with a system that offers only 50 seconds to five minutes of storage? It can certainly be accomplished, but is it practical? If a director has finished shooting a television commercial and has three hours of material, is it practical to work in five-minute sections at full resolution? It depends upon the nature of the finished commercial and the patience of the creative editing team!

The solution lies in the reduction of data for each frame of video. Only when the bandwidth of each frame is reduced do we begin to have additional storage that the editor can gain access to simultaneously. Recall that even the CMX 600 afforded about five minutes per drive for a total of 30 to 35 minutes, even though the picture exhibited many artifacts.

How much resolution should be maintained for each frame of video, and how much digitized footage will be available at any one time? These two key questions will continue to be asked for several years to come. The appropriate resolution depends on the economics of the application, and working at full resolution won't always be necessary or desirable.

The tug of war between full-resolution pictures and large amounts of footage forms the basis for the discussion on the topic of digital video compression.

ANALOG COMPRESSION

Before continuing our discussion of digital video compression, it is necessary to understand those concepts that carry over from analog-based compression (Figure 11–2). Digital video compres-

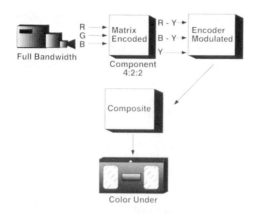

Figure 11–2 The analog compression process for an RGB-originated camera signal shows the degradation inherent during the various compression stages as the signals approach their final distribution form on videocassette. Illustration by Jeffrey Krebs.

sion methods have much in common with the manner in which analog compression of video signals occurs in everyday image origination and distribution. Analog video signals are routinely compressed from the time that they are originated to the time that they are viewed. When we watch a television show, we are watching the effects of analog compression. The original signal that was captured is superior to the image we get to view because of the stages that the signal has encountered. These stages may be long or short, depending upon the signal path that the program must take from origination to distribution.

Consider the compression path that an analog video signal will take when we capture images on a video camera for distribution to our audience on 3/4" videotape. Shown in Figure 11–2 is the original camera signal, whose red, green, and blue signals carry 10 MHz of information per channel. These signals represent the full-bandwidth, un-compressed signal that originates from the camera.

Since the distribution form in this example is a composite VHS videocassette, it is necessary to matrix-encode these individual signals. *Matrix* simply means that the individual signals must be combined, in effect, composited together. This condition is also referred to as *color under*, where luminance and chrominance first are separated while the information per channel is reduced from the original signal, then these signals are combined through a heterodyne process. Matrix encoding results in the original RGB signal being broken down into the following components: R-Y, B-Y, and Y, where R = red, B = blue, and Y = luminance.

These Y, R-Y, and B-Y signals exist in a 4:2:2 ratio; where there are four samples of luminance for every two color difference signals. This process also reduces the amount of information for the individual channels, as luminance is filtered to 4 MHz per channel and each of the color signals is filtered to 2 MHz per channel. We have encountered our first stage of analog compression.

The next stage is to encoder-modulate these component signals into a composite signal. Here, the color information is further wave-shaped down to 1 MHz or less per channel. By the time that the composite signal is introduced to the VHS machine and recorded, the color information has been limited to approximately 1/2 MHz in bandwidth. The luminance channel has been further reduced to 2 MHz.

This analog compression process is a normal affair for images that originate in RGB and must be recorded onto composite, color-under videotape machines. From the original 10 MHz of information per channel, there is quite a decrease in the luminance and chrominance information by the time the signal is recorded to tape. This is a form of compression, and many of these same principles carry over to the digital video compression techniques outlined below.

DIGITAL VIDEO COMPRESSION

The digitizing experiments that scientists at Bell Labs conducted regarding visual recognition yielded certain characteristics and traits for the minimal amount of sampling information necessary to recreate an image. In the early 1980's, Sarnoff Labs, a division of RCA, conducted experiments aimed at providing a new form of entertainment. These experiments eventually led to methods of digital image compression.

The appearance of compact audio discs in 1983 heralded the ability to search and play audio in whatever order the listener wanted. RCA felt that "nonlinear home video" was a promising area for product investigation. The goal was to provide home presentation programming through the use of hardware and software systems that would put the viewer in charge of how the presentation proceeded.

To test whether these interactive programs could be successful, four pilot projects were conducted. One involved the subject of archaeology. A production crew shot every conceivable path around a historic temple, allowing the viewer to see every possible point of view to and from the temple. An archaeological dig was in progress, and the production included footage of the excavation of artifacts, the preservation of these items, and their eventual display in museums. The program combined all this motion material with still pictures, audio, and graphics.

A computer program was written that allowed the user to, in effect, explore the archaeological site at will. The user could choose to learn about the history of the site, view the site, watch the unearthing of an artifact, and then see the artifact displayed in a museum. Or if the user did not want to progress in such a linear fashion, she could start out by seeing the object in its museum setting and then jump to an exploration of the site.

The manner in which the material was laid down to disc didn't matter at all; the power to move through the material was in the hands of the user, so in whatever fashion the user wanted to explore, the computer program allowed her to do so. All this was possible because the medium on which the project was stored allowed such freedom.

How was it done, and how did the user have random access to the material? Since the material was digitized and the pathway to the material was a computer program, on what storage medium was the project recorded? The method involved compressing the huge bandwidth of video that made up the program and storing the results onto a digital medium. A video compression method known as *digital video interactive* (DVI) was used. It searched for and discarded redundancies of visual information. In so doing, the resulting pictures were no longer at full resolution. However, the quality was sufficient to enable users to view the program successfully. The information for the program, now digitized and compressed and existing in digital form, was recorded to a compact disc that included video as well as audio, and played back in a compact disc interactive (CDI) system. This happened in 1987.

Combining all these different media into one medium that could then be easily accessed by the user via the computer required the use of digital video compression. Digital video compression can be achieved by software alone or by software and hardware.

There are several current types of digital image compression methods:

Digital video interactive (DVI)

Joint photographic experts group (JPEG)

Moving picture experts group (MPEG)

Each type of compression scheme is examined in this chapter. Each offers different benefits and liabilities.

As we know, one frame of RGB video requires just under 1 MB of disk space. Considering that disks on personal computers usually range in size from 20 to 40 MB, preciously few video frames can be stored and accessed. When we chose to compress our document, we did so with the intention of saving quite a bit of space; by using a lossless compression technique, we were able to save about 50%, a ratio of 2:1, and what was even more attractive was that we were able to decompress the file and return to the original size without loss of content.

However, at 1 MB or more per frame and with 30 frames per second, we can readily see that if we want to play digital video from computer disks, fantastic ratios are required. Without subsampling, to have even ten seconds of full frame rate video on a 40 MB hard disk would require a compression ratio of 15:1; to get one minute of video on the disk would require a ratio of 90:1; two minutes would require a compression ratio of 180:1!

HARDWARE AND SOFTWARE COMPRESSION METHODS

Compression, reducing a volume of data, can be accomplished using software-only methods or through a combination of hardware and software methods. The advantage of being able to compress in a software-only system is that dedicated hardware that aids the compression process does not have to be designed and implemented. However, the disadvantages far outweigh this economic benefit.

The most powerful compression methods are achieved through a combination of hardware and software. Adding hardware that can process more instructions per second than can a software algorithm helps to avoid some of the consequences of a software-only method. In general, when hardware is introduced, compression methods that are capable of ratios from 8:1 to 150:1 are achieved. Unburdening the CPU also allows more bits for each frame to remain intact; as a result, better images are possible with hardware-assisted compression methods.

SOFTWARE-ONLY COMPRESSION

As we know, it takes time to analyze analog signals and determine how the signal should be sampled and coded. Software-only compression methods can be made to work in the real-time world of video, which moves at 30 frames per second, but the amount of information that they can process and pass—in effect, take in and move out—is limited. Software-only compression schemes can only process frames of video that have a certain number of bits representing that frame of video. When the compression scheme reaches the threshold of data that it can process and pass, more information must be thrown away, the number of frames per second must be reduced, or the process must become slower than real time.

All three consequences can be objectionable. If we can process only a certain number of bits, the visual quality of our digitized video frames will be limited. If the number of frames that we can process per second must be reduced, we are no longer looking at full frame rate video; instead of looking at 30 frames per second, we are looking at something less. If we must slow down the entire process to spend more time analyzing each frame, our process becomes non–real time. Real-time digitizing and compressing methods represent the most viable means of transferring material from the analog world to the digital domain when preparing for editing sessions. If the transfer of one minute of material suddenly has to be accomplished at half of real time, it will take two minutes to process the data. If we have 20 minutes of footage, it will be 40 minutes before we can start editing with the material. Such delays are usually not acceptable.

Software Compression Methods, 1987–1989

Software-only compression methods use the following components: (1) CPU, (2) flash convertor (A/D; the digitizer), (3) framestore (computer memory), (4) D/A convertor, and (5) software algorithms that orchestrate the components and the compression process by determining the amount of data that will be stored for a video frame.

The following compression process uses a Macintosh IIF/X as its computer platform. Certain specifications may differ, depending upon the platform. This method was considered state-of-the-art from 1987 to 1989 before hardware-assisted compression techniques came into use.

Figure 11–3 illustrates the role of the digitizing engine. The process begins when composite analog video is played into the system. The video signal is decoded and broken down into its analog components: red, green, blue, and sync. Sync is composed of stabilizing signals that are present to ensure that the signals being sent to a television screen are in concert with the normal scanning operation of the screen.

The three color components are then processed by three A/D flash convertors. The digitized data (codes now assigned to represent the analog signals) are passed onto a framestore in the computer (the computer's memory) and stored in a new form, as bits. The framestore is capable of handling 24 bit wide samples.

The RGB framestores are connected to the CPU via a 32 bit wide bus. This bus is usually referred to as the *computer back*

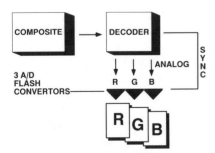

Figure 11–3 The digitizing process begins as composite analog video is decoded into its analog components and flash-converted. Illustration by Jeffrey Krebs.

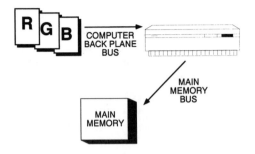

Figure 11–4 The RGB framestores connect to the CPU via the computer back plane bus. Data passing from the framestores through this bus are eventually passed to the CPU's main memory via the main memory bus. Illustration by Jeffrey Krebs.

plane bus. This is a 10 MHz bus (10 million cycles per second). Via this bus, each cycle carries up to 32 bits of data, and the bus can operate at 10 million cycles per second. Since eight bits is one byte, four bytes can be processed (Figure 11–4).

32 bits = 4 bytes × 10 million cycles/sec = 40 MB/sec

This particular bus from framestore to CPU has a capacity of transferring approximately 40 MB of information each second. Without some type of accelerator, no additional data can pass. In actuality, however, even though the bus can sustain 40 MB/sec, it would really only be used for passing 30 MB/sec. Recall that our video frame is providing us with 24 bits of information (8 red, 8 green, 8 blue). Therefore:

24 bits = 3 bytes × 10 million cycles/sec = 30 MB/sec

The CPU is connected to its main memory by another bus called, appropriately enough, the *memory bus.* The width of these buses vary from computer to computer and can range from 8 to 128 bits wide or more.

What we have achieved thus far is the process of digitization: getting analog signals converted to digital signals and stored into a computer's memory.

Once the digitized video frames are resident in the computer, they are available for any additional processing by the computer. It is usually inconvenient to gather people around a single computer display to view an edited program. To disseminate the results more easily, the signals are usually encoded back to composite video and recorded to videotape so they can then be watched on a video monitor or television.

By processing component red, green, and blue and sync through a D/A convertor, these signals are combined and encoded back to composite analog video. Now the signals can be viewed on a television screen.

The entire software-based compression process remained at this stage for some time with few new developments. Since each second of full-resolution video requires over 27 MB of storage, it is easy to note where the potential bottlenecks occurred. Processing the frame was not an issue; we had 24 bits of color coming in and the ability to process 24 bits. Each second of video required passing almost 28 MB of data, and the bus could handle 30 MB. Granted, this affords little margin for error, but so far, so good. The CPU itself, however, was extremely burdened with the entire process of digitizing, and being so taxed, there was little time for the CPU to run software compression algorithms.

The situation was simple. We could digitize in real time. We could even pass the data over the computer bus and store it into computer memory, even though this memory filled very quickly! But with all operations that had to take place, there was little time and little computing power left to run any software compression scheme! Running such schemes was the only way to realize a decrease in the overall amount of data required to process the full-resolution video frames.

The ability to decrease the workload for the computer and free it up to run software compression algorithms came about in 1988.

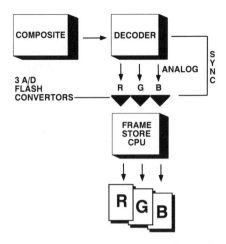

Figure 11–5 Adding a chip-based CPU to the RGB framestore provides the ability to operate on incoming pixels before the main CPU must process them. With this decreased workload, the main CPU can now perform compression schemes. Illustration by Jeffrey Krebs.

Under the original scheme, the computer had to wait until the bits left the framestore and traveled down the computer back plane bus before the CPU could process them with software compression. By the time the bits got to the CPU, processing power was insufficient; as a result, no compression schemes could be run.

The solution was to make an addition to the RGB framestores. A chip-based CPU was added to the digitizing card, and a wide interface from this CPU to the framestores was introduced. Now, by programming this secondary CPU, we gained the ability to operate on the incoming pixels before they got to the framestore. The RGB signals exit the flash convertor, are processed by the secondary CPU, travel to the framestore, and continue to the back plane bus (Figure 11–5).

This breakthrough resulted in decreasing the workload of the main CPU. Suddenly, the CPU was now capable of two important tasks: (1) decreasing the amount of information in each video frame by running software compression programs from the secondary CPU and (2) storing the data to external computer disks.

External Disk Storage

With the CPU connected to external disk drives via a small computer system interface (SCSI 1 type) bus, we gained the ability to process the incoming video frames, store them to computer disk, and play back the images from disk. But before we could store them to the disk, the bandwidth of these frames had to be reduced or else we would still have an age-old problem: the disks would fill too quickly with full-resolution video files!

Software Compression

Software compression techniques were now applied to the frames as they exited the A/D convertor. Software-only methods usually involve two common themes: reducing the size of the matrix for the image by removing pixels (a technique called *scaling*) and reducing the amount of color information (*chroma subsampling*).

Instead of sending the usual 640 × 480 × 3 (921,600) bytes per frame from the framestore to the bus and onto the CPU, the first method of reducing information is to reduce the size of the pixel matrix.

Normally, the RGB frame is 640 × 480. By reducing the size of the matrix, fewer pixels are represented in each frame of video that we have to process. For example, if we use a matrix of 128 × 96, or five times less horizontal and vertical information, we are employing yet another form of subsampling. We are not limited to a pixel matrix of 128 × 96; we could have chosen 256 × 192, 64 × 48, and so on, as long as we remained in a 4 × 3 ratio.

The next form of subsampling that is done is to reduce the overall amount of color that we will pass over the bus. Usually,

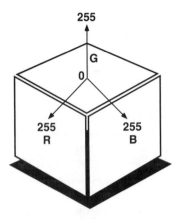

Figure 11–6 The color space occupied by a 24 bit sample. Each color is capable of representing 256 levels on a gray scale. Illustration by Jeffrey Krebs.

we process three colors: red, green, and blue. However, to further reduce our data, we can drastically reduce the amount of color. Again, our normal formula is 640 × 480 × 3 bytes color (8 bits each for RGB). We subsample to a matrix that is represented by 128 × 96 × 1 byte color. Here, instead of representing three colors by remaining in 24 bit mode, we compromise, and attempt to represent three colors by one byte. The question: how do we reduce 24 bits of color (3 bytes) into 8 bits of color (1 byte)?

There are various combinations, of course, but by representing the colors red and green by 3 bits each, and the color blue by 2 bits, we can process all our colors in 1 byte of information.

R (3 bits) + G (3 bits) + B (2 bits) = 1 byte

In Figure 11–6, we look at the color space normally occupied by our 24 bit sample of red, green, and blue. Each color is capable of representing 256 levels on a gray scale. If we multiply 256 × 256 × 256, the total number of colors that this 24 bit sample can represent is about 16.7 million.

In our subsampling example, by reducing the overall number of bits for each color and ending up with just 1 byte for each frame of video, we have lost the ability to see 16.7 million possible colors. Regardless of how complex the color scheme may be in the images being processed, we have only 1 byte to represent that color scheme. Because of this, the most colors we can possibly hope to see represented is limited to 256.

Results of Subsampling

Reducing the combined number of horizontal and vertical pixels and the amount of color information yields significant results. Normally, the framestore would have to pass on 921,600 bytes to the bus:

$640 \times 480 \times 3 = 921,600$ bytes/frame × 30 frames = 27.648 MB/sec

Now, however, the equation looks like this:

$128 \times 96 \times 1 = 12,288$ bytes/frame × 30 frames = 368.6 kB/sec

The savings is on the order of 75:1. The rounded-off 370 kB/sec data stream can easily be processed by the CPU and stored to computer disk. The entire process can occur in real time. With such a large reduction in the amount of data for each video frame, our storage per computer disk increases. As an example, a 100 MB hard disk, storing full-bandwidth video, at just under 1 MB/frame, gives us slightly more than three seconds on the disk. At our new bandwidth of 368 kB, we can fit approximately 272 seconds, or 4.5 minutes, of storage.

However, software-only compression techniques have several limitations. Since they are less powerful than hardware-aided compression methods, fewer analyses can be made on the image that needs to be compressed. Software-only methods employ a brute force approach to compression: the wholesale discarding of pieces of information without the ability to make genuine judgments on the relative importance of the components in the video frame. In effect, the algorithm outlined above directs, "Anytime you see 24 bits, 8 bits each for red, green, and blue, do not let them pass. Instead, give red 3 bits, give green 3 bits, and give blue 2 bits. Repeat."

If we recall our example of the sentence from which we removed letters, certain decisions were made regarding which letters to remove. We did not simply determine that every other time we see the letter *A* it will be removed. We based our decisions on knowledge we possess with regard to how language works and how it is interpreted. But making these judgments takes time. They are based on known information and require additional software programming. In the case of software-only compression methods, there will be little time left over that can be allotted to analyze, judge, and recommend plans of action for each video frame being processed.

HARDWARE-AIDED COMPRESSION

When hardware and software are used together to compress video, several advantages are realized over software only methods. First, hardware-based compression allows more instructions to be run per second. This provides more time to analyze and judge a frame of video before the next frame of video must be processed. Second, the wholesale discarding of information can be relegated as more of a last resort rather than as a given. Instead, more time will be available to examine the video frame, identify redundancies, and assign numeric values. Last, intelligent decisions and plans of action are made based on knowledge regarding how our eyes process information.

As is usually the case with the ever-changing field of computer and silicon technology, the software only compression methods of 1987–1989 have basically run their course. We are currently experiencing the rapid growth of hardware-assisted digital video compression. The extent to which the quality of our original full-resolution video frame can be preserved while packing the data to represent that full-resolution frame into smaller and smaller spaces will preoccupy the manufacturers of systems dependent upon video compression for several years to come.

JPEG COMPRESSION

JPEG compression, which was proposed by the Joint Photographic Experts Group, a subset of the International Telephone and Telegraph Consultative Committee (CCITT) and the International Organization for Standardization (ISO), is a form of hardware-assisted compression (Figure 11–7). The CCITT is the committee that standardized the methods by which facsimile transmissions are sent.

JPEG, it must be noted, is based on still images, also called *continuous-tone images*, whereas MPEG, which is discussed

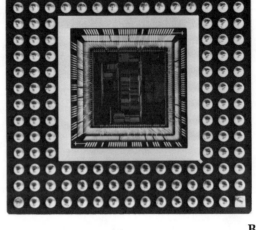

A

B

Figure 11–7 Hardware-assisted digital video compression is accomplished through the use of a computer chip. A variety of JPEG-based computer chips is available from different manufacturers. Courtesy of C-Cube Microsystems, Inc.

Figure 11–8 The RGB signals are converted to YUV, and the color portion undergoes decimation. Next, the signals enter the discrete cosine transform stage. Illustration by Jeffrey Krebs.

later, is based on motion video. Significant and powerful compression methods employ mathematical procedures to accomplish the analysis of the video frame. These mathematical procedures are referred to as *discrete cosine transforms* (DCTs). The DCT is a lossy algorithm, which simply means that when a file is compressed using a JPEG-based processor, information about the original signal is discarded and lost. Which information to discard, which information to keep, and how much information will remain versus the original signal are the key questions that have to be answered intelligently.

In Figure 11–8, the RGB signals are processed by the hardware-based compression "engine," which connects to the RGB framestore. As before, the composite analog video signal is decoded into component RGB, flash-converted, and processed by the secondary CPU. This hardware engine differs from the supplemental CPU added to the framestore because it performs operations based on compression rather than digitization.

Now, however, we introduce compression. Instead of the data leaving the framestore and passing directly over the bus to the CPU, data exit the framestore and enter the compress section of our model. There, the data undergo the JPEG-based compression schemes, travel over the bus, and move on to the CPU.

Adding this compression engine to our model offers additional time to better analyze the video frame. The goals of hardware compression are the same as the goals of any lossy compression scheme: to preserve as much detail as possible in the picture while simultaneously discarding as much information as possible. As we know, there are a variety of concerns, including, how much analysis can be achieved in a limited time, whether the compressed picture will be decipherable, and if there will be enough digitized material to work with on the computer disks.

Inside the Compression Engine

JPEG compression centers around the ability to decode and encode gray scale and color images at full video frame rates. The compression ratio varies depending upon the chip being used, but the top end of the scale would be a ratio of approximately 100:1. Minimal compression is about 8:1. Improvements are to be expected in the performance of JPEG chips, which usually average on the order of ten billion operations per second.

Chroma Subsampling

The component RGB signals from the framestore enter the compression engine. These signals are then converted into YUV: luminance and color difference components. With 8 bits each, we have a 24 bit sample to be processed. While luminance is left untouched, the next step is to decimate color, for it is at this stage that chroma subsampling is applied. It is important to note that, although the digitization process may have introduced quantization artifacts, it is at this stage where we begin a lossy compression process of discarding information that will be irretrievable.

Betacam CTDM Storage Methods

This stage is similar to what happens with analog component signals, such as Betacam's, color time division multiplexing, (CTDM) storage methods. CTDM is a basic technique used in the record and playback processes of Betacam videotape. In CTDM recording techniques, the bandwidth of the color signal is divided by 2. When a composite signal is recorded to Betacam, the signal is broken down into luminance (Y) and the color difference signals, R-Y and B-Y. These color difference signals are then processed through a timebase corrector.

This TBC is a special-purpose device used to provide horizontal management of the color channels. Here, the R-Y signal is shrunk to one-half of its size and placed on the left side of a horizontal matrix. The B-Y signal is also reduced to one-half of its size and then placed on the right side of the same horizontal matrix. This matrix is, in fact, one horizontal line. Both R-Y and B-Y now have only one-half of a line available to them and are, therefore, one-half of their original resolution.

When a CTDM recording is played back, the special-purpose TBC takes R-Y and B-Y and performs a 2× expansion. The luminance signal is delayed to compensate for the delay caused by processing R-Y and B-Y through the TBC. The TBC thus is used to resynchronize Y, R-Y, and B-Y so that they are back in parallel. CTDM signal processing is essentially the reason why it is impossible to purchase a Betacam machine without a timebase corrector. CTDM is an example of the lossy analog signal processing environment.

Continuing with the subsampling model, we process a sample of U while discarding a sample of V. Then we process a sample of V while discarding a sample of U. Alternately discarding samples in this fashion reduces the overall amount of data to be processed. As a result, our 24 bit sample is now reduced to 16 bits: 8 bits for Y and 8 bits for either U or V. In sequence, it would look like this: Y and U, then Y and V, and so on, with each cycle being 16 bits long. Within the 16 bits per pixel, we can represent 256 levels of color (quantization levels). This is akin to what happens with component 4:2:2 color handling.

After decimation, Y and alternating color difference signals are sent to the discrete cosine transform. The main purpose of the DCT is to use frequencies to represent a picture, whereas, prior to this, pixels were used to represent the picture.

It is important to note that compression algorithms vary from manufacturer to manufacturer. Instead of the decimation described above, increased or decreased ratios may be employed. Alternatively, no decimation may occur, and all 24 bits will be preserved. Which algorithm to use and to what degree decimation will occur are choices that will affect the type of picture quality and amount of storage required.

When the YUV components of a frame enter the DCT, the frame is divided into 8 × 8 squares. The entire picture is processed at one time, and the DCT analyzes this whole array. This is the first step in representing the picture as frequencies and not as pixels (Figure 11–9).

Representing a picture's elements via frequencies can be thought of as if we were asking the question, "How much brighter are these elements of the picture than those elements of the picture?" The baseline from which to make these comparisons is zero direct current (0 DC), which, for our purposes, equals the average gray level for the picture being analyzed. These frequency determinations continue until the entire 8 × 8 array has been analyzed. The data that are formed represent information for how the pixels were arrayed. We are able to reconstruct a picture based on interpreting the assigned frequencies that make up the picture.

In general, the overall shape and outline of a picture will be represented by low frequencies, while fine edges will be represented by higher frequencies. Another way of thinking about this is that items that are big and take up a lot of "screen space" are represented by low frequencies. Details for those items, such as fine serifs on a typeface, will be represented by high frequencies. In addition, most of the extraneous and unwanted noise in a video signal is associated with high frequencies.

After analyzing an 8 × 8 array, the process is repeated for the next 8 × 8 array. We started with a picture that had a matrix of 640 × 480 pixels. We first subsampled this to 128 × 96, which gave us a total of 12,288 pixels. We analyze these 8 × 8 blocks (64 pixels) and must do this analysis 192 times to completely analyze the picture. The process repeats itself as the next frame presents itself to the DCT.

The analysis of pixel makeup of a picture by the DCT is a lossless step. We begin with 64 pixels in each 8 × 8 array, and we achieve 64 frequencies as a result of the analysis. No information

Figure 11–9 Using an 8 × 8 array, a picture is analyzed as a series of shades of gray. The purpose is to ascertain levels from one section to another and to use this information to represent the entire picture as frequencies and not as pixels. Illustration by Jeffrey Krebs.

has been lost. Rather, the data have been interpreted in a different and lossless fashion.

In Figure 11–10, luminance (Y) and chrominance (C) exit the DCT and are now represented by frequencies (f). The next stage, quantization, is a lossy step.

QUANTIZATION

It is at this stage where a significant amount of JPEG compression is accomplished. Here, file sizes can be marginally or drastically reduced; on the order of 8:1 to a dramatic 100:1. In any case, this is a lossy process; material will now be discarded that cannot be retrieved. How the resulting pictures will be perceived affects the decisions that must be made regarding how much information should be discarded. It is widely thought that a JPEG compression ratio of 8:1 to 10:1 is relatively difficult to distinguish from the original image.

The quantization table (Q table) and the methods by which it operates are based upon human visual system (HVS) studies, which seek to answer such questions as these:

Which frequencies is the human eye very good at seeing?

Which frequencies is the eye not good at seeing?

How does the eye respond to motion?

How does the eye respond to a static frame?

What luminance information can be removed?

What chrominance information can be removed?

The luminance and chrominance signals that make up a picture are not of equal importance with regard to the human visual system. Digital video compression techniques take into account that when the eye processes a picture, the more important aspect is the picture's luminance content. Color is a much more expendable portion of a picture. Based on some estimates, up to 90% of a picture's color information can be removed without adversely affecting the recognizability of that picture. As a result, most compression schemes take into account that much of a picture's color can be sacrificed. Operating under such principles, digital video compression algorithms seek to discard information in an intelligent way.

These digital video compression algorithms take time to develop. Although the process of integrating JPEG compression has become easier, the decisions regarding what information should be kept and what should be discarded are carried out by instructions given by the compression algorithms. These instructions are designed to exploit the strengths of JPEG-based compression: to preserve the detailed sections of a picture while disregarding sections of a picture in which similarities abound.

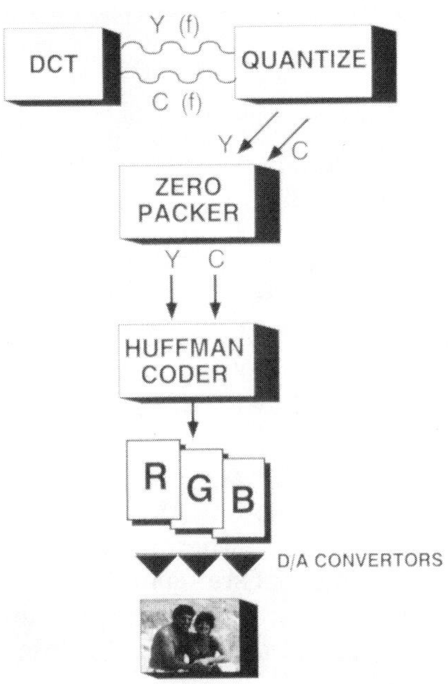

Figure 11–10 The discrete cosine transform processes YUV signals and represents them as frequencies which then enter the quantize section. These frequencies undergo compression and are zero-packed and further coded. The RGB information undergoes a D/A conversion, resulting in a viewable signal. Illustration by Jeffrey Krebs.

Quantization Table Elements, Q Factor, and Quantization Frequency Array

If you think about mathematical ways of reducing a number, one basic method involves dividing that number by a higher number. Simply, the number 50 divided by 5 equals 10. If we increase the denominator to 25, we get 50 divided by 25 equals 2. The Q factor is, in essence, a denominator that reflects how marginally or drastically the frequencies will be changed. The Q factor is the bit/pixel relationship of an image.

A practical guideline is that a Q factor of 50 (Q 50) will lead to little loss of detail (and therefore retains high quality) while up to 95% of the data for that image is removed. When Q 50 is exceeded, deterioration of the image follows rapidly. Q factors of 100 and higher are typically used when image quality is secondary to storage requirements. Therefore, the relationship between the Q factor and kB/frame of images is as follows:

As the Q factor increases, the kB/frame decreases. As a result, a higher Q factor will result in a decrease in picture quality. Conversely, as the Q factor decreases, the kB/frame increases. As a result, a lower Q factor will result in an increase in picture quality.

The goal of JPEG compression at this stage is to represent the 64 frequencies of each 8 × 8 array as smaller and shorter messages to store. For each 8 × 8 array, the frequencies represented by that array are arranged from DC level (0) to 63, which represents the highest frequency in the array.

The quantization factor (Q Factor) is then applied. Here, each of the frequencies for one 8 × 8 array is divided by a set constant:

$$\text{Quantized frequency array} = \frac{\text{Quantized table elements (frequencies)}}{\text{Q factor}}$$

The Q factor chosen has an inverse relationship to the amount of information left intact. Q factors range from 0 to 255 for a total of 256 steps. Recall that we are working in 8 bit samples, equating to 256 levels in gray scale. As the Q factor goes up, more information will be thrown away. As the Q factor goes down, more information will remain.

What is being accomplished during this stage is akin to the same type of process used when long and short codes were assigned to our compacted document. The goal here is to produce a string of as many numbers as possible that are smaller than the original frequencies. By doing this, a series of numbers is created that, when taken in total, will represent transmitting less information for the original number of frequencies. By doing this, we have quantized out a large number of the original frequencies. We have compressed the array and, ultimately, the file.

The quantization process, the Q factor, and the resulting quantization frequency array represent the primary area in which the significant compression steps are taken. The "magic," if you will, with regard to what constants are chosen for the Q table, is the algorithms and formulas based on examining the perceived differences in pictures treated with different Q tables. Recall that we divide the quantized table elements (the frequencies) by a Q factor. The Q factor, in turn, utilizes a table of divisors, the Q tables. These are the tables that further influence the quality of the picture.

The "quality" of a JPEG-compressed image is the ratio of compression (in bits/pixel), which is regulated by the Q factor and the size of the image (in pixels/second).

Our 64 frequencies have now been shortened. When we examine the quantized frequency array, we would expect to see a predominance of zeroes.

Zero Packer

The next step is to process the quantized frequency array. This is done through the use of a *zero packer*. The goal of this stage is to take the resulting zeroes and assign a code that designates how many zeroes there are, which aids in the transmission of less data as the process reaches its end. It is a lossless procedure.

Zero packing uses a process called *run length encoding* (RLE). There is a history of using RLE processing in broadcasting equipment. Early character generators utilized run length encoding to define the shape of characters. Run length encoding determined when to switch a character from black (invisible) to white (visible) over the length of a horizontal line. Each character was defined by zeroes and ones, which would run for a certain length along a horizontal line.

Huffman Coder

The zero packer passes its information on to the Huffman coder. The coder runs a Huffman table. The purpose of this stage is to calculate redundancy in order to store and, therefore, to transmit less data. This is a lossless stage, and it utilizes the process of Huffman encoding.

The Huffman coding stage results in data that is stored to computer disk. When we want to view material, the data is accessed from the computer disk, undergoes decompression, and is reprocessed by the three D/A convertors in the framestore. At the same time, sync is reapplied, resulting in a composite analog video signal. This signal can then be viewed and recorded to videotape.

These are the various stages of decoding, transforming, and encoding that are typically associated with the digitizing and compressing process of JPEG. Different manufacturers and system/software developers may vary or rearrange the order of these processes, but this provides an adequate outline.

Finally, returning to the example of the compressed document, if we were to take a JPEG file and run a software compacting program designed to reduce the size of that JPEG file, we would find that very little, if any, additional space can be saved. Although the compacting program will make attempts, usually the result is 0% saved; the entropy of the message having been reached long before.

Additional aspects of JPEG are of importance with regard to how it is implemented in the compression process.

SYMMETRICAL COMPRESSION VERSUS ASYMMETRICAL COMPRESSION

JPEG is a compression process that is symmetrical in nature. This means that it takes an equal amount of processing power to compress an image as to decompress that image. This is important because in applications designed for editing, the compression of a frame must occur in real time. Decompressing that same frame must also occur in real time. Popular examples of symmetrical methods include teleconferencing and videophones.

Asymmetric compression techniques, on the other hand, require a greater amount of processing power, almost always during the compression stage. Once the material has been compressed, it can be decompressed with fewer processing requirements. Obviously, there are limitations with regard to the editing process. Since asymmetric compression is a non–real time process, it will require a delay while material is transferred from analog to digital, compressed, and readied for playback.

Compact disc—read only memory (CD-ROM) is an example in which asymmetric compression methods are employed. CD-ROM is an optical disc that usually stores 500 MB of data. Transferring information and ordering it onto the disc is an asymmetric process and can require expensive compression systems, but once the disc has been made, the data on it can be read and quickly accessed. CD-ROM is used extensively in industry and education, and the playback systems are quite affordable.

FIXED FRAME SIZE VERSUS VARIABLE FRAME SIZE

JPEG seeks to determine redundancy in a frame while preserving detail. The compression stage, where the Q factor and Q tables are employed, may or may not take into account the complexity of information for each and every frame that must be processed. JPEG is neither a fixed-frame-size or a variable-frame-size compression method; that is the choice of the implementation method utilized.

Fixed-frame-size implementations mean that there is a fixed amount of data that the compression algorithm will allow for each frame. It will not expand this amount of information if the frame contains more data than the algorithm is set to process.

Variable-frame-size approaches, in contrast, are set for a range of data for the frame. If some action occurs from frame to frame, the compression algorithm will, for those frames, expand accordingly and allow more data to pass in order to preserve those elements in the frames related to the temporary increase in data.

For example, let's say we aim a film camera at a boy who is sitting on a bench in a park. Behind him, in the distance is a landscape of trees. The boy is fairly static, and the camera is not moving. A frame of this boy will yield a certain amount of data. Just for purposes of illustration, let's say that each frame will require 10 kB.

Next, we continue to film the boy, and between the bench and the trees in the distance, a woman riding a horse passes through the frame. The amount of data in our frames increases. Let's say that these frames jump from 10 kB to 16 kB because there is more going on in the frame. For the time where the horse and rider are in frame, we will be required to store more data than if only the boy were in frame.

As a frame increases in detail content, the frame will contain more data that will have to be either processed or discarded. Because we have no way of knowing if there will be extraneous and unexpected action in our filming of the boy, how will JPEG compression handle the resulting frames of information?

Fixed-Frame-Size Technique

In the fixed-frame-size approach, a threshold is set for all frames with regard to the total amount of data that can be stored per frame. If the threshold is set at 16 kB, we will preserve all the information in our example with the boy, the horse, and the rider.

However, if the threshold is set lower, to 12 kB per frame, there will be a loss of information (Figure 11–11a). The frames of the boy will be fine, since they only require 10 kB, but the more complex frames will have truncated (missing) data. Missing information in these cases is usually characterized by a loss of data in the coding blocks that make up the picture. When we look at the compressed file as it is being played back, entire sections (blocks) of the picture could be missing; there are no data there to be displayed. Usually what is represented in these sections are random (and incorrect) contents of memory or what can be termed as *pixel confetti*. We do not have the benefit of error masking techniques associated with digital oversampling methods.

Conversely, if the threshold is set too high, to 30 kB per frame, we waste valuable storage space (Figure 11–11b). We continue to use 14 kB more than the 16 kB needed. But we have no choice since we cannot scale down our 30 kB algorithm. It is fixed, and even though the frames coming in only require 16 kB to be stored, we are quite ineffectively using extra storage.

Variable-Frame-Size Technique

In the variable-frame-size approach, no threshold is set for frames being processed with regard to the total amount of data that can be stored per frame (Figure 11–12). By dynamically

A

B

Figure 11–11 (a) An example of fixed-frame-size compression with the threshold set too low. There is a resulting loss of information, often evidenced as a mosaic effect where a picture's pixels cannot be properly drawn. (b) An example of fixed-frame-size compression with the threshold set too high. Here, storage is used ineffectively with regard to the actual file size to be stored. Illustration by Jeffrey Krebs.

Figure 11–12 Variable-frame-size compression accommodates fluctuations in a picture's storage requirements by dynamically allocating the threshold as a picture requires more or less storage. As a result, the captured pictures are less likely to exhibit the artifacts associated with a fixed threshold. Illustration by Jeffrey Krebs.

changing the algorithms to facilitate the variable data size of the frames as the complexity of the image changes, several benefits are realized.

First, storage space is used intelligently. By using only the amount of storage that the individual frame requires, we are able to maximize the effective use of storage. Second, all the data necessary to display the picture properly is captured. In our example, a variable-frame-size algorithm would have accepted data based on the 10 kB frames of the boy. Then, as the horse and rider come into view and the data increases to 16 kB frames, our variable-frame-size approach expands accordingly. When the horse and rider clear frame and only the boy remains, our algorithm contracts to continue to process the less complex 10 kB frames. Last, because we have not failed to store data adequately for the frames, we will not see missing information in the form of pixel confetti.

A variable-frame-size approach involves somewhat more work to design, but the benefits derived are well worth the effort. Despite the clear benefits, it is wise to examine the methods used in the system of choice to judge if variable-frame-size techniques are employed.

Intraframe Coding

When we watch a film or a video, each frame of information is distinct and discrete, representing an entity complete unto itself. Movement is perceived by taking into account that each successive frame has changes that are slightly different from the previous frame. The phenomenon known as *persistence of vision* finds our eyes holding some remnant of information from a previous frame as we process the next frame. Slight, static changes within each frame are blended together to create the semblance of motion.

Intraframe coding of frames under JPEG compression means that each JPEG-compressed frame contains all the information that the frame requires to be displayed (Figure 11–13). While the frame is dependent upon the previous and successive frame to continue to take its role in the appearance of movement, an intraframe-coded frame does not need and is not dependent upon the data contained in any other frame in order to be displayed.

Intraframe coding is important when we think about the editing process. When we are editing, we are always removing and adding frames, joining frames from different sources, and sometimes even removing frames from within the same shot to create jump cuts or to speed up action. This unpredictability of where previous and successive frames will be in our editing program is handled well by intraframe coding. Since each frame that we use during the editing process can be displayed without having to draw information from a previous frame, intraframe coding is currently the method of choice for editing. If we had to restrict our editing choices because we could not reliably display the correct frame at the point that we wanted to make an edit, the creative process of editing would be hampered.

Figure 11–13 Intraframe coding allows each frame to carry its own information and to be drawn independently. Illustration by Jeffrey Krebs.

MPEG COMPRESSION

MPEG compression, which was proposed by the Moving Picture Experts Group, a subset of the International Telephone and Telegraph Consultative Committee (CCITT) and the International Organization for Standardization (ISO), is a form of software compression as well as hardware-assisted compression. The MPEG committee, organized in 1988, began to draft proposals in 1990.

Whereas JPEG is based on still images, MPEG is based on motion. Digital video compression based only on algorithms originally designed for still images is a logical reason why there is a need for both types of compression. There are several important distinctions between JPEG and MPEG.

In MPEG, the basic digitization and compressing processes that have already been outlined in detail are very similar. MPEG also employs 8 × 8 blocks during the DCT stage in which spatial redundancies are analyzed. However, larger coding blocks are employed to examine larger sections of a picture being analyzed for movement. The information regarding redundancy and that regarding movement travel together as they leave the compress engine.

The MPEG video compression algorithm has many of the same applications as other digital video compression techniques. These include digital nonlinear editing, video conferencing, interactive presentations in the home, and electronic publishing, including presentations that combine video and text. Clearly, the applications are wide ranging. MPEG, or MPEG I as it is sometimes called, has a bandwidth limitation of about 150 kB/sec, which gives it just the range needed for CD-ROM applications.

Real-time MPEG compression methods are not yet available, and it is for this reason that MPEG is compared to the non–real time aspects of PLV in DVI compression.

Interframe Coding

The method by which data is stored for MPEG frames is a combination of intraframe and interframe coding. While JPEG is solely an intraframe-coding scheme, MPEG uses both methods. Interframe coding offers a major benefit of MPEG compression: a significant savings of storage over JPEG methods. A ratio of 3:1 is the most usual estimate given for such savings. If we originally could store 30 minutes of JPEG-compressed material, we would realize 90 minutes of MPEG-compressed material. Such benefits seem overwhelming; why use JPEG at all if we can save so much space with MPEG?

As we know, in JPEG, each frame is compressed independently of the previous frame. However, when a series of frames under MPEG compression, each frame is not compressed independently of the others. Instead, MPEG employs both intraframe and interframe coding. Because of this, certain MPEG frames displayed require the presence of codependent frames in order to be drawn (Figure 11–14).

Figure 11–14 With the interframe coding methods of MPEG compression, each frame is not drawn independently. Instead, certain frames are predicted, resulting in a decrease in the amount of storage required. Illustration by Jeffrey Krebs.

In interframe coding techniques, there are a series of frames that are referred to as *I*, *P*, and *B frames*. *I* is an intraframe-coded frame. Data are independent of other frames. It is also called an *intrapicture* and a *standalone frame*. *P* is a predicted frame. Data are predicted from a previous intraframe frame or from a previous predicted frame. *B* is a bidirectional frame. Data are interpolated from the closest I and P frames.

When the MPEG-compression process begins, a frame is coded. This is called an *I (intraframe)* frame. This I frame is the exact type of model used in JPEG. It is a stand-alone frame in that it is completely independent and can be drawn and displayed based solely on the data that it contains.

P frames are created based on predictive coding. For example, if we have a sequence of numbers, 2, 4, 6, 8, we can somewhat reliably predict the continuing sequence: 10, 12, 14, etc. This is *predictive coding*: A routine can be written that attempts to complete the sequence.

It is important to note that I frames and P frames do not follow one another. They are separated by B frames. A way to think about this is that an I frame is created, and then a P frame is predicted. More P frames are predicted until it is time for an I frame to be created. Between the I and P frames are B frames. In general, the I frame is created due to a change in the movement of pixels in the incoming frames to be compressed. However, in MPEG, approximately every one-half second interval will have a new I frame association. This interval may vary, depending upon the scheme that is chosen.

However, with only I and P frames, we have already experienced significant savings since complete data are being stored only for the I frames. P frames are predicted. What is being predicted is movement, and only the data that represent the change in movement need to be stored, not the complete data required for that individual frame.

Bidirectionally Predicted B Frames

B frames separate I and P frames. B frames are bidirectionally interpolated from the closest I and P frames. If we have an I frame with a certain amount of data associated with it and we predict a P frame, we know that a certain amount of time has elapsed between the I and P frames. This elapsed time is represented to us visually by B frames; frames that do not truly exist in terms of their data content. They are interpolated and drawn based on the information contained in frames that do have true data: either the closest pair of I and P frames or the closest pair of P frames.

Using predictive coding (P frames) and bidirectionally interpreting frames (B frames), only storing changes in data (P frames), and only periodically storing all data (I frames), MPEG-compression techniques offer greater storage savings.

Implications For the Digital Nonlinear Editing Process

It is important to discuss some of the aspects of MPEG as they relate to the digital editing process. Under JPEG, since each frame can be drawn and displayed independent of other frames, the editing process can proceed in a normal fashion. However, under MPEG, frames are not independent of one another. P and B frames are dependent upon the I frame. Consider the implications this could have for editing.

If we have two segments of digitized video and we want to edit from a point in segment 1 to a point in segment 2, we will be removing some of the data that precedes the place in segment 2 where we want to make our edit. The editing process cannot proceed if we want to edit from, let's say, a P frame in segment 1 to a B frame in segment 2. The B frame can no longer receive its information from preceding frames; it is now appended to segment 1. A possible solution is to restrict editing only from I frame to I frame, which may be unacceptable to the creative editing process in which access to every frame is usually desired.

In 1991, Didier J. Le Gall, chairman of the MPEG working group, presented a paper outlining the important features of MPEG compression and what it should be capable of offering: random access, fast forward and reverse searches, reverse playback, audio-visual synchronization, robustness to errors, coding/decoding delay, editability, format flexibility (width, height, and frame rate), and affordable decoder systems.

MPEG Asymmetric Compression

The MPEG proposal seeks to address the decoding/encoding delay. MPEG, currently an asymmetric process, requires more time at compression stage than at playback stage. This delay is a consideration, of course, and needs to be rectified as the demands of digital nonlinear editing systems require a symmetric compression model. Some proposals outline a delay period ranging from 150 ms to one second. The shorter delay would achieve real time, while one second would suffice for applications where real time is not as critical, such as electronic desktop publishing.

MPEG Delivering Methods

While there are problems with regard to editing, MPEG can certainly serve as a mechanism to deliver playback-only capability. In this regard, we should not lose track of the possibilities this could lead to, including video kiosks and electronic publishing. While MPEG is not yet ideal for frame-by-frame editing, being able to compress data at a ratio of 3:1 compared to JPEG is a significant attribute.

However, there are intriguing possibilities with regard to the use of MPEG as another facet in the digital nonlinear editing process. If we were editing with JPEG-compressed material and could convert the resulting program to MPEG, a great deal of space would be saved. The decrease in file size would allow the files to be transmitted in far less time than JPEG-based material.

Further intrigue occurs when one considers what the possibilities could be if a JPEG to MPEG to JPEG cycle could be developed. Someone could edit a program with JPEG-based material, convert it to MPEG to decrease the file size and storage requirements, and transmit the program. The receiver could watch the program in MPEG, but if frame-by-frame changes had to be made, the receiver could convert the program back to JPEG to continue editing. If it were a seamless conversion, no further loss of signal would be experienced.

Compressing images without bias and maintaining allegiance to the original compression method is a major step. The ability to convert seamlessly back and forth between and among various compression methods has yet to be realized. In the extremely fast moving compression industry, however, it is perhaps not as far in the distance as one would at first expect.

DIGITAL VIDEO INTERACTIVE COMPRESSION

At the beginning of the discussion about compression technology, the development of DVI technology was briefly outlined . DVI, which began at Sarnoff Labs, is currently being developed by Intel Corporation. DVI, operating much like the process described for JPEG, consists of a programmable chip set and software.

DVI supports both still images and motion video and can be both asymmetric and symmetric in nature. A current implementation can decode and encode in DVI as well as in JPEG. There are two modes of operation: presentation-level video (PLV) and real-time video (RTV).

Presentation-Level Video

PLV mode is an asymmetric process (Figure 11–15). It takes more processing power to compress the image than to decompress and play back the image. While this limits PLV's possibilities for real-time editing applications, PLV is in use in video kiosks and training programs in industry and education. PLV programming is widely distributed on CD-ROM, which offers 70 minutes of full-motion video and audio with random accessibility.

As with any compression technique, the quality of a PLV-compressed image is fairly subjective. One must look at the original picture and the compressed result to make fair comparisons. Early DVI examples, both in PLV and RTV, were characterized as somewhat "soft" in appearance. This has become less of an issue as the compression technologies continue to improve.

Real Time Video

RTV mode is a symmetric process. Compressing and decompressing the image require equal processing power times, and as its name suggests, they occur in real time. RTV, therefore, is the process of choice with regard to digital nonlinear editing.

The current image quality of an RTV frame is characterized by somewhat limited resolution and severely limited chroma information. A great deal of information is discarded, on the order of

Figure 11–15 This is an example of presentation-level video achieved by DVI compression techniques. Global Access is a permanent interactive exhibit at the National Geographic Society in Washington, DC. Photo copyright ©1991 by National Geographic Society. Courtesy of Intel Corporation.

75% of all pixels in the incoming frame. Improvements in chip sets will have a beneficial effect on this characteristic.

PX64

Px64 has been proposed by the International Telephone and Telegraph Consultative Committee. Px64 is a draft of a standard for motion video compression in any transmission applications designed around systems that operate at 64 kbit/second. P refers to the number of channels that could be used in tandem and that would be transmitting the data, each at 64 kbits.

While the main concern thus far has been the compression and editing of motion video, an emerging and rapidly growing emphasis is being placed on the transmission of these compressed files. Px64 seeks to address these transmission issues.

DEVELOPING AND EMERGING COMPRESSION TECHNOLOGIES

With a caveat that the digital video compression movement is evolving rapidly, what follows are descriptions of the developing aspects of compression technology.

DVI

Future chip sets will improve the image quality of the RTV mode. MPEG and Px64 will also be supported in developing chip sets. It is possible that we will also see singular chip sets that simultaneously offer the ability to decode and encode in DVI (PLV and RTV), JPEG, MPEG, Px64, and unnamed methods.

MPEG

Based on its current path, interframe coding and frame-by-frame interframe editing may become a possibility. While MPEG currently offers a 3:1 ratio compared to JPEG, it will most likely be 3 to 5 years before a ratio of 10:1 is achieved. MPEG Extended, commonly referred to as MPEG II, will provide for a bandwidth of about 500 kB/sec, nearing broadcast-quality images. The immediate goal will be to pursue real-time encoding and decoding of MPEG II images.

Fractals

A general overview of how fractal technology has been developing follows. Instead of breaking a picture down into frequencies, fractals utilize fractal patterns to represent every possible pattern that can exist. By dividing a picture into small pieces, in effect, into a fraction of its whole, these smaller sections can be searched and analyzed fairly quickly. On a smaller level, instead of trying to find the pattern for the entire picture, the search is for smaller patterns of pixels.

Figure 11–16   A variation of a Mandlebrot fractal set, the original pattern is magnified to reveal sections that resemble the original pattern. Illustration by Rob Gonsalves.

Fractal compression is an emerging technology. The digital video compression methods that have thus far been discussed—DVI, JPEG, and MPEG—all utilize discrete cosine transform technology in which an image is represented by frequencies.

In Figure 11–16, the original pattern (#1) is broken down into smaller sections (this varies in terms of matrix: one typical pattern consists of 16 × 16 blocks) and magnified. To find where image #4 exists in the expanse of image #1, each subsequent magnified section (#2 to #4) is then analyzed and examined to find something in the fractal set that resembles the original sampled image. Once a match has been found, only the x, y, and z coordinates of where the sampled image (#4) can be found in the fractal set space (#1) are noted.

In this way, pixels do not have to be transmitted for each section of the original picture. Instead, the process continues until the entire picture has been broken down into subsections and analyzed. The result of this is a string of coordinates. To play back an image, these coordinates from the data stream are processed, and then, returning to the fractal set (#1), data are copied based on the coordinate locations needed. The data represented by each of these x, y, and z coordinates are then copied and merged. The result is a recreation of the original picture.

One important aspect of fractal technology is that it is a very asymmetrical process, and thus it takes a long time to compress an image. Conservative predictions see fractal technology in use in digital video compression schemes in 1993.

Wavelets

Wavelets represent a recursive technique that is quite analogous to LZW compression. Wavelets take a picture that is to be compressed and introduce aliases.

Wavelet compression is divided into two processes: a *scaling* function and a *convolving* function. Convolve means to roll or wind together. First, the scaling function is used to take the original picture and squeeze it down to reduce it on the x and y axes (Figure 11–17). Once the image has been scaled, the convolving process is run. This is a set of wavelet functions (transforms) that seek to encode error. For example, the information in the original picture is compared to the differences in the scaled version. The differences represent error. The convolving process can be run approximately five times.

Once the picture has been scaled and analyzed in this manner, the result of this transformation is to achieve information based in "wavelet space." Wavelet space is error. The goal is just to store one of four portions of the picture and then to quantize the other three portions of the picture (Figure 11–18).

Different wavelet scaling functions can be used to form the set of bases that allow a picture to be analyzed. The analog to this is the method by which JPEG compression uses frequencies to analyze a picture.

Wavelet compression offers an advantage over JPEG compression in its treatment of edges to objects in a frame. JPEG analyzes pictures in 8 × 8 coding blocks, and these boxes have edges. An edge represents an infinite number of frequencies, many of which are thrown away when a JPEG-compressed image is quantized. Wavelets, on the other hand, are not block-oriented. As the entire picture is being analyzed, edges are better preserved.

One current characteristic of wavelet compression is that it has difficulty preserving textures. For example, whereas JPEG does a very good job at preserving the detail and textures of a brick wall, wavelets have difficulty coding all the edges of the bricks; one side effect is that the overall texture of the brick wall is softened and compromised. Would you rather have edge artifacts or a loss of texture? This example illustrates that even emerging compression technologies will have artifacts of which we must be conscious.

Figure 11–17 The first step in wavelet compression is the scaling function. The original picture is reduced on the X and Y axes. Courtesy of Aware, Inc.

Figure 11–18 This is the result of the first level of a wavelet transform. The three portions of the picture shown in outline are represented as quantized information. This procedure continues for each cycle of transformation. Courtesy of Aware, Inc.

EVOLUTION OF DIGITAL VIDEO COMPRESSION TECHNIQUES

The history and outlook for digital video compression techniques follows:

Early 1980's	Digital video interactive
1989, 1990	Software compression
1990, 1991	JPEG
1992	JPEG; MPEG I introduced
1993	MPEG II. Various applications using MPEG: compact disc interactive; playback of edited sequences; use of MPEG-compressed video as a means of transmission
	Fractals, wavelets
1994	Unknown

COMPRESSION TECHNIQUES FOR HIGH-DEFINITION TELEVISION

While there not need be a new development in compression technology to compress HDTV pictures, the amount of data required to store one uncompressed frame is quite large, requiring over 3.4 MB. The amount of data in one second is so vast (approximately 1.1 GB) that JPEG, operating at high Q factors, would significantly deteriorate the image.

Compression of HDTV frames requires higher Q factors, but not with the accompanying loss of quality. Acceptable compression and acceptable quality will most likely be addressed by MPEG II and future compression methods.

With regard to the different aspect ratio of HDTV, 16 × 9, it is important to note that JPEG permits for variable aspect ratios, which is to be expected from a system based on still images. One of the proposed characteristics of MPEG is the ability to handle different aspect ratios.

For HDTV-originated material that needs to be edited on digital nonlinear systems, the material is first down-converted, either to NTSC or PAL, and digitized and compressed in the usual fashion. The final edit is done with the original HDTV source material in an HDTV machine-to-machine edit.

Clearly, the ongoing goal to create viable and economical solutions to compressing HDTV images needs to be based on the successful treatment of NTSC and PAL video solutions since working with these lower-resolution formats will continue as a digital nonlinear system is sought for HDTV.

CHARACTERIZING THE RESULTS OF DIGITAL VIDEO COMPRESSION

Perhaps the most difficult questions to answer have to do with attempts to label the results of the digitization and compression processes. Invariably, the comparisons that are sought are in relation to videotape formats. While this is understandable, it is often a futile effort, perhaps as futile as attempting to define *broadcast quality*.

Digital compression yields different types of fidelity and tonality, and therefore, a direct corollary is difficult to explain or express. Clearly, the resulting data that leave the encoder in our digitizing process is compressed video that exits as an analog signal. The digitizing and compressing artifacts that are inevitable as a result of these processes now reside within the material.

The most common question that is asked regarding compressed digital video is, "What is the resolution of the images?" The reason that this question is so difficult to answer is that it assumes that digital video compression techniques operate under the same rules as analog video recording methods. This is not the case. The major difference is that analog methods provide a static allocation of resolution. Digital methods offer a dynamic allocation of resolution.

Analog methods must be prepared to deliver resolution throughout the screen at all times. This is precisely the reason that analog video recording and playback techniques have such a difficult time preserving resolution when signal degradation occurs. Alternatively, it is a straightforward matter to measure the resolution of an analog medium.

Digital video compression techniques dynamically allocate resolution because digital methods provide resolution only when it is needed. It is therefore difficult to measure the resolution of a segment of compressed video. Resolution for the frames in that segment will change as the compression algorithms dynamically allocate and provide varying lines of resolution over the length of the segment.

Film has long been categorized as the image-capturing medium that excels at providing a soft, ethereal quality for its images. Video has always been associated with providing sharpness and detail while excelling at preserving the immediacy of a recording. Whether to shoot an event on film or videotape is influenced by many issues. The leading factor may not have anything to do with establishing a creative look. Rather, a project may be shot on videotape rather than film simply because of economic reasons, or a project may be shot on film to have a negative to cut for foreign distribution.

Digital video compression has not had a long enough history to adequately qualify as having developed "a look." Further, the little history that video compression has is not useful in attempting to define that look. The technologies that affect the quality of digital video images are advancing rapidly enough to thwart attempts at defining what the images look like. In January, the images may look highly quantized. In June, these artifacts could have been completely addressed by improvements in hardware and software. For these reasons, it is important to note that the "compressed digital look" is an erratically moving target that is difficult to assess.

While it is clear that making comparisons between compressed video files and analog or digital videotape formats will continue, it is important to realize that digital video compression creates something new, even though it may share some artifacts with both analog and digital videotape formats. This is a new type of media, even though it may reside in a decoded state on a computer disk or in an encoded state on a videotape.

There will come a time when comparisons to videotape formats will be less overt. This will occur when the compression factors increase, and higher quality images become the norm rather than the exception. When we begin to hear the question, "Is that the digitized image or the original image?" then the attempts to categorize the compressed images in relation to videotape formats will become far less frequent.

Another common request is to associate a kB/frame with a known type of image quality. Working against the figures for one frame of RGB video, at 921.6 kB, it becomes completely subjective if a particular kB/frame results in VHS quality or broadcast quality. There are, of course, many factors, such as the nature of the original material, the digitizing interface, and the quality of encoders and decoders.

Very generally, the following kB/frame outline seems to have a close relationship to a known videotape type. Again, this association can be quite subjective. For example:

1 frame RGB image = 921.6 kB

1 second RGB image = 27,648 kB (27.64 MB)

VHS quality = about 500–550 kB/sec = compression ratio of 50:1 = Q factor of

75–110

1 type C quality = about 1 MB/sec = compression ratio of 28:1 = Q factor of 30–65

Broadcast quality = about 2–2.25 MB/sec = compression ratio of 12:1 = Q factor of 15–30

If pressed to make a comparison, digitizing and compressing test signals will help to provide a fair basis for comparison. When hardware and software are ready to be tested at a specific Q factor, the material to be digitized should be any test pattern that shows horizontal and vertical lines of resolution. The resulting compressed files can be played back, and the number of horizontal and vertical lines of resolution will be displayed in clear detail for that Q factor. This method may help in attempting to qualify an example of digital video compression as fitting into a particular tape format category.

The following table shows horizontal lines of resolution for each visual format:

Format	Lines of Resolution	Format	Lines of Resolution
VHS	200	35mm Film	6x resolution of average "broadcast picture"
SVHS	272		
3/4"	240	2"	336
Laserdisc	336	1" type C	336
Hi8	320 (luminance)	D1	336
Betacam	288	D2	336
Betacam SP	336	D3	255

The issue of technical versus observable is a point that must be recognized. Let's say we run an experiment and create a JPEG-compressed file that shows horizontal and vertical of resolution in excess of U-matic (3/4") videotape. Many factors can affect the resulting picture, making it look superior or inferior to 3/4". Technically, the horizontal and vertical resolution of a compressed digital image may be equal to a specific tape format. However, when the file is viewed, the pictures may appear to be inferior or superior to that tape format. This is one reason why a degree of care must be taken during the digitizing process. Although we can measure a compressed file to compare it to the videotape formats, the question should be, "Is it of acceptable quality for your purposes?"

Whether those purposes are digital nonlinear offline editing, digital nonlinear online editing, total program creation and delivery, archival and retrieval, or broadcast on cable or over the air, the two issues are whether the compressed images are indistinguishable from the original and, if they are not, whether the images are adequate when judged against demands for the immediacy of the message being delivered.

STATE-OF-THE-ART DIGITAL VIDEO COMPRESSION 1989–1992

The examples shown in Figure 11–19 represent the evolution of digital video compression in both software-only and hardware-assisted compression techniques. All images shown were shot with a charged-coupled device (CCD) video camera and recorded to Betacam videotape. The images were then digitized and compressed. Finally, the images were captured in a graphics program for display here.

Software-Only Compression, 1989–1991

Example 1
In late 1988 through the end of 1990, software-only compression techniques yielded results similar to examples 1 and 2. The data size for part a is 6.144 kB per frame at 4 bits per pixel. Note the significant pixillation and coding block artifacts within the image. Depending upon the images being compressed, editorial decisions based on what can be seen in the frame may be difficult to make.

Example 2
The data size for example 2 is 12.413 kB per frame at 8 bits per pixel. With more than double the data compared to 1, this example shows an increase in detail preservation and a decrease in the coding block artifacts.

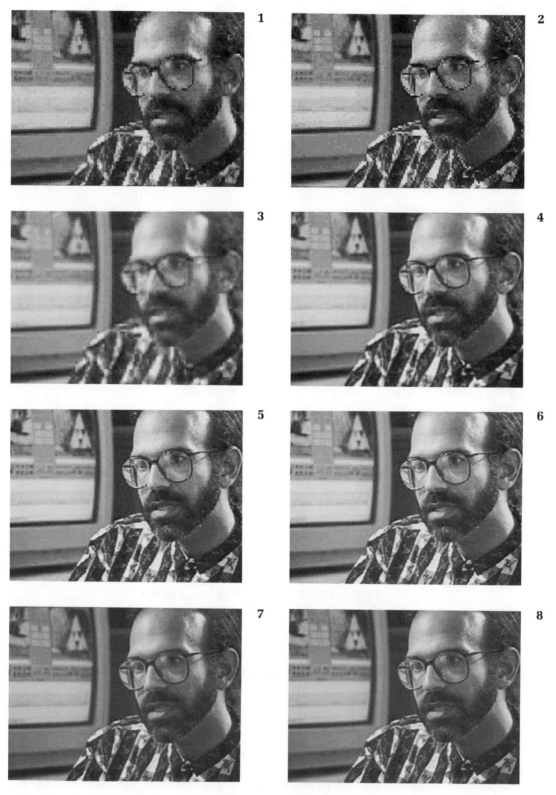

Figure 11–19 These examples of digital video compression show the original uncompressed image and the results of software and hardware compression at various kB-frame rates.

9

10

11

12

Figure 11–19 *(continued)*

Hardware-Assisted JPEG Compression, 1991–1992

Example 3

Hardware-assisted JPEG compression began to surface toward the end of 1990. Digital nonlinear systems began to offer either JPEG- or DVI-compressed images by the first quarter of 1991. Shown under this category are four examples. The data size for example 3 is 4.206 kB per frame at 15 bits per pixel. Although, as a still, the image blurred, this artifact is not as apparent when the image is in motion.

Example 4

This example, at 6.483 kB per frame and 15 bits per pixel, clearly shows the advantage of JPEG-assisted compression over the software-only methods. If we compare examples 1 and 4, there is a dramatic difference. While 1 is at 6.144 kB, 4, at 6.483 kB, is not significantly greater in terms of its storage per frame. The difference in the two images, however, is significant. Notice the decrease in coding block artifacts and the preservation of detail, especially in edge areas, such as the computer monitor and the eyeglasses.

Example 5

This example, at 8.209 kB per frame and 15 bits per pixel, shows increased detail being preserved in the image. This is especially noticeable in the rims of the eyeglasses where there is an increase in definition. There are also noticeable high-frequency artifacts shown in these areas.

Example 6

This example, at 18.109 kB per frame and 15 bits per pixel, represented the state of the art of JPEG digital video compression by the end of 1991. At this data size, much of the pixillation has been removed from the image, while a high level of image detail remains. For many applications, this image resolution is acceptable for direct output from the digital nonlinear system.

Second-Generation Hardware-Assisted JPEG Compression, 1992

Example 7

At the start of 1992, improvements were realized in JPEG compression chips. As a result, larger frame sizes became possible as increased data rates were supported. The data size for example 7 is 7.914 kB per frame, 24 bits per pixel. By increasing the number of bits per pixel from 15 to 24, an improvement in both image resolution and storage capacity was realized. Comparison examples 7 and 4.

Example 8

The data size for example 8 is 9.615 kB per frame, 24 bits per pixel. At this size, an excellent comparison can be made between examples 8 and 5. With a slightly larger size, at just over 1.4 kB, there is noticeably less ringing in the high-frequency areas such as the eyeglasses.

Example 9

The data size for example 9 is 17.509 kB per frame, 24 bits per pixel. This image, at less kB per frame than 6, shows improvements compared to 6, especially in the areas of the eyeglasses.

Example 10

The data size for example 10 is 23.007 kB per frame, 24 bits per pixel. This image, while very similar to 9, improves dramatically when examples 9 and 10 are expanded to fill the computer screen. With more than 6 kB per frame more than 9, example 10 shows less artifact and degradation at full screen size.

Example 11

The data size for example 11 is 37.889 kB per frame, 24 bits per pixel, slightly less than a 25:1 compression scheme. The improvement over example 10 is particularly evident in the lower and upper left sections of the eyeglasses. In 10, there remains a great deal of high-frequency ringing. In 11, this distortion is not present.

Example 12

This is the uncompressed image at its original resolution of 900 kB per frame, 24 bits per pixel. Compare this uncompressed image with the other compression levels to help determine which kB per frame may be suitable for your needs.

One of the most common issues with hardware-assisted digital video compression, whether the method employed is JPEG or DVI, is the extent to which image improvement can be realized. As these examples show, there will continue to be some form of compromise with regard to what can be achieved in image quality versus the effects on storage capacity as more kB per frame are stored for each frame of information. However, images that rival the original can be achieved as long as a decrease in the overall amount of storage capacity is acceptable. Improvements in video compression schemes and the alternatives to JPEG and DVI methods discussed earlier will provide improved image resolution with lower storage demands.

TIMEBASE CORRECTORS AND THE IMPORTANCE OF A PROPER INPUT SIGNAL

Prior to the digitization and compression stages, several steps can be taken to ensure that the detail content in a picture to be compressed is not extraneous detail. Video noise, caused by many troublesome sources, such as generation loss, can have significant and detrimental effects on JPEG compression. Reducing video noise through the use of first-generation footage and digital noise reducers helps ensure that JPEG compression schemes will not attempt to process this unwanted noise as elements of detail that should be preserved.

When we need to create a copy of a videotape, attention must be paid to the proper setting of video playback levels. The same care must be taken when information is being digitized. For example, if we have a videotape with a black level that is higher than usual (with normal setup at 7.5 IRE units), information will be present due to the fact that the data that should be set to play back at 7.5 IRE is now playing back at slightly higher levels. The JPEG algorithms will attempt to preserve information in these portions of the picture. Instead of uniformly coding all sections of a frame the same wherever 7.5 IRE black levels should have been, the deficiencies in the tape, as presented by the higher black levels, will represent, to the JPEG algorithms, more information and more detail, which JPEG will seek to preserve.

To avoid using more storage space than necessary in coding portions of a picture where detail is a result of generation loss or incorrect video playback levels, a timebase corrector (TBC) can be employed. By using a TBC, a waveform monitor, and a vectorscope, the videotape can be set to its proper playback level prior to digitization.

12

Storage Devices for Digital Editing Systems

COMPUTER STORAGE DEVICES

Thus far, we have been concerned with the digitization and compression process. In real time, using symmetric compression techniques, footage must be transferred to disk such that the entire process continues uninterrupted.

As we know, there are bandwidth issues that the CPU must address. If a series of frames is being processed and there is more information associated with the frames than the CPU can process in a given amount of time, there will be difficulties. Either the process cannot continue (in effect, the CPU times out on the activity) or a smaller portion of each frame is processed or perhaps every third frame is processed.

The same bandwidth issues are integrally linked to the storage mechanism being used. The digital density and the image resolution of our compressed pictures are linked to the storage system. If the computer disk cannot hold as much data as we would like, we must compromise somewhere: either choose a more rigorous compression scheme or store less material on disk.

Disk Characteristics

Disks have three characteristics: capacity, data transfer rate, and access time. They are discussed below.

Capacity
Capacity is measured in kilobytes (kB), megabytes (MB), gigabytes (GB), and terabytes (TB).

The amount of footage that can be stored on the digital nonlinear system is affected by the capacity of each disk drive and the number of drives that can linked together.

Data Transfer Rate
Data transfer rate can also be termed: *read/write speed* or *transfer speed*. Data transfer rate is measured in kB/sec and MB/sec.

The amount of data that a computer disk can write and read in a certain amount of time is the disk's data transfer rate. Let's return to the issue of bandwidth. We can decide that we would like to store pictures at their full resolution or, in the case of HDTV frames, pictures that require more than three times the storage of an RGB picture. However, if the disk is not capable of reliably accepting this data and returning it to us, we have another decision to make. We must compromise on the image resolution that the disk is capable of supporting. If the disk cannot keep up with accepting the data being sent to it, we must send less data. The disk's data transfer rate represents the bandwidth limitation of the disk drive.

Access Time

Access Time is measured in milliseconds (ms). A millisecond is 1/1000 of a second.

Once frames have been transferred to disk and we request that the frames be played back, there is a time delay between the moment that we ask for the frames until the data are available to the computer. This time delay represents the access time of the disk drive. Whereas the disk's data transfer rate is the amount of data that the disk can reliably read/write in the span of one second, in this example, once the material is on the disk, the issue is how quickly the disk can locate the material, process it via the CPU, and pass it on to the monitor for display. There are many delays between disk and computer screen; disk seek time is only one component.

All three characteristics—capacity, data transfer rate, and access time—must be considered not only to understand fully the present methods of digital video compression, but also to judge how improvements in disk technology will affect the picture quality and the amount of information that can be stored. Although we tend to think of the capacity of the disk as the most important concern, it is not. The disk's data transfer rate is the enabling/disabling characteristic that is the most important for our purposes.

Data Rate

To determine which disk type will be needed, we need to understand first the roles that the disk will be asked to play. The only way to do this is to ask these questions: How much material do I need? What image quality will be acceptable once the footage is compressed? How quickly will the material be accessed from the disk? When we answer these questions, the appropriate choice of disk type will become quite apparent.

kB/Frame

As shown in our discussion about the large amount of information contained in an uncompressed video frame, we use the term kB to refer to the number of kilobytes of information. This is the amount of digital data used to represent each frame. For example, we know that an uncompressed RGB frame requires 921,600 kB. However, after choosing a particular compression scheme, we wind up with 9 kB/frame, a compression ratio of about 100:1. We will now identify this file as 9 kB/frame.

To determine the amount of data that the disk needs to be capable of transferring, we calculate the amount of data that will be digitized and compressed. Let's say that we judge that a 100:1 compression scheme gives us picture quality that is suitable for our purposes. We also decide that we require two channels of 22 kHz sound. For each second of material, how much data must the disk be capable of processing?

Formula for Determining Data Rate and Storage Capacity

The following formula will be used to determine the required data rates as well as the disk's capacity to store information:

t = disk capacity kB ÷ {60 × [(kB/frame × fps) + (audio K × audio chans)] }

where t = length of material in minutes; disk capacity kB = the disk's capacity in kilobytes; 60 = the number of seconds in one minute; kB/frame = kilobytes per frame for the compressed image; fps = the digitized frame rate (30, 25, or a down-captured rate such as 15, 10, 5, etc.); audio K = audio kilobytes per second; audio chans = the number of audio channels.

We must take into account the number of bits per sample of audio; this could vary. For example, we could have 22 kHz audio at 8 bits per sample or 22 kHz audio at 16 bits per sample. The 8 bits per sample audio file will require less storage than the 16 bits per sample file.

Down captured refers to the ability to digitize at some fraction of the normal playback speed. For example, if footage is normally played back at 30 fps, we may triple our storage by digitizing the footage at 10 fps.

To determine the data rates required for each second of our 9 kB/frame and our two channels of 22 kHz sound, we calculate as follows:

(kB/frame × fps) + (audio K × audio chans)

9 kB/frame × 30 fps = 270 kB/sec + 22 kB/sec × 1 byte

270 kB/sec + 22 kB/sec × 2 =

270 kB/sec + 44 kB/sec = 314 kB/sec

This material requires that we be able to transfer consistently 314 kB each second. The disk type we choose must be capable of that data transfer rate. This is precisely the reason why image quality is dependent upon the storage mechanism's capability to transfer data.

Additionally, to determine how much of this material will fit on a 1 GB disk drive, we complete the formula:

t = disk capacity kB ÷ {60 × [(kB/frame × fps) + (audio K × audio chans)] }

t = 1,000,000 kB ÷ (60 × 270 kB/sec + 44 kB)

t = 1,000,000 kB ÷ (60 × 314 kB/sec)

t = 1,000,000 ÷ 18840 kB/min

t ≅ 53.07 minutes

At these settings for image and sound resolution, 53 minutes can be stored to a 1 GB disk.

As we are aware, there are two ways to increase the quality of a compressed image: either store more kilobytes per frame or develop better compression methods. The first option is under our control today. By devoting a higher kB/frame to the material, less information is discarded. As a result, the kB/frame is increased, picture quality is enhanced, the demand for a higher data transfer rate from the computer disk increases, and overall storage time decreases. Clearly, a complete understanding of the quality of picture and sound that you need to work with and that will be adequate for your purpose will greatly influence the type and number of disks that are required.

DISK TYPES

Computer disks can be magnetic or optical in terms of the actual recording medium. The process by which data is written and read to the disk can be magnetic, optical, or both. Each disk type has a unique set of characteristics with regard to data transfer rate, capacity, and access time. Which type of disk to choose will be a decision that is made when one weighs the amount of footage required (capacity), the quality of that footage (data transfer rates), and the speed with which the footage must be available (access time).

Floppy Disks

Floppy disks offer a data transfer rate of 80 kB/sec, a capacity of 1 to 2 MB, and an access time of 180 ms. Floppy disks consist of a thin, flexible magnetic medium that is encased in a protective plastic shell. They are erasable and can be used many times. They come in several sizes, ranging from the older 8" to 5.25" disks. The newer 3.5" disk is increasing in popularity and in disk capacity. Floppy disks are used primarily to store and transport data for small files, typically text-intensive files such as documents and spreadsheets.

The 8" floppies originally had about 100 kB capacity. The 5.25" disks now average 360 kB. The 3.5" floppies averaged 600 to 800 kB for some time and now range to about 1.4 MB. Newer 3.5" disks have in excess of 2 MB. More data capacity in a smaller amount of space is being achieved.

If we take the example of our 314 kB/sec of material, we would be able to store just over four seconds of material on a 1.4 MB floppy disk. It is important to note that computer disks are data disks: they write, read, and store data. On this floppy disk, we are able to store almost four seconds of our digitally compressed video and audio. Our files are digital data, and a computer disk accepts data.

When we look at the characteristics of the floppy disk, we can see that it does not provide the data transfer rates that would allow us to digitize to or read from the disk in real time. But as a storage medium, floppies can accept our data files. This may appear humorous now (after all, how many floppy disks would we need for a fifteen-minute program?), the fact that all computer disks can be used to store digital data will be important later when archiving issues are discussed.

Magnetic Disks

Magnetic Disks offer a data transfer rate of 0.5 to 3 MB/sec (1.5 to 1.8 MB/sec for SCSI 1 type; 3.0 to 4.0 MB/sec for SCSI 2 type), a capacity of 2.4 GB, and an access time of 8 to 16 ms.

Magnetic hard disks have a long history of development and use. They are the most reliable disk technology from that point of view. They are erasable and are usually rated in terms of hours of use, on the order of about 40,000 mean time between failure (MTBF). Today, most personal computers have internal hard disks. The size of these drives usually ranges from 20 to 40 MB, and the drives are contained in a 3" cartridge. However, for digital nonlinear editing, drives are external to the computer, and disk capacities of 20 to 40 MB are insufficient for our needs.

Inside the hard disk is usually a number of disk platters (the number varies but is usually no more than ten) that are able to store data on both sides. The reading and writing mechanisms of the disk drive consist of multiple heads on parallel stacks that align with the disk platters. These heads read and write information from tracks that are in the form of concentric circles. Hard disks spin quickly, about 3,600 r.p.m for SCSI 1 disks and 5,400 r.p.m. for SCSI 2 disks. SCSI, pronounced "scuzzy," is short for small computer systems interface. The read/write heads come into close contact with the surface, some within the distance represented by the size of a particle of dust, but they do not touch. Hard disks can be fixed in place or can be removable (removable magnetics, or RMAGs). In the case of removable disks, the disk housing remains in place while the disk cartridge is removed and another is inserted. Although the packaging size of a 40 MB hard disk is 3", the package that contains a 1.5 GB hard disk is 5 1/4".

Logically enough, hard disks are used to store data-intensive files. They are the storage system of choice for digital video and audio files. While the specifications for hard disks vary slightly from manufacturer to manufacturer, in general they offer an excellent combination of characteristics with regard to the roles that are important for digital video compression: data transfer rates, capacity, and access time.

With increased data transfer rates that are able to transfer, on average, about 1.5 MB/sec (considerably more for SCSI 2 type disks), hard disks allow us to increase the kB/frame for material we want to digitize and compress. Higher quality images, regardless of the compression method used, will always be associated with the type of disk that can handle the highest data transfer rates. When we want to combine higher quality pictures with higher quality audio or multiple channels of audio, we require hard disks. For some time to come, hard disks will have an advantage over all other types of digital disk storage devices in view of the demands of nonlinearly editing digital video and audio files.

Returning to the example above, if we decide that 9 kB/frame is unsuitable and that we require 25 kB/frame pictures and that instead of 22 kHz audio we require 44.1 kHz audio, 16 bits per sample audio, our calculations are as follows:

(kB/frame × fps) + (audio K × audio chans)

(25 kB/frame × 30 fps) = 750 kB/sec + 44.1 kB/sec × 2 bytes =

750 kB/sec + 88.2 kB/sec × 2 =

750 kB/sec + 176.4 kB/sec = 926.4 kB/sec

This image and sound resolution requires that we can consistently transfer 926.4 kB, almost 1 MB, each second. Since the hard disk can transfer up to 1.5 MB/sec, the amount of data that must be transferred is well within the disk's capacity. As long as our compression model allows us to compress to 25 kB/frame, we are armed with the correct type of disk to store the material.

To determine how much of this material will fit on a 1 GB disk drive, we complete the formula:

t = disk capacity kB ÷ {60 × [(kB/frame × fps) + (audio K × audio chans)] }

t = 1,000,000 kB ÷ (60 × 750 kB/sec +176.4 kB)

t = 1,000,000 kB ÷ (60 × 926.4 kB/sec)

t = 1,000,000 ÷ 55,584 kB/min

t ≅ 18.00 minutes

The need for a removable form of magnetic disk becomes evident when one considers situations that occur when we do not have simultaneous access to enough footage because of the capacity of the disk drives linked together. We may have an overall disk capacity of eight hours and ten hours of footage. The use of removable magnetic disk drives can help to address constraints on overall storage capacity. Editing based on sections or acts of a program can be accomplished with one set of disk drives for program act 1, another set of disks for program act 2, etc.

When the requirements for image and audio quality are high, editors often decide to use magnetic disks. To solve the problem of having access to all footage at one time is a major reason why removable magnetic disk drives are used.

Another logical thought would be to increase the number of disks with which the CPU can communicate simultaneously. These developments are already underway and are beginning to appear in the marketplace. Different computers have different computer buses. In the PC environment, both 16 bit ISA (industry standard architecture) and 32 bit EISA (extended industry standard architecture) buses are used. In the Apple Macintosh, a 32 bit bus, the NuBus, is used. Sun computers use the S-Bus, and so on. These buses communicate with peripheral devices in different ways and can communicate with a variety of different devices: disk drives, printers, etc. In the case of Apple Macintosh computers, the CPU communicates with disk drives via the SCSI chain. SCSI consists of a 50 pin cable and a protocol and format for sending and receiving commands and data. SCSI is not used solely by Apple computers.

SCSI 1 offers a data transfer rate of up to 1.5 MB/sec, but there is a limitation on the overall length of the SCSI chain when totaled among the disk drives. This occurs to ensure that data is sent and received properly. The maximum length of this SCSI chain does not vary, and for both SCSI 1 and SCSI 2 chains, the length will not exceed six meters. Up to seven external disk drives can be attached to one SCSI chain.

SCSI 2, on the other hand, represents improvements made in the semiconductor industry and takes advantage of increasing bandwidth and transfer rates. SCSI 2 offers two modes: fast, which is twice the bus speed, and wide, which has twice as many bits per transfer. SCSI 2 offers a data transfer rate of about 4 MB/sec, with some rates reaching as high as 6 MB/sec. Fast and wide modes can be used separately or together.

By adding a SCSI 2 card to one of the slots in the computer, an additional seven drives can be attached to the CPU. This provides 14 disk drives that are available for digitizing and playback, in effect doubling the amount of footage that can be accessed at a given time. For projects in which having access to a greater amount of footage is important and reducing image and audio resolutions is not desirable, adding additional disk drives is now an option. The addition of more cards allows more disks on the system.

Disk Arrays

Another method of adding storage is to employ disk arrays, which in effect are multiple hard disks that appear as one disk to the CPU. If the CPU can only communicate with seven disk drives, by utilizing disk arrays, we are able to have seven disks that, appear as one disk to the CPU. Thus, there is the potential to link 49 disks! Such disk arrays are currently being introduced, and staggering capacities are being discussed on the order of tens of gigabytes. The development of hard disk technology is progressing quite rapidly. The process of packing bits closer together is being refined, and the recording medium is increasing in density.

Evolution of Capacity and Cost of Hard Disks

Increasing the size of hard disks has been a constant activity since the 1960's. As a general rule, capacity doubles every two years. Prior to 1989, hard disks in excess of 400 MB were uncommon. In early 1989, 600 MB disks began to appear. These disks averaged about $6,000 each. In 1991, the disk capacity doubled to 1.2 GB, while the average price dropped to about $5,000 each. In late 1991, the disk capacity was increased to 1.5 GB, and the average price dropped again, to about $4,000. In 1992, 2.0 to 2.8 GB disks appeared, and based on this history, 6 GB disks will be available within three years.

The rapidly evolving hard disk technology is a promising sign with regard to a bottleneck for digital nonlinear editing systems. The goal is to provide better quality pictures, more capacity, and leverage economic outlay with regard to the number of hard disks that are purchased. The kB/frame will continue to be raised as system users continually move toward working with a digital version that shows no degradation from the original full-resolution image.

Hard disks offer extremely fast access times. These range from 8 to 16 milliseconds, but the more usual average is 14 to 16 ms. Fast access time means lower latency rates (the time spent waiting for the material to be shown). Fast access time means that when we are editing, we can have a series of short edits one after another. It is possible to experience trafficking problems when editing in a videotape- or laserdisc-based system; editing from magnetic disks means a greater chance of being able to play through a series of faster cuts without experiencing situations in which the magnetic disks cannot keep up with the editing pace. If faster paced work is important to the editing of the program, magnetic disks will offer advantages over optical discs.

Depending upon the kB/frame and the audio resolution, it is not unusual for one hard disk (access time 16 ms) to permit the following cutting scenarios. Note that the terminology is defined as follows:

V   Video only

A1   Audio channel 1

A2   Audio channel 2

Therefore, VA1A2 indicates that the sequence is composed of edits on the video and the two audio tracks. This example is based on digitized video at 7 kB/frame and 22 kHz audio.

Fifteen-frame cuts composed of video and audio channel 1 along with continuous playback of audio channel 2. A 60-second sequence should play back without reverting to non–real time playback or stopping.

Four-frame cuts composed of video and audio channels 1 and 2. The disk should be able to play back at least seven seconds of material.

One-frame cuts composed of video and audio channels 1 and 2. The disk should be able to play back at least 27 frames in succession.

Magnetic disks, again, are like any other computer disk with regard to storing digital data. They can be used to store not only digital video and audio files but also high-quality graphics, which are also quite data-intensive. However, long-term storage on hard disk for archival purposes is costly. In these cases, an optical means of storage should be used. By transferring data from the hard disk to the optical disc, the hard disk then becomes available for new material, while the original files are on optical disc. Although data can be recorded to or from any of these media, the drawback may be that large files may or may not play back from disk types that have lower data transfer rates than the files may require. If we have high-quality pictures playing back from hard disk and we transfer this file to an optical disc, the file may or may not play back from the optical; it will depend upon the optical disc's data transfer rate and whether it can read the information in real time.

Optical Discs

Optical discs used for nonlinear systems are 5 1/4" in diameter. However, unlike the laserdiscs used in the second-wave systems, these optical discs store data in digital form rather than in analog form. The optical discs used in digital nonlinear editing systems are all of the write many, read many category and are erasable. *Worm* discs, many of which are used in the second-wave systems, can only be written once.

Compact disc—read only memory discs are actually different from WORM discs in that CD-ROM discs are manufactured with their data on them and are never written. CD-ROM is used for many purposes, including motion digital video for DVI presentation level video programming.

Not all optical discs are similar, and they currently fall into two categories: magneto-optical discs and phase-change optical discs. The specifications of each type of disc are somewhat different, depending upon manufacturer. However, overall, their capacities and data transfer rates are similar. The media for both types of disc are reliable, the discs are estimated to be good for 1 million uses, and they have an archival life estimated to be ten years.

Magneto-Optical Discs
Magneto-Optical (MO) discs combine two technologies: magnetic and optical recording methods. While the basis for this marriage is decades old, lasers that could perform the optical function are a much more recent development. The first MO discs began to appear in the mid-1980's. MO discs are usually two-sided, although one-sided discs can be obtained. MO disc drives, however, only afford the ability to read data from one disc side at a time. The disc must be turned over to access files stored on the second side.

Magneto and *optical* refer to the method by which data is erased, written, and read. An MO disc consists of a protective polycarbon (or glass) layer. Beneath is the recording medium,

Figure 12–1 The read and write process for a magneto-optical disc. A laser (right) erases spots on the disc by heating these sections while the disc is held in a strong magnetic field. The magnet is then used to change the orientation of the spot. The same laser (shown in the illustration as another beam, left), operating at a different power level, determines the orientation of the disc spots to read information. Illustration by Jeffrey Krebs and Ryan Murphy.

which is metal-based. This layer cannot be easily magnetized and therefore is not susceptible to accidental erasure. The recording layer has two possible states: positive or negative. In digital terms, this would equate to zeroes or ones. The recording surface of a new MO disc consists of spots that have the same orientation.

Before we are able to write information to the disc, whether for the first time or the 100th time, we must first erase the sections of the disc where we want to record new information. Erasing the disc is accomplished by heating the sections with a laser while the disc is in a strong magnetic field. The spot is heated such that the magnetic material loses its orientation. This is known as its *curie point*. Curie point refers to temperature. One laser, running at two different power levels, is used to heat spots and to read the orientation of these spots. While a spot is being heated, the magnet is used to change the orientation of the spot. The recording medium rapidly cools, and the spot has a new magnetic orientation. This is how material is recorded to a magneto-optical disc.

To read information from the disc, a laser beam is used to read the orientation of the spots on the disc. If the light from the laser beam is reflected off the surface in a clockwise direction, the spot has a certain orientation; if it is reflected counter-clockwise, the spot has the opposite orientation. This is known as Kerr Rotation (Figure 12–1).

When we compare the data transfer rates of MO discs to those of hard disks, we can instantly see that MO discs will both read and write data far less rapidly than hard disks. Data transfer rates, capacity, and access times for both magneto-optical discs and phase-change discs are far less than those of hard disks.

In the case of MO discs, before data can be written, space must be made for the new material. The process of erasing and writing information to an MO disc is done in three passes (revolutions) of the disc. These passes add time to the process, and MO technology does not permit for the direct overwriting of old material by new material. This is a significant consideration when time and data rates are critical to providing the best quality picture and sound.

Whether we use a magneto-optical disc or a phase-change disc, it is also important to note that the data transfer rates are not equal over the surface of the disc. When a new disc is used, it begins to fill from the outer portion to the inner portion. The outer ring of optical discs is considered the "fast" portion, and the inner ring is considered the "slow" portion.

If we are attempting to store data that is well within the overall bandwidth limitations of the disc, both its fast and slow portion, no difficulties will be experienced. However, if the bandwidth that we require is close to the limitations of the disc, there may be problems. For example, let's say that we want to digitize two channels of 44.1 kHz audio to an MO disc. We calculate the bandwidth requirements as follows:

44.1 kB/sec × 2 bytes/sample × 2 (audio channels)

88.2 kB/sec × 2 = 176.4 kB/sec

The disc must be capable of reliably writing and reading 176.4 kB of data. When we look at the measured data transfer rates of the typical MO disc, we see data rates of 180 kB/sec to write and 360 kB/sec to read from the inner portion and 280 kB/sec to write and 560 kB/sec to read from the outer portion. We are within the bandwidth capabilities of 180 kB/sec to write on the inner portion.

Note that the actual measured data transfer rates will often differ from the official specifications and claims of the manufacturer, which are often "best case" examples. It is critical that these claims be substantiated.

Now, let's say that we want to digitize 7 kB/frame pictures as well as two channels of 44.1 kHz audio to the disc. We calculate the bandwidth requirements for one second of material as follows:

7 kB/frame × 30 fps = 210 kB/sec + 176.4 kB/sec (for audio) = 386.4 kB/sec

Clearly, over the entire span of the disc, the 386.4 kB/sec of material cannot reliably be written or played back. Error messages, such as "Disc I/O error," will occur or the digitizing process will halt during write time or the play back will halt during read time as the slower portions of the disc are accessed.

Again, the solution to this situation is to decrease the bandwidth requirements or to move to disks that have faster data rates. If we decrease the kB per frame of our pictures from 7 kB to 4 kB, decrease the audio quality from 44.1 kHz to 22 kHz, and digitize only one channel of audio, our calculations change as follows:

4 kB/frame × 30 fps = 120 kB/sec + (22 kB × 1 Byte/sec) = 120 kB/sec + 22 kB/sec = 142 kB/sec

This data rate is well within the bandwidth that the disc can provide over its entire surface. The manner in which these compromises are made are subjective; most manufacturers offer modes in which various combinations of image resolutions and audio resolutions will work for different types of disk drives, both magnetic and optical.

Access times for MO drives are slower than phase-change optical drives and a major influence is the size and weight of the magneto-optical pickup head. Since the laser and magnetic portions of the device add to its overall weight, picking up and moving this head over the surface of the disc takes time. Improvements in laser design will help to some degree. There are no magneto-optical disc drives which are fitted with dual lasers to read simultaneously from both the A and B sides of the disc.

Although we know that access time is less important than the data transfer rates in determining the quality of pictures or sounds that can be read from the disc, the slower access times of MO discs in relation to hard disks is important if the job to be done requires faster paced editing. Again, an examination of the project's requirements in terms of both overall storage as well as cutting pace must be considered.

Access time and cutting pace have sometimes been overstated with regard to both magneto-optical and phase-change discs. This has led to an uninformed and often cursory dismissal of optical discs. Faced with the opinion that optical discs are "too slow" for fast cutting, one must always counter, "Too slow compared to what?" Both types of optical discs are faster than shuttling linearly through a roll of film or through a videotape cassette. Both types of optical discs are faster than laserdiscs. Optical discs are only slower than hard disks, but if the application requires optical discs because of storage requirements and the cost of that storage, access time becomes a less important factor.

MO discs have joined hard disks in more widespread use for digital audio workstations. Although the majority of main editing functions for digital audio work is done from hard disk, often additional backup work is archived to optical discs. For example, it may be desirable to have a library of sound effects stored on optical disc that can then be transferred to hard disk when required or played back directly from the optical disc.

Phase-Change Optical Discs

Phase-change optical (PCO) discs became available in 1990. They represent a technology based on the ability of a material to exhibit two properties: amorphous (without shape; not crystalline) and crystalline (a solid having a characteristic internal structure) properties.

PCO discs consist of a protective layer made of polycarbon or glass. Beneath are several layers of the recording medium, which is composed of nonmetallic substances that have an electrical resistance that can vary with the application of light. PCOs, like MO discs, are quite durable, and the data stored on a PCO cannot be easily erased by stray magnetic fields. They can also be dual-sided.

Writing data to a PCO disc involves changing the state of the recording medium from amorphous to crystalline. All recording spots on a new disc have the same property: They are amorphous. To record data to the disc, a laser beam is used to change the spot on the disc from amorphous to crystalline (Figure 12–2). The reflective properties of a crystalline spot are different from those of an amorphous state. This difference will eventually be interpreted by a read laser.

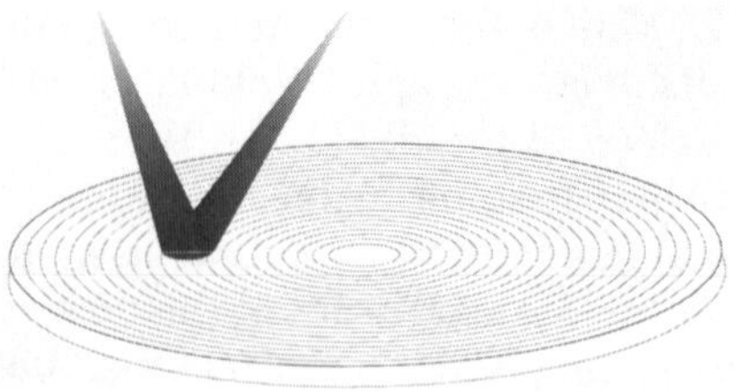

When we want to change a spot from a crystalline state to an amorphous state, we must heat the spot in the same fashion as we heated the magneto-optical disc, in effect reaching the melting point of the recording medium. The same laser, running at a more powerful rate, is used to heat the spot, and it turns from its crystalline state to an amorphous state.

The important aspect of the medium that makes up a PCO disc is that it can be changed from an amorphous state to a crystalline state and back to an amorphous state simply by varying the amount of light applied. Because of this nature, writing data to a PCO disc can proceed more efficiently than writing data to a magneto-optical disc.

Whereas MOs require that the spot first be erased and then magnetized, PCOs change states depending upon whether the data to be stored will be represented by an amorphous or a crystalline state. This capability is referred to as a one-step read/write process, and it represents a significant time savings over

Figure 12–2 Changing a spot on a phase-change optical disc from its original state to either an amorphous or a crystalline state results in a read/write cycle that is more efficient than magneto-optical disc technology. One laser, which is shown twice in the illustration to indicate two separate power levels, serves as the read/write mechanism. Illustration by Jeffrey Krebs and Ryan Murphy.

the write process of MO discs. This can effectively double or triple the writing speed. Reading speed is not affected.

Information is read from the PCO disc by measuring the intensity of light reflected from a spot in an amorphous state versus light reflected from a spot in a crystalline state. A read laser is used for this purpose. Depending upon the level of reflected light, data is interpreted as either a 0 or a 1.

By being able to determine whether data should be interpreted as either a 0 or a 1 through an examination of the different reflected values, the PCO holds distinct advantages over the MO disc. Recall that the MO must determine the Kerr rotation to read the orientation of the recording spots. Because the laser pickup head on the PCO has a less critical task to determine what a spot's orientation is, the device can be designed to be smaller and lighter. It therefore takes less time for the head to move along the surface of the disc.

The most significant result of this improvement in technology is an increased data transfer rate from PCO discs. If we compare the current wave of optical disc drives, the write times of PCO discs are approximately 2.5 times greater than MOs. This increase in bandwidth capacity can be utilized by increasing the quality of image and audio that are stored to the disc.

Returning to the previous example in which we wanted to digitize 7 kB/frame pictures as well as two channels of 44.1 kHz audio to a disc, our calculations were 386.4 kB/sec. Whereas with MOs this bandwidth pushed the limits of the disc, with the 430 to 450 kB/sec, worst-case scenario (inner portion of the disc), the bandwidth requirements now can be met, and the material can be reliably written to and read from the disc.

In this, the first generation of PCO discs, there is virtually no difference in the access speed of the disc versus that of certain MO discs. In fact, depending upon the manufacturer, some MO discs offer access times almost three times as fast as those of PCOs.

Phase-change optical discs are on a rapid improvement cycle, as are magneto-opticals. Because of their advantage over magneto-optical discs in terms of the capabilities of the recording medium, they promise to evolve for some time to come. Larger disc capacities, faster seek mechanisms, and more accurate lasers that will facilitate higher data transfer rates seem likely for this technology's future. However, it is quite unlikely that PCO discs will overtake magnetic disks.

Economics plays a major role in the decision to use optical discs, whether they are magneto-optical or phase-change. The price of a 1 GB magneto-optical disc or 1 GB phase-change disc averages $200 versus approximately $3,000 for a 1 GB magnetic disk. While all optical discs have limitations versus magnetic disks, price is not one of them. However, it should be noted that a 1 GB optical disc has 500 MB per side available to the user at any given time, whereas with the magnetic hard disk, the entire 1 GB of storage is available. A decisive factor in the further acceptance of optical discs for digital editing will be the appearance of discs with larger capacities per side, on the order of at least 1 GB.

13

Transmitting Video Data

The discussion on digital video compression began with the example of a document that we compressed and sent over a phone line with the use of a modem. The process of moving data from one place to another has been transformed from luxury to a necessity for communicating faster, wider, and more efficiently. Today, the fax machine is enjoying a stunning popularity; color fax machines (using JPEG no less!), portable computers with internal modems, and videophone service are now being introduced. The reasons for and methods of transmitting information that is more data-intensive than text documents are continuing to grow and emerge. Business teleconferencing, educational seminars via satellite, national and international offices, companies linked by common voice mail and electronic mail systems, neighborhood libraries tied into computers at a central library, personal computer owners dialing into neighborhood bulletin boards, and so on, are all ultimately concerned with one item: sending and receiving information that arrives intact or in adequate enough form to be interpreted.

The applications and reasons for sending digital video and audio files from one location to another, whether within the same office building or from one city to another, will grow until it is common to be able to monitor progress on the editing of a project without being in the same room as the editor who is cutting the show. However, until that day, there are details to consider regarding the method of transport for these signals.

Video compression has done a great deal to make an unwieldy signal more manageable. The signal can now be stored and edited. Sending motion video and sound from one location to another, on the other hand, is dependent upon having the right combination of hardware, a realistic schedule for delivery, and the willingness to compromise on the quality of signals being sent as represented by the data for those signals.

There are many different methods of transmitting data. Certain modes require specialized hardware, but all methods are currently available and in use. Table 13–1 lists various communications methods. The following section describes several methods.

Table 13-1   Transmission Methods

Method Data Transfer Rate

Method	Data Transfer Rate
Modem 2400 baud	2.4 kbits/sec
Normal phone link	15 kbits/sec
Leased-line phone service	56 kbits/sec
ISDN	64 kbits/sec per channel
Computer Network:	
AppleTalk (Macintosh)	230 kbits/sec
Ethernet	10 Mbit/sec
	2 Mbit/sec
T1	1.5 Mbit/sec
T3	45 Mbit/sec
FDDI	100 Mbit/sec (Option 1)
	200 Mbit/sec (Option 2)

METHODS OF TRANSMITTING DATA

Modem

A modem is used to transmit and to receive digital data. The channel that the data can travel through is small, at 1.2 kB/sec. The bandwidth capability of an ordinary phone line is approximately 1 kB/sec, leaving some room for overhead and control codes that must be sent with the message.

Modems are used to turn digital information into analog information. The analog signals are in turn represented as sounds, which can travel over a phone line. The receiving modem turns these analog signals back into their original digital form. Modems operate at different baud rates, which is a measurement of the transmitting and receiving capability of the modem. Baud rate does not mean how many bits can be sent per second, but this is often thought to be the case. Modems are available in baud rates of 300, 1200, 2400, 4800, 9600 bits per second (bps) and higher.

A 9600 baud modem can transmit approximately 9.6 kbits/sec. We then divide 9.6 kbits by 10 (9600 kbps ÷ 10 bits = approximately 1 kB/sec) to get 1 kB that the modem can pass each second. It would be logical to assume that we should have divided by 8 bits since 8 bits = 1 byte, but there are actually other signals being sent in the span of each second, such as control codes, stop bits, and so on, so we increase the divisor slightly to adjust accordingly.

Integrated Services Digital Network

Integrated services digital network (ISDN) is a transmission method that is designed to combine various sorts of information, from telephone transmissions to fax to images and sounds, into one digital network. ISDN is used for many purposes, and a typical example would be a large corporation that has to communicate with its branches on a regular basis; for example, an insurance company that must regularly receive and send large database files.

ISDN requires that a communications link be placed at the transmitting and receiving sites. These installations are offered by telephone companies, who apply a use fee to such installations. At 64 kbits/sec, the bandwidth is not that much greater than dial-up service. However, up to 24 channels of ISDN can be combined, depending upon how many are needed. If 24 channels are used, the data transmission rate totals 1.5 Mbit/sec, a respectable data rate. For companies that regularly must transmit many forms of digital information but do not require very large bandwidths, ISDN is a very acceptable solution.

Computer Networks

A great number of computers in business and education are networked, or linked. In an office setting or educational institution, computer terminals on many different desks that can access a central computer have become quite commonplace. The central computers may contain databases or information that many office workers or students can tap into, query, and obtain materials from. These central stores of information are often referred to as *servers* because they serve the demands placed upon them by those requiring data stored on the server. Networking offers many possibilities, including transferring data files from one computer to another and actually controlling a computer on one floor of an office with a computer on a different floor.

Most of the material that users have heretofore moved back and forth has been in the form of documents, spreadsheets, graphic presentations, and files that are not particularly data-intensive. However, transferring digital motion video and audio files from computer to computer will eventually become more commonplace. As video compression techniques evolve and improve, data rates for pictures will decrease. As a result, sending and receiving this information will take less time.

Whereas we have had servers for documents and databases, servers for different forms of media will begin to surface. Some of these will be private servers, and some will be public servers in which digital media, whether film, video, audio, or graphics, can be down-loaded from the server in exchange for a usage fee.

Computer networks can be very sophisticated and have different bandwidths. The network protocol for the Macintosh computer is AppleTalk, at 230 kbits/sec. Its bandwidth serves as a general benchmark for personal computers.

Ethernet

Ethernet is a form of local area network (LAN). It consists of a coaxial cable that can extend to approximately 1.25 miles and can offer up to 1,000 nodes (computers and peripheral devices).

Ethernet operates on a principle known as *carrier sense multiple access* (CSMA). While all LANs require computers to "wait their turn" before sending information, CSMA allows a computer to send information, and when the cable is not in use, another computer can begin sending its information.

Ethernet is rated at a bandwidth capability of 10 Mbit/sec, which is what the system is capable of delivering. However, because multiple computers can be waiting to transmit, the more practical throughput (the amount of data that can regularly be processed) is about 2 Mbit/sec (about 200 kB/sec) to any given computer. Networking computers via Ethernet affords an adequate throughput size for transmission and is a popular choice of linking systems that must transfer data-intensive files. The cabling used, however, is not inexpensive. There are also variants of Ethernet that are much cheaper because they run on "twisted pair" telephone wire instead of coaxial cable.

T1 and T3

T1 and T3 are best referred to as common carriers of signal that require dedicated systems installed at two sites that will be in communication with each other. Often misinterpreted as terrestrial, T1 and T3 links are not solely terrestrial in nature; they can involve communication using satellites as well as over land. T1 and T3 service, available from phone companies, is arranged in advance of use and involves hardware that, in the case of T3, offers greater bandwidth capabilities than, say, ISDN. Coaxial cables are used.

T1 links have a bandwidth of approximately 1.5 Mbit/sec (about 150 kB/sec), which is about one-half greater than a network using Ethernet, but equal to 24 channels of ISDN. T3 links, on the other hand, have a very significant bandwidth of 45 Mbit/sec (about 4.5 MB/sec). Although T3 service is expensive to install and utilize, transferring data over these links is an extremely efficient method when large files are routinely sent and received.

Fiber Digital Data Interconnect

Fiber digital data interconnect (FDDI) is a transmission method that uses fiber optics to transmit and receive signals. These fibers are transparent and are usually constructed of either glass or plastic. Fiber optics transmit energy by directing light instead of electrical signals. Electrical signals are encoded into light waves, transmitted, and received and decoded back into electrical signals.

While fiber optics are still considered a somewhat fragile mechanism, these interconnects have steadily grown in popularity. It may not always be economical to create a network based on copper conductive wire when fiber optics can be used. FDDI is expected to be offered in two configurations: bandwidths of 100 Mbit/sec and 200 Mbit/sec. However, just as there are trafficking issues in networks using Ethernet, there are similar throughput constraints for FDDI networks. Because of this, a 100 Mbit/sec FDDI network will most likely deliver about 2 MB/sec.

ISO Synchronous FDDI

ISO Synchronous FDDI is an emerging method of fiber optic transmission. Initial data indicate that this mode may reach 100 Mbit/sec as practical throughput, which would be quite an achievement.

TRANSMITTING PICTURES AND SOUNDS AMONG EDITING SYSTEMS

Consider the following scenario. Although somewhat unusual, it is not at all uncommon for requests like these to be heard. Let's say that editing is progressing in New York on a documentary. A digital nonlinear editing system is being used. In Los Angeles, a different section of the documentary is being edited. The same nonlinear system model is being used. The editor in New York finishes a 15-second prologue and wants to send it to Los Angeles. Crew members in Los Angeles need to see it to judge how to start their portion of the show, which immediately follows this introduction. Normally, a videotape is sent via an overnight delivery service.

What if the actual files for New York sequence could be sent to Los Angeles? The files could be played back on the Los Angeles system, and since the files are completely compatible, the Los Angeles crew could make adjustments to the sequence if required. Such flexibility speeds up the process of viewing and interpreting the sequence, directing changes to be made, waiting for the changes to be made, receiving the changes, and completing the process.

This scenario, while it may appear somewhat unusual due to its unfamiliarity, represents only one of the many options revolving around the transmission of digital video and audio files. First, consider the bandwidth requirements of the 15-second sequence to be transmitted. The 15-second sequence consists of JPEG-compressed video at 5 kB/frame and one channel of 22 kHz sound. Our calculations show the following:

5 kB/frame × 30 fps = 150 kB/sec + 22 kB/sec × 1 byte = 172 kB/sec × 15 sec = 2580 kB (2.58 MB).

We must transmit 2580 kB, or 2.58 MB of information.

Next, we must consider the choice of transmission method. We would logically seek the most rapid form of transmission, but we may not have access to the needed equipment. First, we use a 9600 baud modem and begin transmitting the files. A 9600 baud modem transmits at approximately 1 kB/sec (9600 bps ÷ 10 = approximately 1 kB/sec). We have 2580 kB to be transmitted. To find out how long it will take to transmit our 15-second sequence using a 9600 baud modem, we calculate as follows:

2580 kB ÷ 1 kB/sec = 2580 seconds ÷ 60 seconds/minute = 43 minutes

Unarguably, taking 43 minutes to transmit 15 seconds of motion video and one track of audio is not a very effective use of time. However, consider the alternative: waiting for an overnight package. Perspective, of course, is always important. If an overriding benefit justifies the 43 minutes spent transmitting and receiving the files, the option will be utilized.

Consider how the transmission time of our 15-second sequence is affected when a communication system that provides for greater throughput is used. If both sites have dedicated T3 links, the transfer time is extremely rapid. T3 links transmit on the order of 45 Mbit/sec. We have 2580 kB to transfer. First, we represent the megabits as kilobytes:

45 Mbit/sec ÷ 10 bits = 4.5 Mbit/sec = 4500 kB/sec

2580 kB ÷ 4500 kB/sec = 0.5 second.

Clearly, depending upon the nature of the files that will routinely be sent from one location to another and the amount of money that can be invested in the network, it is possible to realize considerable improvements in transfer time. If a post-production facility considers offering a service of transmitting compressed digital video files from one location to another, investing in the most economical network is necessary.

TRANSMITTING VIDEO DATA FOR THE BROADCAST INDUSTRY

There are many developing requirements for being able to send pictures and sounds over regularly available communications networks. Broadcasters throughout the world routinely consider alternative methods of communicating news stories from locations where traditional forms of transmission are not available.

For example, consider the reporter tasked with providing a story from a country where there is no method of getting news camera video footage sent to the broadcasting country. Faced with no satellite uplink, what are the alternatives? The reporter can call in the story, and the news station can, and usually does, broadcast a still slide of the reporter or a graphic of the location while the reporter's voice-over (VO) is heard.

But what about the actual video of the scene? One solution is to use still video cameras, which store still frames in analog form on floppy disks. These disks typically hold 25 frames of color or black and white video. It is possible to transmit the contents of these disks in much the same way that fax machines transmit text. For many news events that cannot be aired with full-motion video, broadcasters have substituted a location graphic with a still video slide. Often, the slides are cycled to provide some semblance of movement. The news reporter often speaks over a normal telephone line as the still video images are broadcast.

When a communications ban is enforced in a country and a

reporter is in the country with a camera crew, how can they get the moving pictures out of the country and to the broadcast station? Interesting solutions to this dilemma are being considered. Digital video compression is at the fore of these proposals.

Our 15-second sequence took about 43 minutes to transmit over a telephone line. Clearly, we cannot utilize T3 links, since they most likely won't be available to the camera crew. However, compromises can be made. Instead of sending the normal number of frames per second, we can send fewer frames. Motion will be more staccato, but the data requirements will be reduced. Instead of sending audio with the pictures, we can send no audio, and the reporter can simply describe the scene over the telephone line. Other considerations are available. Instead of sending color pictures, black and white pictures can be sent.

Consider this possible scenario: War breaks out in a country. Military law is declared, and a ban is placed on all communication out of the country. Journalists from around the world have footage that has been shot showing the first hours of the war, but there is no way to transmit the motion video. However, there is a way to make phone calls and to send documents via modem. Some reporters use the phone to call in their reports. Others use modems in their computers to send stories. Others send still video images.

The race is on. Which network will be the first to transmit motion video from the ravaged country? Consider digital video compression. A small digitizing and editing station is set up in a hotel room. The video footage of the war is digitized and stored in the computer. A lower-than-normal frame rate is chosen along with a monochrome picture setting. Instead of digitizing at 30 frames per second, we compromise and digitize at a fraction of that: five frames per second. At five frames per second, the motion will be stepped in appearance, but we will still have a good semblance of movement before the next updated frame appears. We also remove all color during the digitization stage and work in black and white to achieve additional space savings. We choose a Q factor that gives us acceptable results to make out the action transpiring within the frame; this gives us a 4 kB/frame sample, and we have video that lasts 1.5 minutes.

The data that must be sent is calculated as follows:

4 kB/frame × 5 frames/sec = 20 kB/sec × 90 seconds = 1800 kB

We use a 9600 baud modem, transmitting 1 kB/sec:

1800 kB ÷ 60 seconds/minute = 30 minutes

Is it worth taking 30 minutes to send 90 seconds of material that still is not at normal play rate? It may very well be when no other motion video has been transmitted out of the besieged country and there is a broadcast war in progress.

Certainly, improvements in compression methods will further facilitate being able to provide motion video with less transmission time involved. If MPEG compression were used in the above situation, assuming roughly a 3:1 savings of storage space, our 90-second sequence would require the following:

$$1800 \text{ kB} \div 3 = 600 \text{ kB}$$

$$600 \text{ kB} \div 60 \text{ seconds/minute} = 10 \text{ minutes}$$

IMPROVEMENTS AND STANDARDIZATION REQUIREMENTS OF DIGITAL VIDEO COMPRESSION METHODS

Improvements to a vast array of technology upon which the digital video compression process relies are steadily appearing. Future compression methods will permit greater compression ratios without increasing quantization artifacts. It is inevitable that a compression method—be it fractals, wavelets, or some yet-to-be-unveiled method—will appear that will offer an increase over conventional JPEG or DVI by a factor of, say, 10:1 and that can be accomplished in real time. As risky as predictions are, the timetable for this will be within the next three to five years.

Improvements will continue to be made at a dizzying rate to the hardware associated with the digitization and compression process. This will reduce artifacts associated with material as it becomes decoded and encoded. While digitizing boards are currently limited to accepting only analog video signals, digitizers that can accept digital video data, such as component D1 or composite D2 signals, are imminent. These will serve to preserve better the integrity of the original signal.

While these scenarios certainly sound exciting and promising, care must be taken to ensure that progress is made concerning the further standardization of methods involved in digital video compression. If one considers almost any new technology that appears in the marketplace, the introductory price of that technology to the end user is typically quite high. For example, fax machines were introduced before a standard methodology for transmission and reception was established. Some of these machines were quite expensive, often costing thousands of dollars. When fax was standardized, prices dropped dramatically. Today, perfectly acceptable fax machines can be purchased for approximately $500. As prices dropped, other models were created, based on emerging technologies, such as the color fax machine. Users benefited from the decreased price of the monochrome fax, and manufacturers benefited from technological advances by being able to create new products that, in turn, could be sold for higher prices.

Standardization is the key to leveraging existing equipment for the user and better ensuring that the investment will be protected. Although JPEG is a compression standard, in no way should we conclude that there is a standard of file compatibility between or among digital video compression methods. This is precisely the perplexing concern: If we two manufacturers of digital nonlinear editing systems use JPEG-based compression, even if the two systems use the same JPEG-compression chip set, the actual JPEG files will probably be incompatible. Files from System A will not play back on system B due to the coding methods of each manufacturer.

Is it important that the files are compatible? It is important for a variety of reasons. First, the user is able to take files created on one system and use them on another system. This enables a company to be built around various types of machines: high-end capability machines and lower-end capability machines. Second, the work flow of the organization improves as files can be exchanged and work does not have to be duplicated. Third, other manufacturers gain the opportunity to create peripheral systems that can work in conjunction with the digital video files; as a result, a greater variety of tools becomes available.

Often there is resistance to standardization until it can be made clear to manufacturers that there are additional economic benefits to be realized that justify their taking part in a standardization effort. Standardization means file interchange possibilities. The ability to take a file from one system to another system and have the file instantly recognized is a major benefit to the user.

At this early stage in its evolution, digital video compression has yet to undergo these growing pains of standardization. However, file compatibility will be a distinguishing factor when one considers how the digital nonlinear editing system fits into the current method of working and the profile of products and systems in the user's environment.

14

Logging for Digital Nonlinear Editing Systems

Logging footage, whether thousands of feet of film or tiny snippets of audio, and being able to find the needed footage quickly and painlessly has always been one of the most important and least-liked tasks associated with editing film or videotape. Maintaining good records and creating a successful filing system on paper means cross-referencing a variety of logs.

Consider the editor's position: For an action-adventure feature film, approximately 350,000 feet of 35mm film, or about 64 hours of material, may be shot, and this is a conservative estimate. In addition, wild sound and temporary audio, such as music and sound effects, must be factored into the amount of material that the editor needs access to at any time. It is not at all unusual to see much higher figures for footage shot in the feature film category. Figures from 600,000 feet to 1 million feet are often possible.

Much of the editorial task involves searching for the right take to create the best possible performance. It is often the case that dialogue passages are created by mixing lines of dialogue from several takes. To the editor, 60-plus hours of material is the pool of footage from which a story is to be molded and performances created. It is undeniably the worst and best of all worlds: so much footage to move through, but potentially so many alternatives. Recall that editing is as much about deciding what footage not to use as it is about deciding what footage to use.

It is the editor's job to create the right feeling and to find the best shots and sounds available to reach that goal. This is not a denigration of the acting talents of the cast; the greatest possible acting performance always requires honing during the editing process. The trimming of just a few frames here or the addition of a few frames there will always be necessary to make one scene flow more easily and successfully into another scene.

Finding the right material means good logging techniques and the ability to quickly let the editor find a shot when it is needed. If the editor is cutting a dialogue scene and has to replace an actor's sentence by using the line from another take, it should be an easy routine of finding the different printed takes of the scene, locating the line of dialogue, and trying the replacement. The ability of the editor to find items easily, regardless of whether there are 20,000 or 1 million feet of film, is the purpose of bringing the logging and cataloging processes under the feature set of the nonlinear editing system.

THE VIDEO SCRIPT

Projects often begin with a script that has been carefully written and carefully reviewed while it is prepared for shooting and post-production. There are several script forms; typically, the two most common formats are for film and for video. In a video script format, there are two columns: one for video and one for audio. In the video column, there will be a description of the visual on the screen, while the corresponding audio will be written in the audio column. Figure 14–1 is an example of the two-column video script.

The Log Sheet

When a video script is used, it is accompanied by a video log sheet. The log sheet usually includes entries for the following:

Production	Name of project
Date	Day of shooting
Page	Page number and total pages
Reel	Videotape reel for logged footage
Scene	Scene number
Take	Take number
Timecode Start	Rough start of action: not exact, and often slightly late
Timecode End	Rough end of action: not exact, and after action is complete
Description	Notes usually include director's comments and some notation of preference

When a scripted project is shot on video, the editor usually receives a copy of the script, and each videotape box includes a log sheet describing the footage on the reel. One hundred reels of videotape usually mean 100 sets of log sheets that the editor has to refer to when material must be found. Still, the information contained on the log sheets is crucial, especially if the time to edit the project prevents the editor from reviewing all the footage that has been shot.

The process of logging is often viewed as a thankless task, but it is a form of preediting, and its importance cannot be overemphasized. In the example above, consider what will happen if the editor has limited time to cut the show. Initially, he will scan the logs and cue the tapes to the circled takes. If the performance and action in a circled take are acceptable, the alternative take may not even be reviewed. The editor will use the circled take and move on to the next piece of material as the show is edited. There may be no time to accomplish more than this.

Log sheets help to ensure that the best possible footage is selected, and they also help the editor to get to the best possible

PRODUCTION: "SWIMMING CHAMPIONS"   Page 3 of 15
WRITER: John A. Seta
CLIENT: Sampson Pool Company

<u>VIDEO</u>

Slow Motion:
Shot of swimmer, seen from rear.
Walks away from camera, moves
into shadows.

Camera looking down at pool surface,
the water is completely still.

Low angle underwater shot, looking
from pool up to the deck. The swimmer
approaches the edge of the pool.

Side angle, swimmer holding a pair
of goggles.

Swimmer's eyes: the goggles are
snapped into place. We change from
slow motion to normal motion the minute
they are in place.

Back underwater: the swimmer dives in
and blurs out camera view.

<u>AUDIO</u>

Music Up.

Music continues as we
hear exaggerated snap
sound effect.

Music continues as
natural sound of splash is
added.

Figure 14–1 The two-column video script separates the picture and sound portions of a program.

footage in the shortest amount of time. Just by looking at the log, the editor can instantly see which takes are probables, which are possibles, and which should not even be considered.

The logging process can be very subjective. If the person who creates the log sheets introduces his or her own opinion to the descriptions, it is possible that, when faced with minimal time, the editor will scan the log for the circled takes and proceed to use a take which that not be indicative of the best footage available. It is up to the editor to know when to look at alternative takes when the circled take just doesn't meet the requirements of the scene, but it is still important to note that the log may be the editor's first view of the material. The person responsible for logging the material, usually an assistant editor, needs to concentrate more on accurately describing and organizing the footage than on making subjective judgments.

On a project where footage arrives daily from the shooting location, the editor usually plans on setting aside time each day to review the dailies that have arrived. In this way, the editor will be better able to judge scenes that have already been edited with scenes about to be edited. Having the time to review material as it arrives for editing does not remove the necessity of good logging, but it is of invaluable assistance to the editor in knowing what more of the pieces of the whole look like and how they can all be fitted together. Unfortunately, having the opportunity to screen daily footage is usually limited to the schedules afforded to longer works or to situations in which there is enough time to screen footage before the editing process itself must begin: specifically, offline situations.

Video log sheets are not really any different from film log sheets. The process of logging, however, is quite different. The methodology of logging for video is usually concerned with noting everything that is on the individual videotape reel, but cross-referencing one shot to another shot is not typically done. This means that if an actor delivers a line of dialogue in a medium shot, and the same line is covered in a wide shot, the video log usually does not specifically show that the dialogue is covered in this way. Instead, the log usually shows the following:

Reel	Scene	Take	Start	End	Description
2	41	3	2:00:12:05	2:00:33:10	MS John
3	41A	1	3:12:09:15	3:12:35:02	WS John

If we examine this log, we see that the editor will find scene descriptions, so it will be clear that there is both medium and wide shot coverage of the same scene. However, note that the reel column shows that 41A is on a different videotape reel. The editor will have two sets of log sheets, one for each videotape reel. The script that the editor is working from may or may not indicate that there is a wide shot for John's line of dialogue. If the script was marked accordingly, the editor may have indications written directly on the editing script that show where footage is for the lines of dialogue. In this way, the editor will know that wide shot coverage exists for the scene. However, if the script was not prepared for the editor in this way, the editor must rifle through several different log sheets to ascertain whether there is alternative coverage for the scene.

THE FILM SCRIPT

The logging methods for a film script are usually quite different than the logging methods for a video script. There are many different types of logs that the assistant editor and the editor pore over during the course of the project. The methods of cross-referencing vary, but an extremely large amount of information and paperwork are generated as part of the identifying and logging process when, for example, a theatrical motion picture is created. The editor's first look at information will be a script, in the form shown in Figure 14–2.

This is one page of a script. It is in its finished form and has been approved so that the production can begin. A rough method

13.

22   CONTINUED: 22

                         NICKY (V.0)
          There wasn't that much time left until I had to
          leave for the meeting of my life. The only thing
          I didn't know was if I was ready for it or not. I'd
          soon find out.

23   <u>MONTAGE</u> - INT. OFFICE COMPLEX - DAY   23

      Nicky and Russell walk through the office areas,
      past desks, etc. JACKIE RANDOLPH, a thick man in
      his 50's looks at them briefly and walks past.

                         NICKY
          Did you see that? That's Randolph. Nice of
          him to stop by and say hello, huh? What a
          wise guy; I should stop him and pop him one.

      Antonio Botelli glides past.

                         NICKY
          Tony Botelli is just as bad. He got to where he is
          the easy way; he had money. Doesn't matter, but
          do they acknowledge me now? They don't even
          stop and say hello.

      Harrison Frederick passes Nicky and Russell. He looks up
      but doesn't say a word.

                         RUSSELL
          Good old Harry. He lived fast, made money for
          all of us and then forgot who his friends are. He'll
          get his, I promise you that, Nicky.

24   INT. MEETING ROOM - BOARDROOM MEETING - DAY   24

      Twelve MEN who make up the Board of Trustees are in the
      midst of a meeting. In addition to Nicky, Russell, there is
      Michael Johns, at one time one of the nation's most wanted
      for insider trading. Despite the noise, business is being done.

                            (CONTINUED)

Figure 14-2   The traditional layout
of a film script.

of assessing the overall length of a project is to apportion one minute of screen time for each script page. A two-hour motion picture will be approximately 120 script pages. From the script are generated a series of work products for the production crew: storyboards that serve as a guide during shooting, shooting schedules, talent call sheets outlining times when actors will be required, and so on. The editor will not have to process most of this information, although copies of storyboards may make their way to the editing room, especially for a particularly well-planned action sequence.

Marking up the Script

Figure 14–3 shows the same page of the script as in Figure 14–2, but it is now marked using a traditional method that serves as an outline to the editor. During shooting, notes are made directly on the script, and these notes indicate for the editor how a specific scene was covered.

The person responsible for indicating on the script how a scene was covered is the script supervisor. The straight lines on the script indicate that a particular actor was shot, or covered, for the material indicated by the straight line. The zigzag lines indicate the sections where the actor was not covered. By looking at the marked script, the editor can easily tell whether or not there is footage for a line and how the footage was covered, for example, as a master shot or a close-up.

The Continuity Sheet

Another piece of paperwork is the continuity sheet (Figure 14–4). These notes indicate the specific action taking place in a shot. This is particularly important in matching material from one day's shooting of a scene to a different day's shooting of the scene. For example, if an actor enters a building wearing a brown jacket and the editor cuts to a shot of the inside of the building showing the actor entering wearing a blue jacket, there is clearly a loss of continuity. While this is an example of an egregious mistake, it shows how keeping track of continuity is tied to the process of keeping good script notes.

Even though continuity is more an issue during the shooting stage, the editor may have to refer to continuity notes when assembling a scene. If the editor knows that a particular take cannot be used completely because there are continuity problems at the end of the take, this will allow her to be on the lookout for alternatives.

The continuity notes are also referred to as *left page notes* since they are usually on the left side of the script. The script supervisor marks the script on the right side and makes additional continuity notes on the left side of the script. The editor can refer to either page easily enough. There are always copious notes describing how characters enter and exit the camera frame so that the action is not duplicated or violated in ensuing shots.

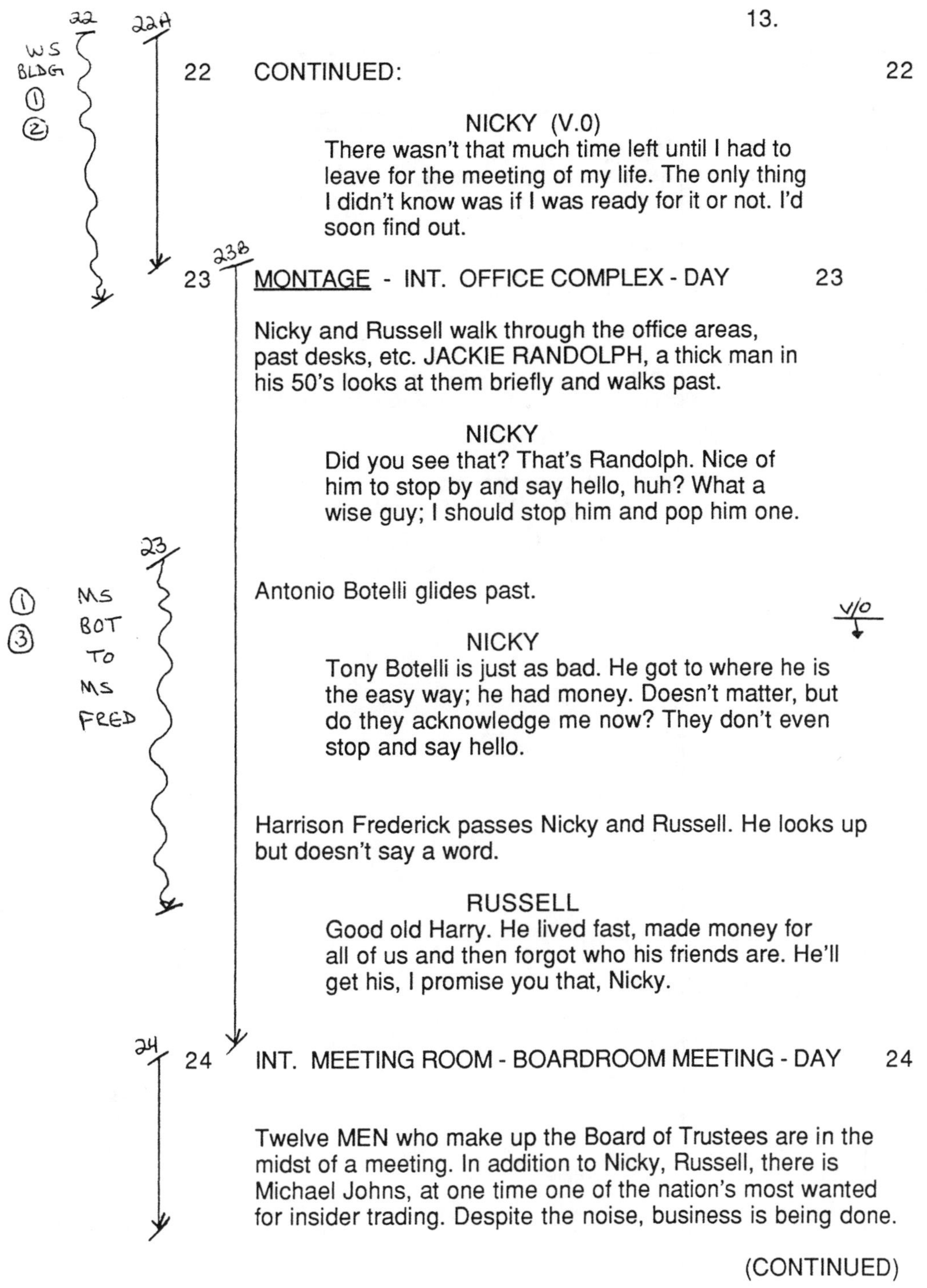

13.

22 CONTINUED: 22

> NICKY (V.0)
> There wasn't that much time left until I had to
> leave for the meeting of my life. The only thing
> I didn't know was if I was ready for it or not. I'd
> soon find out.

23 MONTAGE - INT. OFFICE COMPLEX - DAY   23

Nicky and Russell walk through the office areas,
past desks, etc. JACKIE RANDOLPH, a thick man in
his 50's looks at them briefly and walks past.

> NICKY
> Did you see that? That's Randolph. Nice of
> him to stop by and say hello, huh? What a
> wise guy; I should stop him and pop him one.

Antonio Botelli glides past.

> NICKY
> Tony Botelli is just as bad. He got to where he is
> the easy way; he had money. Doesn't matter, but
> do they acknowledge me now? They don't even
> stop and say hello.

Harrison Frederick passes Nicky and Russell. He looks up
but doesn't say a word.

> RUSSELL
> Good old Harry. He lived fast, made money for
> all of us and then forgot who his friends are. He'll
> get his, I promise you that, Nicky.

24 INT. MEETING ROOM - BOARDROOM MEETING - DAY   24

Twelve MEN who make up the Board of Trustees are in the
midst of a meeting. In addition to Nicky, Russell, there is
Michael Johns, at one time one of the nation's most wanted
for insider trading. Despite the noise, business is being done.

(CONTINUED)

Figure 14–3 The marked-up film script graphically displays to the editor how a scene was covered during the shooting stage.

The Lab (Telecine) Report

A log that is indispensable to the editorial staff is the report that accompanies the footage. Whether the project is being edited on an electronic nonlinear system or on film, knowing what pieces of film have been developed, printed, and transferred to videotape is critical information to the editor. The lab report may be automatically created, or it may contain data that have been entered by hand.

SCRIPT PAGE __13__   DATE __7·31__

CAM ROLL	SND ROLL	SCN NUM	TAKE	TIME	COMMENTS	ACTION
B7	#8	39E	①	:32		40mm/None: CU: Alexis in mirror
			2	:43	2nd STX	Michael enters BG CL [out when
			③	:33		she stands] She looks off [in
						mirror] to him. Exits
B7	#8	39F	1	1:37	(overlap) NGS 6.5	50mm/None: MED/OTS: Oscar walks
			②	1:37	GREAT!	sits on porch step. Looks off CR
						to Alexis. She enters CRS ≈ OTS
B7	#9	39H	1	1:56	Good	54mm/None: MED/OTS: Alexis turns
			②	1:42	Loved It!	clockwise and looks off CR to
						Oscar. Xs R→L to sit next to
						ⓄO = OTS.

Figure 14–4 The continuity sheet indicates the action that takes place in a shot and describes other important items.

The Film Transfer Log Sheet

Shown in Table 14–1 is a typical lab report. It consists of the information shown on the next page.

Camera Roll (CR)

The camera roll from which the footage comes. The camera roll is the individual roll of film that is loaded into the camera. Each camera roll for 35mm four-perf film is approximately ten minutes.

Table 14–1   Film Transfer Log Sheet

Roll: 47 Prod: Frame of Mind   Date: 12/14   Page 2 of 5

CR	SC	TK	Video TC	Key Numbers	Pull	Audio TC
47	41	1	08:00:01:21	KJ 289027-5073+00	D	08:43:10:18
47	41	3	08:00:39:00	KJ 289027-5162+00	C	08:47:04:19
47	44	2	08:01:12:26	KJ 289027-5254+00	D	11:52:18:26
47	44	3	08:01:40:20	KJ 289027-5296+00	A	11:54:02:10
47	44A	1	08:02:09:27	KJ 289027-5340+00	B	13:29:21:21
47	44A	3	08:02:33:18	KJ 289027-5414+00	A	13:31:47:27
47	44A	4	08:03:01:09	KJ 289027-5456+00	A	13:35:28:19

Scene (SC) and Take (TK)
Scene and take numbers for the piece of film.

Video Timecode (TC)
The timecode of the videotape to which the film was transferred.

Key Numbers
The edge number of the film.

Pull
Shows the location of a specific frame within the pulldown sequence.

Audio Timecode (TC)
The audio timecode of the audio tape playback.
By referring to the lab report, the editorial staff can determine the location of footage for a particular scene and take. The lab report is yet another crucial piece of paperwork in the editorial process.

Camera and Sound Reports

Additional technical notes made by the camera and audio crew are reflected in the camera and sound reports. They are of value to the editorial staff because they indicate items that the editor may require. A common example is a scene that has been covered by multiple cameras or a scene in which multiple microphones have been used. By looking at the audio report, for example, the editor will easily be able to tell which audio channels represent primary tracks and which channels represent ambient tracks.

Additional Notes

Additional notes may have been made by any number of individuals, including the writer, producer, and director of photography. These comments may be extremely brief or quite lengthy.

They advise the editor regarding specific concerns about a camera setup or take. If shooting is taking place at a location distant from the editing area, the director and editor will exchange notes. The director's preferences are yet another part of the logging process.

THE COMPUTER DATABASE

Keeping track of all the various logs, reports, and notes that are generated requires time, patience, and extreme organization. Locating a piece of footage that has been misplaced involves cross-referencing several logs. These methods traditionally have all been manual. In fact, the room adjacent to the main cutting room is usually replete with logs of all types affixed to clipboards that adorn the walls of the room.

The ability of a nonlinear system to create and search through a database to catalog footage is therefore extremely important. If the log is bad or if the log is nonexistent, the editor will spend much more time locating the footage to be used. Once footage has been loaded into the nonlinear system, what is the editor's view of the footage? Does the footage appear as text descriptions or as pictures that can be seen on the system's display screens? Further, what happens to the log information? If the editor is looking at footage on the screen, is it still necessary to rifle through many different paper logs? The database functions of the nonlinear system can be rudimentary or quite extensive and should be investigated. Using computers to associate logged information with the actual pictures and sounds contained on the nonlinear system results in the creation of computer databases.

Creating different views or windows to the stored footage can be thought of as combining information from the various paper logs. Shown in Figures 14–5 to 14–7 are three different ways of looking at information that has been cataloged on a digital nonlinear system.

Figure 14–5 shows a descriptive log for the footage that includes columns for scene, take (under the "name" heading), description, director's comments, and script lines. The entries for these columns can be made manually or, depending upon the nonlinear system's ability to import ASCII information from

	NAME	Description	Director Comments	Script Lines
	34	Good	*	Get out of here...
	34/2	-	-	"
	34A/1	Boom in Shot	-	"The money was here all the time...
	34A/2	Nice	*	"
	35/2	Good Take	-	
	35A/1	Statue	-	Voice Over...
	35A/2	Best Take	*	"
	38	Logo Draw	-	
	38A/1	-	-	
	39	Eyes / Keeper	**	Voice Over...
	39	Eagle	-	"
	40/3	Stereo	-	
	40A/1	Beach	-	Music Under
	40A/2	Beach	*	
	42	Logo Expands	-	"Tough driving through fog...
	42/1	-	*	"Kim couldn't have predicted...

Figure 14–5 This view of a computer database for footage contained on the digital nonlinear editing system shows customized information that further describes the footage.

	NAME	Tracks	Start	End	Duration	Disk	Mark IN	Mark OUT
	34	V	18:34:27:01	18:34:31:02	4:01	1	18:34:27:01	18:34:31:02
	34/2	V	18:34:21:26	18:34:24:09	2:13	1	18:34:21:26	18:34:24:09
	34A/1	V	18:09:09:05	18:09:11:24	2:19	1	18:09:09:05	18:09:11:24
	34A/2	V	18:08:49:16	18:08:57:06	7:20	1	18:08:49:16	18:08:57:06
	35/2	V	00:15:40:23	00:15:49:27	9:04	1	00:15:40:23	00:15:49:27
	35A/1	V	00:12:43:01	00:13:15:10	32:09	1	00:12:43:01	00:13:15:10
	35A/2	VA1A2	00:12:38:25	00:13:16:16	37:21	1	00:12:38:25	00:13:16:16
	38	V	00:20:23:08	00:20:30:08	7:00	1	00:20:29:26	00:20:30:08
	38A/1	V	00:19:57:11	00:20:03:23	6:12	1	00:19:57:11	00:20:03:23
	39	V	00:18:08:16	00:18:11:16	3:00	1	00:18:09:07	00:18:11:16
	39	VA1A2	00:12:38:25	00:13:16:16	37:21	1	00:12:38:25	00:13:16:16
	40/3	V	02:22:25:11	02:22:29:25	4:14	1		
	40A/1	V	02:17:35:01	02:17:41:05	6:04	1		
	40A/2	V	02:17:24:25	02:17:29:15	4:20	1		
	42	V	17:00:40:00	17:00:48:11	8:11	1	17:00:40:00	17:00:48:11
	42/1	V	15:24:35:08	15:24:40:16	5:08	1		

Figure 14–6 A different view of the database exhibits statistical information about the footage.

other programs, could be created automatically from the most important document: the script. If the system is capable of importing such information, a great deal of manual data entry can be avoided.

Figure 14–6 shows technical information about the footage. If the editor needs to know what videotape the footage is on, there may be a column labeled "tape name." Statistical information about the footage can be kept separate from descriptive views of the footage. Often, the technical database can be built automatically since many nonlinear systems will identify many of the technical attributes for the footage while the material is being transferred into the system.

Figure 14–7 shows pictorial representations of the footage. One or several frames that serve to represent an individual take are displayed on the screen. These views are digitized frames that have been converted from their analog form into digital data. Regardless of whether the nonlinear system is tape-, disc-, or digital-based, these frames are all examples of digitized pictures. If such a view is included in either the tape- or disc-based systems, there will be some form of digitizing board within the system to provide this feature. Some systems only display these frames in black and white, while others provide a color display range of 256 to millions of colors. High-resolution screens, such as 1280 × 1024, may also be available, increasing the detail of the pictures displayed on the screen.

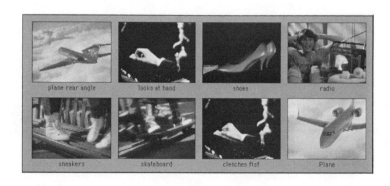

Figure 14–7 This view of the database shows the digitized footage in pictorial form.

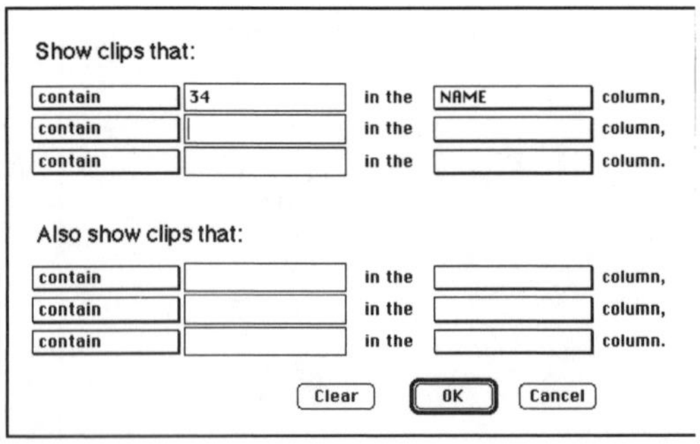

Figure 14–8 Once information has been entered into the database, the required footage can be automatically located by outlining specific search conditions.

Searching with the Computer

Once descriptive data is associated with the picture and sound files, there should be a way to use the power afforded by the computer to perform operations on the database. One of the more common tools that an editor can draw upon from a database, of course, is to find material in the database based on specific search conditions. Searching, sorting, sifting, and operations that are loosely referred to as *global actions* are all activities that benefit greatly from the fact that a computer is replacing the manual task of rifling through the various logs that are associated with the production.

Consider our database for footage. There are a variety of columns that describe the material in some fashion. If many shots are resident on the system, the editor may find that searching for only specific shots in the database can be very useful in quickly getting to and displaying just the shots that fit the search criteria. Instead of wading through all the footage, the editor may want to create certain search conditions. For example, consider Figure 14–5. If the editor only wanted to see shots for scene 34, the searching conditions would be as shown in Figure 14–8. Based on these conditions, all footage contained on the system would be searched and sifted. Only those shots meeting the conditions would be displayed (Figure 14–9). Last, the shots can be viewed in any mode that the system offers: either text-only or pictorial representation.

Searching and sifting the database for specific shots is enormously useful when large amounts of material are associated with the footage. If a feature film is being created and a million feet of film have been shot, there will be about 185 hours of material to wade through. Looking for a specific shot becomes vastly easier when a computer is enlisted to pore through the data. Similarly, documentaries are often accompanied by very large amounts of descriptive information. On large documentary projects that draw upon hundreds of hours of footage, being able to find shots easily is extremely important.

In addition, on large projects that employ a team of editors, knowing what footage each editor is working with is an important piece of information. If a documentary is being made about the history of a person's life, the database plays a major role in being able to find and cross-reference footage quickly. It may also be important that the same piece of footage not be used twice, and if a shot has already been used by one editor, this can be noted in the database.

Clearly, being able to search a database rapidly for a specific item of interest is dependent upon how the database was initially created. If a column for close-ups has not been created, how can

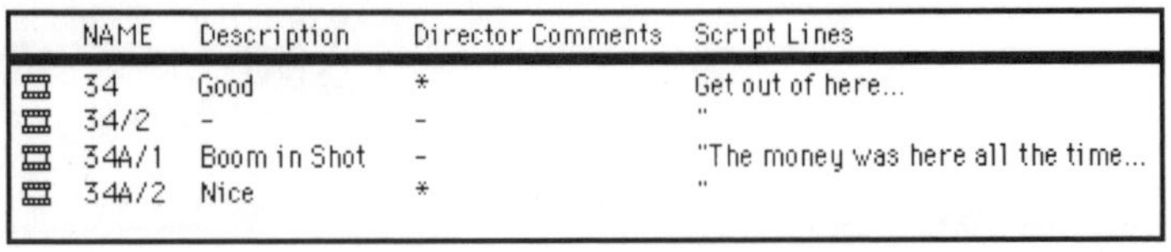

Figure 14–9 As a result of the sift conditions, the editor is only presented with material relating to scene 34.

	NAME	Description	Director Comments	Script Lines
	34	Good	*	Get out of here...
	34/2	–	–	"
	34A/1	Boom in Shot	–	"The money was here all the time...
	34A/2	Nice	*	"

the database be searched for this condition? A database should offer a great amount of flexibility to the user. If the database is created and the editor decides that an additional field should include close-ups, it should be easy to add this field at any time.

An additional benefit of the database is that it can be used as a scratch pad for information. If the editor is looking through footage and finds a shot that she likes but does not yet have a place for, she can make a note in the database that the shot has some attribute and should not be forgotten. Most editors make notes on scripts or on scraps of paper regarding the possible use of a shot somewhere in the project. Putting these notes in the database creates information that, unlike tiny scraps of paper, does not get lost. Later, when the editor reaches a section where the shot can be used, the database can be searched to locate the shot based on the note that the editor made, even if the comment made was as vague as, "Save it for somewhere."

Replacing the paper log sheets with the electronic log sheet is useful for more than creating one large database that the editor can manipulate. Getting the data from many different types of logs into one common location lessens the time that is required to flip through the pages of several log books. If the database available is comprehensive and allows the user to customize different views—for example, combining descriptions, technical data, and pictorial representations—then the result will be as if the different types of log books have become simultaneously available.

Script Integration

A number of commercially available word processing programs are specifically designed to create both video and film scripts. Creating databases on the nonlinear system can begin early on in the scripting stage of production. If the script has been word processed, it may be possible to import the text files as ASCII data into the nonlinear system. In this way, several benefits may be available. First, it may be possible to have the script reside within the nonlinear system and be displayed for the editor. If script integration is an important feature that is desired, the nonlinear system should be evaluated as to whether it provides this feature.

An interesting feature is to correlate the coverage for a scene with the footage that has been transferred into the nonlinear system. For example, by importing the script, the editor would be able to mark the script according to the script supervisor's original notes.

Logging During or After Shooting

When a videotape recordist makes notes regarding the material being shot and recorded to tape or when the script supervisor marks up a script and continuity sheet, a form of logging is occurring. These observations and notes represent valuable information to the editor. If notes made on set become associated with the footage, these notes can travel with the footage as it makes its way through the telecine process. The additional information from the telecine stage now becomes part of the database. When the footage is loaded into the nonlinear system, additional notes can be added. Associating information with each piece of footage means that the footage takes on a history of its journey through the production and post-production process.

Is this history important? It certainly can be. For example, let's say that an editor is putting together a commercial. Various takes are tried for one particularly difficult section. The editor uses the multiple versions feature that the nonlinear edit system provides to create different versions for the client. Finally, a version is tentatively approved, and the project lies dormant for a short amount of time. Then the project returns for reediting, but it returns to another editor. If the second editor can ascertain what other shots have already been tried by checking an entry in the footage bin that shows the different takes that were tried and rejected, these takes can be initially bypassed while the editor looks for alternative footage to use. Decisions were made regarding whether footage was acceptable in a certain section of the commercial; these decisions represent information to the second editor, and it can be useful to retain this information. Electronic databases can be utilized to provide this ability.

Preparing for Editing

The logging process and the preparation for the offline session are usually associated with paper logs, burn-in videocassettes, and paper edit lists. The purpose of using computers to log footage is to remove much of the manual process of creating these preediting work products. The logging process can be easily accomplished via a computer and videotape player.

A number of programs are available that link the serial port from a computer to the serial port of a videotape machine. In this way, the tape machine is controlled by the computer, and as footage is played back from the tape, timecode start and end times can be automatically grabbed by the computer, and a log can be created in the computer. Where this information goes and whether the nonlinear system is capable of importing this data must be evaluated. Clearly, the desired path would be to bring the data from the computer logging stage directly to the nonlinear system. The nonlinear system would automatically load the logged footage without requiring human intervention.

Several nonlinear systems work in this manner, allowing the editor to log footage at home or on a very streamlined system and arrive at work with a computer floppy disk that contains the information necessary to create an automatic "batch load" process to be run.

Avoiding Duplicative Work

The goal of the entire logging and cataloging stage is to allow editing to proceed quickly, directly, and efficiently. Logging done in one area is often done in another area, and this duplication of work takes time and effort. Having to type telecine notes into the database of the electronic nonlinear system means that the original source of information, the lab report, is not compatible with the database structure. It is this one-way form of information passage that must be corrected. Nonlinear systems that are able to import and export data easily over the many paths that footage takes can result in much less work for the editorial staff in logging, cataloging, and cross-referencing.

EVOLUTION OF THE LOGGING PROCESS

Planning a production with the understanding that the edit will be on a nonlinear system is affected by the method of logging that will be done. In some situations, logging is done on set with a laptop computer that is connected to a videotape recorder and shares the same timecode source. While the action is being taped, the logs that the electronic nonlinear editing system utilizes to batch-load footage are being created. At the same time, takes can be further described and comments can be added.

Knowing what takes will be considered can significantly lessen the amount of time spent logging footage prior to editing. The editor can receive a floppy disk that identifies the material to load into the nonlinear system as a first pass at cutting the material. While most editors insist on watching the footage if time permits, the alternative becomes significant when a client has logged material and decided what will be used and loaded into the system. When this is the case, the reliance on burn-in videotapes lessens.

The benefit of more and more information traveling with and becoming part of the identity of the footage will continue as the logging and preediting process becomes easier to accomplish either on the set or at home. Smaller and less functional editing systems based on personal computers will allow both the editor and the client to prepare better for the editing session with more information attached to each piece of footage.

TALKING WITH EDITORS

How Has Logging Software Affected the Editorial Process?

Paul Dougherty, editor, New York:

It has made a profound difference. Logs want to be computerized, notes are made to be copied, searched and sorted—this is what computers do best. Having this integrated into an editing system represents the "final frontier" of power editing. When I'm old and gray, I'll still be trying to find ways of better organizing my footage.

Tony Black, ACE, Washington, DC:

I spend a fraction of the time I used to spend finding shots while editing. Any shot in a system which has a computerized database can be located and played within a few seconds. This translates into an uninterrupted creative flow and a better edited product.

Basil Pappas, editor, New York:

Many projects need a great deal of databasing power. It is important for the database of the nonlinear system to talk to other programs since external and dedicated database programs will have more functionality. I have found that most clients are happy to log in their offices. A great database means that I have to spend less time looking for footage and more time editing.

Alan Miller, editor, New York:

Logging has always been a part of editing but the big difference now is that a database is created. This makes retrieval of footage a simple matter. The days of searching notes or index cards have been replaced by random-access retrieval of information. This makes editing quicker and easier.

15

The Digital Media Manager

Combining pictures and sounds is a task that has always been associated with a variety of equipment and a very procedural way of working. If the project was being edited on film, the editor would most likely be sequestered in an editing room and, after some time, would emerge with the editor's cut. Revisions would be made, and the steps associated with finishing the film would be taken. If the project was being edited on videotape, the editor would decide whether to edit in either an offline or an online environment. The finishing steps, including audio sweetening and audio layback to the finished master tape, would be made as the project is readied for tape duplication.

The personnel associated with the many crafts involved in making a finished film or videotape have certain areas of expertise. There are areas of specialization in the film and television post-production industries, including audio sweetening, digital video effects, electronic paint systems, film color correction, and so on. Editing is further broken down into specialties: Some people specialize in commercials, others specialize in film documentaries, and so on. It is the rare individual who can move easily among the various specialties.

What the film and video editing processes have in common is a very procedural way of working. A project is taken to a certain stage by one specialist, and other specialists are brought in for the next series of steps. For some projects, one person can accomplish everything that the film or video requires, but this is quite rare since the steps and the equipment used in finishing a film or a video are usually disparate.

There is a range of programming that does not fit into the "for broadcast" category. Corporate films, industrial training tapes, and product promotional tapes are all examples of presentations that must be created. These "not for broadcast" film and videos form the majority of work that is done throughout the world. Of course, the most visible projects will perhaps always be television commercials and feature films. For example, in the United States, about 400 feature films are made each year but significantly more industrial videotapes are made.

It is important to remember these types of programming because the personnel who are responsible for creating these presentations may not be classically trained film and videotape editors who have served as apprentices and assistants. These audio-visual professionals may in fact be concerned with more than finding a way to put pictures and sounds together to form the presentation. They often have to combine many more elements in creating the program. Computer graphics, character generation, music, sound effects, material from 35mm slides, and consumer videotape may all have to be combined to create the program. Managing all the different forms of media to create the finished program can become a difficult undertaking, especially when the corporate audio-visual department is limited in terms of equipment and personnel.

Why are there so many different pieces of equipment in the online video editing suite? If we take stock of the contents in a traditional online room, we will usually find the following:

Videotape machines (three to four machines)

Character generator (either dedicated or based on a personal computer)

Digital video effects unit to manipulate and move images

Digital disk recorder to combine layers of images

Edit controller to orchestrate the different machines

Audio console to route and mix audio signals

Video switcher to route and mix video signals

Monitors (separate monitors for the different systems in the room)

Title cameras to shoot graphic cards

Audio tape player (usually 1/4" or DAT)

Cabling (not visible, but of considerable length) to connect the different systems

The reason for having so many different pieces of equipment is that the online room must be able to accomplish a wide variety of tasks for any one project. If we were asked, "What task is more complicated: editing a two-hour feature film on film or editing a 30-minute industrial training tape on video?" there would be a variety of ways to consider the question. For the feature film, most of the editing will involve basic visual effects: cuts, dissolves, and fades. A 30-minute industrial film could involve every electronic technique available. As a result, the equipment list necessary to finish the industrial will be far longer than the equipment needed for the film. Alternatively, the industrial tape may have far less footage that the editor must wade through to determine which takes to use than the feature film has.

File Incompatibility

Whenever a variety of equipment is used, file compatibility issues will arise. If a videotape is being offlined at one facility and titles are being composed on a character generator and stored to floppy disk, when the project goes to online in a different facility where a different character generator is being used, the chances of the character generator files being compatible are usually quite poor.

Manufacturers have traditionally been suspicious of file compatibility, which they have viewed as a potential threat. Few companies have been forthcoming in promoting a common interchange format. If there is any doubt at all about this, simply consider the situation of videotape edit decision lists. There are a variety of formats, such as Ampex, CMX, Grass Valley, Paltex, and Sony. None of these lists can be recognized by another editing system. For example, an Ampex list will not load into a CMX editing system without modification. The most common format that most manufacturers try to provide as a form of interchange is the CMX 3400 list.

File incompatibility is not only a result of proprietary manufacturer formats. Without a common standard, sharing files between different electronic paint systems or between different digital nonlinear systems cannot be accomplished. Another aspect is the operating platform itself. Returning to the subject of edit decision lists, there are three disk types and three disk sizes. When the offline editing stage is completed, one work product is the video EDL. The online system may require the EDL on either Macintosh, DOS, or RT-11 formatted disks. In addition, the online system may require either 3.5", 5.25", or 8" disks. Many circuitous paths may need to be taken to get from one system to another system.

File incompatibility remains a leading cause of additional time to transfer data from one form to another form. Dedicated machines that offer one form of expertise—for example, the digital video effects unit, which is solely designed to manipulate images—represent the traditional way that machines in the film and video industry are approached in terms of design and manufacturing. This is changing.

Dedicated Systems versus Software Modules

Linear videotape editing systems have usually acted as sophisticated controllers of the different peripheral devices in the editing suite, directing these devices through the common element of SMPTE timecode. The videotape machines, digital effects units, and switchers can all be triggered automatically and repetitively from the edit controller. However, linear edit controllers have progressed only in the number of serial devices that they can control and the list management tools that they offer.

Nonlinear systems are a combination of analog device control and software tools. Where the waves significantly diverge is in how the integration of additional features is accomplished. The analog signals that are associated with the tape-based and videodisc-based systems can only be manipulated so much by software. Instead, hardware is used to provide the desired feature. For example, tape-based and laserdisc-based systems create dissolves through the use of an internal switcher, which is a hardware component that facilitates the transition from the A source to the B source.

Digital systems, however, manipulate the digital data to create a blend of the A and B sources. There is no requirement for an additional piece of hardware; the optical effect is accomplished through a combination of the software and the digital video compression hardware that are the normal components of the system. While this may appear to be a subtle point, a question is raised: "Can nonlinear editing systems begin to offer the features and capabilities normally reserved to other dedicated peripheral devices?"

Rather than being the "junction box" or routing system for the signals from the dedicated character generator, digital video effects unit, electronic paint systems, and so on, the tools themselves become incorporated into the editing system. As a result, the system becomes much more than just a nonlinear editing tool. Instead, the editing system truly becomes a device that manipulates the many forms of media typically associated with the editing process, whether they are still photographs, slides, film, video, computer animations, or multiple channels of digital audio.

The digital media manager (DMM) is concerned with creating systems that manipulate and orchestrate data. Rather than furthering the creation of systems that operate by manipulating analog devices or the analog signal, the goal of the DMM is to treat all information as data that can be manipulated and to offer the feature sets of different analog and digital machines within one digital system. It is a concept very similar to that of the Turing machine: Computers have the ability to do all tasks, if the software and hardware are extensible to the set of features necessary to perform the desired tasks. This is the path that digital nonlinear editing systems could take.

The DMM concept exists in two stages. The first stage is to incorporate the features and functions of dedicated workstations within the domain of the digital nonlinear editing system. The second stage represents the evolution of the DMM, where data files, whether they are associated with pictures, sounds, stills, or three-dimensional graphics, are manipulated as data file types. The DMM therefore evolves into a manager of digital signals rather than remain as a digital nonlinear editing system.

While it is questionable whether the manufacturers of digital nonlinear editing systems will take this route, an outline of the development path that videotape-, laserdisc-, and digital-based nonlinear editing systems are currently on should be examined. Regardless of whether systems that operate under the DMM scheme arrive from the offline, online, or nonlinear worlds, the DMM is going to appear.

DEVELOPMENT OF THE TAPE- AND LASERDISC-BASED WAVES

The first two waves of nonlinear systems are not on a developmental path that is associated with complete manipulation of visual and aural data in the digital domain. Instead, the path of these systems is relatively fixed. These systems are based on the manipulation of analog signals and achieve the majority of their feature set from serial machine control. Additional features that manipulate the analog video and audio signals will be available through the use of additional hardware.

For example, if the capability of providing digital video effects on these systems is desired, the analog video signal must be manipulated digitally; there are only two ways of accomplishing this: digitize the video signal and then manipulate it or route the analog video signal through a peripheral digital video effects unit. In either case, the nature of the signal must be changed. This will represent incremental hardware and software that must be built into the system.

As a result, to offer additional features that are normally accomplished through the use of other dedicated systems will require the incorporation of additional hardware in the tape and laserdisc systems.

We need only to view the product announcements from manufacturers of the tape- and laserdisc-based waves of nonlinear systems to ascertain how future developments are viewed. Many of these manufacturers are now pursuing the transformation of their operating systems toward a digital nonlinear offering.

DEVELOPMENT OF THE DIGITAL-BASED WAVE

Digital nonlinear systems have already transformed the analog video and audio signal. The data now exists as digital information that can be manipulated. Not only can the data be manipulated, but it can also be augmented with data from different sources. In effect, the digital data that represents those original analog video and audio signals can be changed. It is this ability to change the inherent nature of a signal that gives digital nonlinear editing systems the possibility of achieving greater capabilities through software than by incorporating hardware with each additional feature.

The developmental path of the digital nonlinear system can be twofold, depending upon the philosophy of the system's designers. The digital system can perform offline editing tasks, or it can migrate toward what have traditionally been online tasks: that is, finished results directly from the system. Incorporating more tools into the digital nonlinear system is necessary to create presentations that can be considered finished works. After all, if the program requires titles and multiple audio tracks that must be mixed together to be a deliverable program, the project cannot be finished if the tools required to perform these operations are not available.

Digital nonlinear editing systems will vary in the features that are available, but the trend is most definitely toward incorporating more tools that heretofore would have required dedicated hardware systems. Following are some examples.

Figure 15–1 The background video and foreground title (with key channel intact) are combined in the digital nonlinear system to allow the editor to previsualize keyed effects. Illustration by Rob Gonsalves.

Keying

The ability to do visual keys usually requires the use of a video switcher, a background element, a foreground element, and a key source (Figure 15–1). By using the capabilities of the digitizing board in the digital system, an object can be keyed over the background video. The foreground title is keyed into the background video by using a matte. The matte channel, referred to as an *alpha channel*, defines the keyhole. This definition results in a fully anti-aliased key that ensures that the title has no edging artifacts. This keying capability can be luminance- or chroma-based.

The process, which previously required one videotape machine for the background video, a still store device for the graphic and its matte shape, a switcher, and a videotape machine to record the effect, is easily accomplished in the digital domain. Among the benefits are less reliance on other dedicated machines that would have been necessary to perform this key, less tendency for capital outlay for these additional devices, and increased capabilities within the nonlinear system.

Advanced Audio Editing

Recreating the techniques and methods of the multiple audio track editing and mixing environment is another feature that can easily be incorporated into the digital nonlinear system. Since the audio already exists in a digital form, it can be manipulated by the digital audio software module of the nonlinear system. Several digital nonlinear systems have already incorporated sophisticated audio editing and mixing capabilities (Figure 15–2).

Multiple audio tracks, audio waveform editing, and audio mixing capabilities in the digital nonlinear system represent additional functions that are brought into the domain of one system. These features include the ability to create multiple tracks of audio, simultaneously monitor several channels of sound, edit the audio by manipulating its waveform shape, and mix the different audio tracks to create a finished digital stereo mix. This complement of tools represents capabilities that have never been addressed in the nonlinear *picture* editing system category. These capabilities, long the domain of dedicated digital audio workstations, are now finding their way into the picture-dominant digital nonlinear systems.

The benefits, again, range from less reliance on other machines to a reduced need to go to the audio sweetening facility and recreate all the audio work done in the editing session. Even if the audio sweetening session is still required, there is a greater chance that having these digital audio editing tools available will mean that better audio work products are taken from the digital nonlinear system.

Digital Video Effects

Digital video effects units offer sophisticated manipulation of the shape and trajectory of an object. DVEs have always been periph-

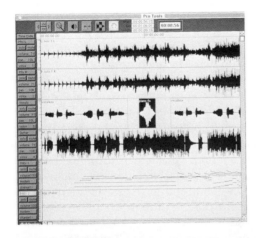

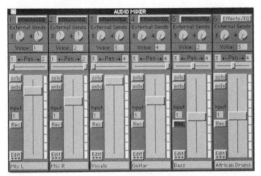

Figure 15–2 Audio waveform editing tools and the audio mixing console are represented here in software. Courtesy of Digidesign, Inc.

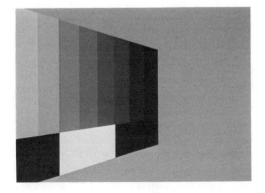

Figure 15–3 Digital video effects capabilities accomplished within the digital nonlinear editing system. Illustration by Rob Gonsalves.

eral devices that are used in conjunction with an edit controller. DVEs can also be thought of as dedicated picture manipulators. However, being able to bend or move a picture along a path defined by a number of key frame positions is not dependent upon having a dedicated peripheral device to perform these functions.

Under the realm of the digital media manager, truly digital video effects will be created not with dedicated workstations, but within the tool set of the digital nonlinear editing system. The incorporation of digital effects has already been accomplished by several manufacturers of digital nonlinear systems. While these early applications do not offer the sophisticated imaging techniques that can be found in dedicated DVEs, they do offer many of the basic aspects of size, position, and trajectory (Figure 15–3).

Compositing

The area of compositing, or layering many images together, is usually associated with digital video devices that are a combination of component digital videotape machines, component digital disk recorders, and component digital switchers. All of these devices are used to maintain the highest possible integrity of the original images. In compositing the materials together, projects are created that combine hundreds of visual layers without noticeable signal loss, especially in the area of chroma distortion.

The use of digital component or digital composite devices to accomplish such sophisticated layering of images usually costs more per hour than conventional A/B roll online editing due to the cost of these specialized devices. While precompositing plans can be made on paper, it is very difficult to preedit for a compositing session. Until the elements are all orchestrated, it is quite difficult to judge what is and is not working.

Digital nonlinear systems are beginning to offer compositing capabilities. While the quality is not even close to the quality of the D1 component layering devices that now dominate, the goal of the digital nonlinear system offering compositing features is simply to bridge the gap between the design of the layered effect on paper and the execution of the effect during the online compositing session. If there is a way to preedit the layered effect and to know that the elements that will be used together are the right elements to choose, time can be saved during the online session. There has been a tremendous increase in interest in creating a digital offline compositing system.

For example, if children have been shot against a chroma key green background and the desired effect is to have the children playing while puffy clouds roll in the background sky, one of the elements that must be chosen is the cloud backgrounds. It is unusual for filmmakers to shoot the clouds because there is commercially available stock footage from special effect libraries that offer many types of specialized backgrounds. There may be 50 or more types of cloud backgrounds from which to choose.

Why choose the appropriate background of clouds—the clouds that work well with regard to what the children are doing in the foreground—during the expensive compositing session? Even if the digital nonlinear system does not accomplish the layering at nearly the quality of the dedicated D1 component system, the benefit is that the visual selections and the timing relationships among the items can all be worked out in advance.

Further, as the digital video compression methods continue to improve, the goal will be to take finished results directly from the digital nonlinear editing system, which will ultimately evolve into the DMM. Treating data as digital data means not only that pixels of two sources can be combined, as in the case of dissolves, but also that pixels from multiple sets of sources can be combined to create complex layered visual effects. If the quality is good enough the results can be used to work out all the relationships before hand, or the results can be used as a finished item.

Journaling

A developing subset of digital nonlinear systems is the concept of a creation and playback journal. If we examine the editing process on a digital nonlinear system, we find a series of commands that result in an edited sequence. These commands, in effect, computer routines, are similar to the commands found in any editing system, linear or nonlinear. The difference is the manner in which these commands are stored.

For example, in a linear online edit room, using an edit controller involves typing in a series of timecode numbers, changing reel numbers, changing cuts to dissolves, typing in dissolve durations, and so on. Many of the repetitive keystrokes and operations are streamlined through the use of programming macro keys, which allow several keystrokes and routines to be programmed into one keystroke. As a result, there is less work for the editor, and there is a time saving. Depending upon the edit controller, macro keys can be saved to floppy disk and reloaded when necessary. Many editors religiously protect and guard their personalized macro files; some share their macros, but others do not.

The journaling method of the online edit session is an EDL. The EDL contains the information necessary to reedit the show if it becomes necessary to make changes after the online is completed but before final approval is given for dubs to be made. The EDL does not show how the program was created; instead, it has a series of numbers, which, if rerun, will create a program similar to the original program. The important word used in the last sentence is *similar*. Even if the EDL is rerun, the resulting program may not be identical to the original program.

There are variables that may or may not appear in the EDL. All of the various items that the online editor must be conscious of during the online session, such as video and audio playback levels, how a DVE move was programmed, and how a fader bar on a switcher was moved manually to create a certain effect, will not be presented in the EDL. Although several manufacturers of linear editing controllers have begun to integrate more of this disparate information into a unique and unifying EDL structure, most of the information that describes how a program was made is not evident when we look at an EDL for that program.

The journaling feature means that everything done during the editing session can be recreated simply by rerunning the journal for that edit session. Computer journals are lines and lines of computer code that track the commands that the editor made while the sequence was being edited. Rerunning these commands will result in editing a sequence exactly as it was done the first time. A journaling method becomes a significant feature as the digital nonlinear system continues to provide finished results.

Journals can be a significant asset to the person putting together a presentation if the journal allows the editor to write a routine for the journal to run or if the editor is able to change an existing journal, in effect, to create a subroutine, to make changes to already edited sequences.

Since a journal can track the exact commands that were done during the session, one can think of the playing back of a journal as if an automation file were playing back. The automatic recreation of the sequence that was made from the journal is accomplished. Consider a particularly intricate sequence that has been edited. If several things were done to the sequence, such as the editor changing audio volume settings, manipulating the fader positions on an audio board or a video switcher, or changing playback video settings gradually during the course of a take being used in the sequence, these operations can be difficult to document in the conventional EDL. Even if these operations can be captured as documentable steps, they may not be easily recreated in an automatic fashion. However, running the journal for this sequence in the digital nonlinear system would produce the exact results as when the sequence was originally created.

Journals borrow from musical instrument digital interface (MIDI), which, at its most basic form, is a music *journal*. Journaling, while not used nearly to the extent possible in the digital nonlinear systems, can also be thought of as a very advanced form of MIDI automation techniques.

Musical Instrument Digital Interface

MIDI, used to describe the translation of musical information into digital terms, has much in common with the concept of digital nonlinear editing. Linking musical instruments digitally to computers relies on a simple given: that musical information, notes, and how the notes are created, can be expressed by numbers.

The MIDI keyboard is at the center of the link. When notes are played on a MIDI keyboard, signals are sent serially via a MIDI cable to a computer. These signals describe each note by a variety of characteristics: note on, note off, key pressure, attack velocity, duration, and so on. With a set of such data and an understanding of how the information should be deciphered, sounds can be interpreted and exactly played back as digital data. Sixteen

Figure 15–4 Audio effects and audio sequencer panels, along with a MIDI sequencer, are represented in software. Courtesy of Digidesign, Inc.

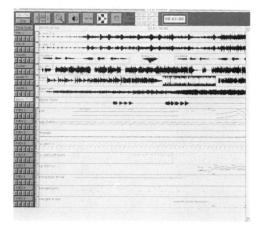

Figure 15–5 MIDI instrument data assigned to separate tracks by the MIDI sequencer. Courtesy of Digidesign, Inc.

separate channels of MIDI data can travel down one MIDI cable.

The software that captures these incoming signals is the MIDI sequencer portion of the computer program (Figures 15–4 and 15–5). As notes are played on the MIDI keyboard, the signals are sent to the sequencing program and are assigned to one track in the sequencer. Different instruments can be incorporated, such as MIDI drum machines and MIDI samplers; there are even special acoustic-to-MIDI convertors available if acoustic instruments, such as guitars or wind instruments, are required. When each instrument is played, a separate track for that instrument is created and displayed by the MIDI sequencer.

The sequencer, in effect, is similar to the concept of the playlist for the nonlinear editing system. In many ways, the MIDI sequencer represents a log to play music. Once the digital information that makes up a track of MIDI is resident in the computer, it can be changed through a variety of means, such as audio waveform editing and equalization. MIDI is an integral component in the digital audio workstation system.

When the tracks have been prepared, this "sequencing log" can be manipulated and reordered. In the same way that the nonlinear editing system achieves multiple versions, multiple MIDI automation files can be created to play back the tracks in the sequencer in different ways. When a playback file is run from the computer, the digital data is sent back to the musical instruments. If multiple instruments have been used, these instruments will then play back simultaneously, emulating a multitrack recording. At any point, changes can be made to any of the parameters that make up the sounds being played back, and timing adjustments can be made. The output of the sequenced instruments is recorded and then laid back to the master videotape.

MIDI has had a significant impact on how finished audio for pictures is created. The concept of nonlinearly accessing audio files, sequencing them, and creating multitrack playback of instruments paved the way for refinements to the analog nonlinear editing waves and the appearance of digital nonlinear editing systems.

MIDI, like digital video effects capability, represents a form of specialization. The goal of the digital media manager is to offer more of these capabilities within its domain. This is especially critical when one system capable of providing tools for horizontal and vertical editing techniques is desired.

HORIZONTAL AND VERTICAL EDITING CONCEPTS

The editing process is at a clear juncture that separates horizontal and vertical editing techniques. Consider the television commercial. Television commercials have traditionally belonged to the form of editing that can be called horizontal-based storytelling. A commercial portrays people, situations, and products and uses dialogue, music, and sound effects to tell a story. Horizontal editing means telling a story over time and relying primarily on dialogue to do so.

The idea of the "traditional" commercial has changed dramatically as digital video compositing and layering techniques have been popularized. The commercial has evolved and relies heavily on graphics and animation, instead of cuts and dissolves, to sell products. Vertical editing means making finished presentations and finished images from composites, layers, and visual and aural special effects.

Horizontal and vertical forms of editing require different pieces of hardware to achieve results that maintain the integrity of the original signal quality. A major reason why there are so many different devices in the online suite is because this editing environment must be ready to assume all types of editing operations, horizontal as well as vertical.

While editing is at a juncture between these two types of techniques, it is also at a crossroad. Being able to accomplish all types of editing without regard to whether the technique is horizontal or vertical means being able to manipulate information regardless of its origin. The most common form for this data is digital.

FILE COMPATIBILITY

File incompatibility is affected as analog systems try to communicate with digital systems. If each system is treating or processing the signal in a different way, the files from one system may not be usable in another system. Files are not usually severely affected in this way when an analog-to-analog link is made. However, when analog systems must coexist with digital systems, file incompatibility is a matter of course, and special translation programs and bridging hardware between the analog and digital worlds must be employed. When digital-to-digital links must be made between systems, file incompatibility issues similarly surface.

The role played by the DMM is twofold. The first goal is to achieve compatibility across different types of media integration. In the evolution of the DMM, there will be various stages where data from one system will be taken and recognized by another system. For example, if an analog nonlinear system is used to cut a program, one common technique is to use an audio-only EDL from the editing system, transfer it to a dedicated digital audio workstation, and batch-load the audio from the original audio recordings to manipulate further these newly redigitized audio files.

This file compatibility is very basic: An edit list from one system is recognized by another system. It should be noted that almost all of these transfers are accomplished via the most commonly used form of EDL interchange: the CMX 3400 edit list. The next goal is to achieve a higher level of file compatibility where the data files from one system are recognized by another system. For example, in the digital audio environment, audio files can be recognized from one system to another if they belong to several standard file formats, such as audio interchange file format (AIFF) format or SoundDesigner™ 2 (SD2) format.

If a digital nonlinear editing system is being used and the audio files are AIFF- or SD2-compatible with a dedicated digital audio workstation, the digital data of those files can be taken from one system to another. This avoids the additional step of batch-loading the material as a result of file incompatibility. Having file compatibility of this nature is not very common, but it is extremely important as more previsualization and finishing capabilities are built into the digital nonlinear editing system.

The second goal of the DMM is to treat information as digital data that can be searched, manipulated, combined, transmitted, and archived. The goal is not merely to translate pictures, sounds, graphics, and so on, from analog to digital so that they can be manipulated in a common format.

Always having to treat signals according to their origin results in a very constrained and constraining set of circumstances. Because the production and post-production processes must look ahead at how everything will be put together into a final presentation, the manner in which pictures and sounds originate yields a number of concerns that must be addressed. If we have an analog piece of audio that we want to use in the digital nonlinear system, the audio must first be digitized. If we want to incorporate three-dimensional computer graphics of a dragonfly into a previously filmed background of children, all elements must first be brought into a common format in order to exit eventually in a common format: a finished film element to be incorporated into the cut negative.

The digital media manager's ultimate evolution is to treat the various forms of media, whether they be video, audio, computer-generated, and so on, as data without regard to the media's origin. A dizzying set of products and capabilities will arise. This is how editing constraints and analog and digital mismatches are removed, yielding more capable Turing machines that allow editors to be more creative within one system environment.

ARE DIGITAL NONLINEAR SYSTEMS OFFLINE OR ONLINE?

This is an excellent question, and here's a controversial answer. Whether a digital nonlinear editing system is labeled offline or online depends to a great extent on the finished results that the system is capable of delivering and how the user perceives the direct output from the digital system.

For example, consider an industrial training tape. The tape consists of video, graphic titles, digital video effects, and mixed stereo audio, and the usual presentation form is a Betacam videotape master. If the digital nonlinear system can provide all the tools necessary to create the requirements of this presentation and the direct video and audio output are not discernible from

Betacam quality, why should there be reluctance to use the digital system's "video and audio print-outs"?

Alternatively, if the client is used to producing programming on D1 videotape, a Betacam-quality output from the digital system will not be acceptable. Consider the broadcaster who is reluctant to air compressed digital video; if the only video available of a newsworthy event is compressed digital video, the broadcaster must decide whether to air the moving footage or to allow another broadcaster the opportunity to provide the first glimpses of the event. Chances are good that the quality of the pictures will not take priority over their content.

In addition, the increasing number of features that are being incorporated into digital nonlinear editing systems is yet another factor in blurring the distinction between offline and online editing systems. An offline system that offers dissolves, wipes, graphic titling, digital video effects, and multitrack digital audio mixing provides quite a different view of the traditional offline video room. As digital nonlinear systems evolve, it will become very difficult to distinguish whether they are offline or online systems.

The digital media manager, although only a concept, represents the equivalent of the Turing machine in the world of media production and manipulation. Progressive manufacturers will create *digital media managers* in the future, not *editing systems*.

16

The Film Transfer Process

Since the mid-1980's, the necessity of being able to move images through various signal processing paths has increased, regardless of the origin of the images. Images and sounds are often treated with some unique processing method which frequently involves the use of digital video techniques for special effects. In addition, the economic requirement of being able to provide a video release of programming originally shot on film has contributed to the necessity of being able to easily move within and among the worlds of film, videotape, and computers.

Feature films, television shows, and even commercials have world markets. Programming that originates in one country can be remarkably successful in other countries; the United States film industry has a proven record that attests to the international popularity of its films.

Consider the path that a network television show in the United States could take. The show is a one-hour episodic which is shot on film. The footage is transferred to videotape, and an electronic nonlinear editing system is used to edit the program. The EDL is conformed during online, and a finished videotape master is created. Dupes are made of this master and delivered to the network for broadcasting. At a later date, the international syndication rights to the show are sold, and it is necessary to provide a videotape master of each show in the series in PAL format. What is the recourse for the show's owner? If only an NTSC videotape master for each show was created originally, how does the seller of the show provide the PAL masters?

One method is to take each videotape master and perform a standards conversion from NTSC to PAL. Depending upon the scan conversion process used, there may be noticeable visual artifacts whenever videotape images in one transmission scheme are translated into another scheme. The most common detriments are degrees of temporal aliasing and an overall softness to the images. Post-production facilities offer different methods of scan conversion. Real-time scan conversion, converting a one-hour show in one hour's time, offers varying results. Excellent results can be achieved with techniques that take more time to process the individual frames. These non–real time scan conversion methods are usually more expensive than real-time methods, however.

Being able to move easily between film and videotape formats and interject computer technology into these formats requires that there be methods whereby projects started in a particular format can be "matched back" to the original form regardless of the processing done to the images during the post-production process.

CREATING VIDEOTAPE MASTERS FROM AN ASSEMBLED FILM NEGATIVE

In the example of the television episodic, the problem is that a PAL videotape master must be realized from an NTSC videotape master. There is no common ground for the formats, and the recourse is to scan-convert from NTSC to PAL. The solution to this dilemma would have been to create the PAL videotape master from the correctly assembled film negative.

This is precisely the benefit of using the nonlinear editing system to create a film cut list. The selected film negative is transferred to videotape and edited on the nonlinear system. The resulting film cut list is used to indicate to the negative cutter which pieces of negative should be assembled and in what order. The final negative assembly can be thought of as a neutral element that can then be transferred to the required format and is thus used to create both NTSC and PAL masters. The benefit is that no scan conversion method has to be used to translate from NTSC to PAL or from PAL to NTSC. The film is simply projected at an appropriate rate to create either a PAL or an NTSC videotape master.

Another emerging requirement where film, videotape, and computers must coexist involves the area of digital imaging. Film images that are manipulated digitally by computers must originate on film, be transferred to digital data, be manipulated digitally, and then be transferred back to film.

As an example, consider a motion picture that is aimed at a family audience. One scene involves two children who are first filmed in a dark room. The boy waves his hands and talks to a character who is not visible. Later, a three-dimensional computer-generated dragonfly is designed from individual frames and rendered into an animation stack. This element is then composited with the background plate of the boy. The compositing process involves taking the original film and the animation stack and creating a new piece of film, which is then used in the finished motion picture.

Special visual effects done strictly in the film domain, the videotape domain, or the digital domain, are being replaced by methods that include a combination of all three. In the same scene, it is quite commonplace that some effects will have been achieved by a combination of all three techniques. Data must be manipulated, and this data is visual information that can be affected in any way but that most often must wind up in the same form as it originated.

HOW FILM, VIDEOTAPE, AND COMPUTERS COEXIST

There are several paths in which film, videotape, and computers intertwine. This chapter describes three: film to tape, film to tape to film, and film to tape to film to tape.

FILM TO TAPE

Film to tape (FT) is a transfer process. Footage is shot on film and transferred to videotape. The machine that accomplishes this is called a *telecine*. The videotape is then edited, and either the direct output of the nonlinear system becomes the master, or the EDL is conformed to create the final videotape master. Film to tape can be viewed as uni-directional since once the original film is shot and transferred to videotape, the film does not have an additional role in the process. Film was used to lend a specific look to the production, but the program does not need to be delivered as a film release.

Where would film to tape be used? Film to tape is commonly used for television commercials. Original footage is shot on film, transferred to a high-quality videotape format, such as D1, D2, or 1", edited nonlinearly through the use of lower-quality videotape dubs that are loaded into the nonlinear system, and finished on videotape, using the high-quality videotape transfer as source material. This is perhaps the most common situation in which film to tape is used. Throughout the world, the majority of television commercials are shot on film.

Again, for every example there are exceptions. Footage shot for commercials may not necessarily have a final destination on the television screen. Often, theatrical versions of a television commercial are required. It is common for a commercial to be shown on television in one country while a film version of the commercial is being shown in a motion picture theatre in another country. The material is still a commercial, but its distribution form is different.

FILM AND VIDEOTAPE SPEEDS

Film and videotape can move at different speeds. Normal play speeds for film and videotape are different, and normal play speeds for videotape itself are different, depending upon the transmission method used.

Format	Normal Speed (fps)
35mm film	24
NTSC videotape	30
PAL videotape	25

Fast Motion (Undercranking)

It is necessary to understand how slow and fast motion on film is achieved. When we want to achieve fast motion, we shoot the film at less than 24 fps. When film is shot this way, a technique known as undercranking, movement appears to be faster than normal. The term *cranking* harkens back to the time when film cameras were powered by a hand crank and the skill of the cameraman who had, with time and practice, learned what normal cranking cadence should be.

Since the normal shooting speed of film is 24 fps and the normal projection rate of film is 24 fps, even when there are fewer frames of an action, the material will still be projected at 24 fps. If the film is undercranked to 6 fps, the action will be projected on the screen in one-quarter the time that it would normally take. Another way of stating this is that the action occurs four times faster than normal.

Slow Motion (Overcranking)

When slow motion is desired, we shoot the film at more than 24 fps. This is a technique known as *overcranking*, or cranking the film faster than normal. If we expose the film at 48 fps and project it at 24 fps, the action takes twice as long to complete. (The projection speed is never altered because that would affect the speed at which the sound is resolved to the listener.) Another way of stating this is that the action occurs two times slower than normal, at 48 fps, or, in videotape terms, at 50% normal speed.

CORRELATION OF FILM AND VIDEOTAPE

When more or less film is shot per second and the film remains in the same medium and is not transferred to videotape, little attention needs to be paid to the correlation of film to video. If we are told that some footage of a basketball player has been shot at both 6 fps and 72 fps, we know that we will have alternatives when we start editing the piece; there will be action that is four times faster (6 fps) and three times slower (72 fps) than normal speed.

Since the projection of film is steady at 24 fps, the correlation of exposed film frames to projected film frames is quite easy to calculate; a fraction or multiple of 24 is used to determine how the action will occur. However, when film is transferred to video for nonlinear editing or for distribution on video, the correlation of film frames to video frames depends upon what type of video rate is used.

There is usually no easy correlation among various playback rates between film and videotape. For example, when transferring film (24 fps) to PAL (25 fps), one extra video frame must be generated from every 24 frames of film. Similarly, there is no 1:1 correlation between film and NTSC videotape; in this situation, six extra film frames must be created as the transfer takes place between 24 fps film and 30 fps videotape.

One way to avoid the necessity of creating video frames to achieve a 1:1 correlation is to alter the shooting speed. If the eventual goal is to achieve a PAL videotape master, the film is shot at 25 fps. Twenty-five frames of film transferred to 25 video frames creates a 1:1 relationship and ensures that for each frame of film, there is a corresponding video frame. This is similarly achieved in NTSC, if film is shot at 30 fps. Thirty film frames transferred to 30 video frames again results in a 1:1 relationship.

However, these alternatives are rarely taken. Shooting film at speeds greater than 24 fps increases the amount of capital required in a budget without adding significantly to the perception of "normal" motion. As a result, shooting at 1:1 correlation rates is rarely done. After all, why shoot more film per second unless a specific effect or benefit is realized?

Direct Correlation of Film Frames to Video Frames

There are situations in which a direct correlation of film frame to video frame is beneficial. When the FTF situation involves using electronic techniques to perform some type of digital imaging, a 1:1 correlation is necessary.

For example, if a segment of film must be manipulated electronically, such as with computer painting, each frame is transferred to the digital device, painted, and returned, frame for frame, back to film. For visual effects that must be created in the world of videotape, which runs other than normal film rates, a 1:1 correlation often lessens the frequency of aliasing and motion artifacts. (Recall that interpolating fewer frames into a medium that requires more frames than the original requires interpretation; in essence, this is yet another form of sampling except that instead of information being discarded, information must be generated: redundancy must be introduced from fewer samples.) Being able to capture one film frame, process it, and return it back to one film frame without interpreting whether the film frame consists of two or more fields is very advantageous.

FILM TO TAPE TO FILM

When a film is shot and the goal is to edit on an electronic nonlinear system to produce both a video EDL and a negative cut list, the path that the project takes will be something like this: The film being shot is a 60-second commercial. It is shot at 24 fps and is edited on an electronic nonlinear editing system running video at 30 fps. The goal is to load the selected footage into the system, create a correlation between film and video frames, edit the commercial, and arrive at a video EDL that will be used to online the commercial for its television run. The next work product that must be created is a negative cut list because the commercial will be delivered on film for a showing in theatres in another country.

Telecine

The telecine process involves the use of a sophisticated film projection machine that ensures that a certain number of film frames are projected each second. The machine does not waver and delivers the exact number that the user requires. The speeds at which the telecine machine can run vary. Some can only run the film at 24 and 30 fps, while others can run the film at slower and faster rates.

To put it simply, in the telecine process, film is threaded between a feed reel and a take-up reel and run through a projection system. The film is pulled down from the feed reel and held for a time, during which the image is actually transferred from its optical state to an electrical state. These electrical signals are then recorded to the videotape. Telecine systems work in conjunction with color-correction systems that affect the images being transferred to videotape. The color of the film frames can be altered, or special effects such as enhancing or reducing film grain artifacts can be introduced.

The sync sound for the program is played off a 1/4" audio tape or DAT, and both the picture and the sound are transferred to videotape. This source videotape is then used to create dubs that will be used for the nonlinear edit.

EDGE NUMBERS

It is easy to locate videotape frames because a timecode number is associated with each frame. Locating a specific frame of film within a roll is also easily accomplished. This is due to the fact that there are film edge numbers that provide a method of locating specific footage. These edge numbers are placed on the edge of the film by the manufacturer. These latent edge numbers are struck directly into the film negative. Whenever the film is transferred, a unique identifying number for each frame of film becomes part of the transfer. This data is useful and will be used to successfully match back to the original film negative when the editing stage is completed. Edge numbers are also called *key numbers*.

Edge numbers appear at specific intervals on a strip of film. In 35mm 4-perf film, an edge number appears every 16 frames, or once every foot. In 16 mm film, there are 40 frames for each foot of film; edge numbers appear every 20 frames, or every half foot. Every 1,000 foot 35mm film roll translates to approximately 11 minutes.

In Figure 16–1, a strip of 35mm film is shown. The edge number is an alphanumeric identifier. In this case, KeyKode™, from Eastman Kodak, is used to identify the individual film frames. The edge number, KJ 29 1234 5678, has two parts. KJ 29 1234 is the reel identifier; this code reflects the film manufacturer, film type, and reel number. 5678 is the footage counter; this reflects the location of the frame in terms of feet of film.

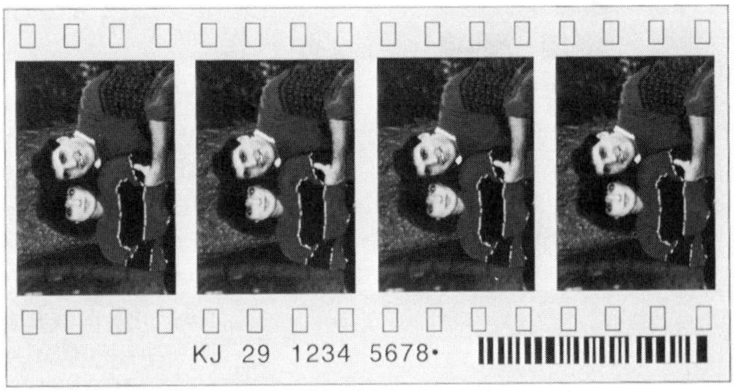

Figure 16–1 This piece of 35mm 4-perf film shows the KeyKode™ edge number and the accompanying machine-readable bar code which is used to identify each film frame on the camera roll. Illustration by Jeffrey Krebs.

A zero frame indicates where the edge number begins. Since the edge number requires the span of several film frames to be displayed, there is an indication of where the 16 frame counting process should begin. A filled-in circle indicates whether the zero frame is at the beginning or at the end of the edge number.

In this example, there is no frame offset because shown is a zero frame reference edge number. If we want to label a frame within the 16 frames-per-edge number, we identify that frame by using an offset to the previous edge number. This offset would appear as "+00"; this reflects the frame offset from the edge number. The frame offset number is important because there is one edge number for every 16 frames of film. For example, if the chosen frame showed, for example, KJ 29 1234 5678 +12, the offset indicates that the selected frame is plus twelve frames from the edge number. Edge numbers, like timecode numbers, ensure that we can find a particular film frame. By using frame offsets, we can find frames within the boundaries of two edge numbers.

PULLDOWN

As shown in Figure 16–2, when transferring film to videotape, a method must be used to create more video frames than there are film frames. *Pulldown* refers to the method by which additional video frames are created from their film counterparts. Since we start out with 24 film frames for each second of action, we must arrive at 30 video frames for that one second. How do 24 frames become 30 frames? The answer lies in the act of "pulling down" the film and holding it for a specific amount of time while it is projected and transferred to videotape.

A 24-to-30 conversion provides us with a ratio of 4:5; for every four frames of film, there will be five frames of video. Six sets of four film frames and six sets of five video frames represent each second of material. For PAL transfers of material shot at 24 fps, the transfer proceeds at 24 fps while one additional film frame is created, for a total of 25 frames of video from 24 film frames.

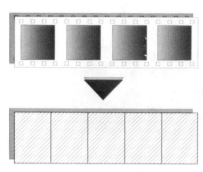

Figure 16–2 This is the pulldown sequence when transferring 35mm film to NTSC videotape. Six extra frames must be "created" when 24 fps film is transferred to 30 fps videotape. This results in a ratio of four film frames to five video frames. Illustration by Jeffrey Krebs.

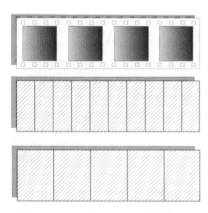

Figure 16–3 In a 2-3 pulldown conversion, four film frames are represented by ten video fields, which in turn create five video frames. Illustration by Jeffrey Krebs.

2-3 and 3-2 Pulldown

Creating five video frames from four film frames does not involve duplicating frames. Instead, the process of pulldown creates duplicate fields. At specific times, a frame of film is recorded as either two video fields or as three video fields.

Since each video frame consists of two fields of information, the extra six frames of video are created by duplicating fields. The telecine machine ensures that a frame of film is pulled down into the projection gate and held for a specific amount of time while a specific number of fields are recorded onto videotape. It is for these reasons that a telecine must be used. Another type of transfer device is a film chain, which is similar to the projection methods described but which cannot ensure specific pulldown. Film chains cannot be used when a negative cut list is required because the pulldown cannot be guaranteed to take place in the same specific 2-3 or 3-2 sequence.

In a 2-3 pulldown, the first film frame is transferred for two fields to the videotape. The next film frame is transferred for three fields to the videotape. The process continues, and the result is that for every four film frames, ten video fields, or five video frames, will be artificially created. This ratio of four film frames to five video frames provides us with the 24-to-30 frame conversion that is required (Figure 16–3).

Although duplicate video fields are created, the action represented in the film frame is not altered. All that is happening is that one frame of film is transferred to videotape as either two fields or three fields.

There may be confusion regarding whether to use 2-3 or 3-2 pulldown. It is often thought that one method is better than another. In actuality, the only difference between the two methods is where the duplication of fields begins. When we use the term *pulldown* we are merely describing the location where the duplication begins. Whether a transfer is in 2-3 or 3-2 pulldown sequence is of little consequence; the result of both is that the additional video fields are created. However, there is usually a preference for the 2-3 sequence because in a 2-3 transfer, the first film frame is associated with two video fields; therefore, there is a direct correlation of one frame of film equaling one frame of video. 2-3 pulldown is also referred to as the *SMPTE-A* transfer method. 3-2 pulldown is referred to as the *SMPTE-B* transfer method.

Pulldown Mode

In Figure 16–4, there are four film frames labeled *A*, *B*, *C*, and *D*. The A frame is transferred, and two video fields are created. The B frame creates three video fields. The C frame creates two video fields, and so on. The transfer of four film frames yields the following video fields:

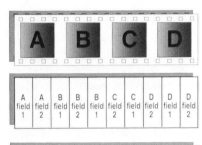

Figure 16–4 Pulldown mode refers to the location of a particular video frame within the pulldown sequence. This diagram shows a 2-3 pulldown sequence and the pulldown mode. Illustration by Jeffrey Krebs.

Video Frame	Film Frame	Video Fields
1	A	A + A
2	B	B + B
3	C	B + C
4	D	C + D
5	None	D + D

When we edit with film footage that has been transferred to videotape, it may be possible to edit on frames that do not have a direct correlation to a piece of film. For example, video frame 3 (:02) consists of a B frame and a C frame. The B frame is a remnant of the 2-3 pulldown sequence from video frame 2 (:01). Some nonlinear systems compensate for this by displaying only video frames that have a direct correlation to the film. In this case, the B frame would not be displayed. Instead, the C frame would be displayed. In this way, the editor would be able to see the direct relation of the video frame being chosen to the film frame that exists on the negative.

Pulldown Mode Identification

The pulldown sequence refers to how duplicate video fields were created: either the first duplicate consists of two fields in a 2-3 sequence or three fields in a 3-2 sequence. The pulldown mode allows us to determine what position a frame will take in the pulldown sequence. If we are given a videocassette that was transferred from film in a 2-3 sequence and we are asked to determine what the pulldown mode of a particular frame is, we can find the answer by looking at specific information on the videocassette.

In Figure 16–4, we see that five video frames are created from the four film frames. If the zero frame consists of two fields, it must be either an A frame or a C frame because only A and C frames take part in creating two video fields. When we jog the videotape through the two fields, if the timecode does not change, the zero frame is an A frame. If the timecode does change between fields, the zero frame is a C frame.

If the zero frame consists of three fields, it must either a B frame or a D frame because only B and D frames take part in creating three video fields. Now, we jog through the videocassette and find the zero frame. If the timecode changes between fields 2 and 3, it is a B frame. If the timecode changes between fields 1 and 2, it is a D frame.

SYNC POINT RELATIONSHIPS

When the film for the commercial is being transferred to videotape, a database needs to be created. The film transfer will most likely be a one-lite process in which only cursory attempts are made at color correction. The goal of the one-lite is to provide an overall exposure to the film to make editorial judgments. When it is finally determined which pieces of film will be used in the commercial, just those pieces of film will be retransferred, and more time will be spent color-correcting each scene. The database created during the one-lite consists of sync points, or common relationships among the various media that are associated with every strip of film being transferred.

Sync points include the following:

Edge numbers, which relate to the sync point of film footage

Timecode, which relates to the film's position on the videotape

Pulldown mode, which relates to the position of the video frame in the pulldown sequence

These three sync relationships are the minimum needed to provide a negative cut list. There may be additional sync relationships; if a sound cut list is required, the sync points will expand to include these:

Audio timecode, which relates the sync point of the original sound recording to the film frame

Ink number, which relates the sync point of the sound transfer to the sync point of the film frame

For each sync point, the scene and take number can be noted.

Punching the Printed Film

One method of quickly identifying the sync point is to make a physical punch to the printed film. By looking at the resulting videotape, the punch hole will indicate that a sync relationship has been established. The database for the sync point illustrated in Figure 16–5 shows the following:

Figure 16–5 Making a physical hole directly on the film permits the easy identification of the sync point, which becomes known as the *punch frame*. Also shown is the "smart slate," which includes a visual reference for the audio tape timecode. Illustration by Rob Gonsalves.

Video reference frame (04:09:28:00.1). This shows the timecode that is being recorded to the source videotape. This tape will be used to perform the final assembly for a videotape delivery. The ".1" designates field 1.

Videotape timecode (04:09:28:00). This shows the timecode that is being recorded onto a second videotape, usually a cassette format. This tape will be used to input material into the nonlinear system. The timecode numbers must agree with the timecode on the source tape that will be used during the final conform.

Film edge number (111590+00A). This indicator shows the edge numbers in feet and frames. The "A" indicates the pulldown mode.

Audio timecode number (08:04:27:01). This indicates the timecode for the audio recording. This format is usually center track 1/4" reel-to-reel audio tape or digital audio tape with timecode capability.

Once these relationships have been established, we can determine the correct profile for each ensuing frame since the rules of counting have been established. When a new sync relationship has been made, that is, when any of the sync points is changed, as is the case when a new punch frame is made, a new database entry must be created. We cannot simply continue to count through a different set of sync points because at least one of those sync points has now changed.

The largest criticism of manual methods of database entry is that mistakes can occur. If a video timecode number is given for a particular key number and the timecode is incorrect, when we create the negative cut list, the incorrect frames of film will be used from the negative. All it takes is one manual entry error for the wrong piece of film to be used.

Automatic Key Number Readers

Removing the possibility of human error in creating the database of sync relationships is the goal of machine-readable key numbers. The essential method of these systems is similar to the bar code information contained on various products. By placing machine readable edge numbers or timecodes directly on the film, edge readers can be placed on the telecine machine to capture sync points automatically. One example is Eastman Kodak's KeyKode™, which consists of the usual edge number information but also includes a bar code alternative. This bar code is deciphered by a reader that is housed on the telecine (Figures 16–6 and 16–7).

Telecine machines pass on information about the items being synchronized. The controlling device can be a lap-top computer that runs software capable of orchestrating the various machines involved in the telecine process. If the film footage has machine-readable edge numbers and the telecine machine is outfitted with

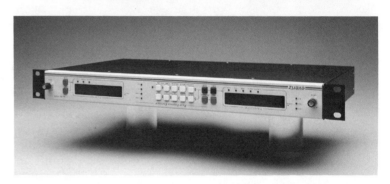

Figure 16–6 The KeyKode™ reader processes the bar code information contained along the edge of the film. Courtesy of Evertz Microsystems, Inc.

Figure 16–7 The 4015 timecode reader/generator from Evertz Microsystems, Inc., processes timecode sync points for reference to film sync points. Courtesy of Evertz Microsystems, Inc.; installed at Mainway Studio, Burlington, Ontario, Canada.

a bar code reader, the relationship of items in the transfer can be automatically captured. Software and hardware to control the telecine include a means of remotely operating videotape and audio machines. The purpose of machine-controllable telecine transfer and bar code information on the film is to provide a complete database for all film transferred to videotape, thus avoiding any manual entry of information to the film database.

When the footage for our commercial is transferred to videotape, if the film has edge numbers in bar code, a database will be created for every item involved at a sync point. This database is created automatically and will consist of the following:

Film edge number. The film edge number is captured via the bar code reader.

Video reference frame. The videotape timecode for that sync point is captured by the timecode reader connected to the telecine and the videotape recorder.

Pulldown mode: A frame. The pulldown mode for that sync point is established; depending upon the system being used, a sync point may consistently be the same type of frame, usually an A frame. If not, the pulldown for the sync point will be calculated by referencing the position of the frame within the pulldown sequence.

Pulldown sequence. The method of transfer is established as either a 2-3 or a 3-2 sequence.

Audio timecode. The audio timecode for that sync point is captured by the timecode reader connected to the audio tape player.

Additional. The scene, take, and description entries for the sync point are entered manually. Since these entries are not critical to the creation of an accurate negative cut list, errors here can be tolerated.

The automatic creation of such a database is a great step forward in reducing errors in cataloging how film frames relate to video frames. Once the methods of automatic capture are in place, the steps are as follows: Footage for our commercial is placed on the telecine, and the audio tape to be synched to the picture is loaded onto an audio tape player; this will be either 1/4" or DAT playback. A videotape is placed in the videotape recorder. All machines are then under the control of the telecine transfer system. The telecine operator cues the film to the first sync point. If the film has been punched, the film is stopped on the punch. This sync point may include clapsticks.

Next, the proper piece of audio for the sync point is located. If the commercial footage we have shot includes sync sound, such as two people talking, we may elect to sync the sound to the picture during the telecine process. Locating the right sound for the sync point may be a very easy process if the shooting stage included the use of clapsticks that displayed the timecode of the audio recorder. These clapsticks are called *smart slates* or *digi slates*.

When the operator parks on the sync point of the film frame, a visual timecode is displayed on the smart slate. This timecode number is then typed into the telecine controller, which remotely operates the audio tape player. The result is that the correct sound for this visual sync point is found. When the film is run forward, the sound is in sync with picture. Both picture and sound are then recorded to the videotape.

When all the film for our commercial has been synced with the corresponding audio and transferred to tape, the work products of the telecine session will include one or both of the following:

Computer printout of sync relationships. A hard copy printout of all the sync relationships involved is generated. The relationships are correlations: This edge number equals this videotape timecode equals this pulldown mode equals this audio timecode, and so on.

Software file of sync relationships. In addition to the hard copy, database information in digital form may be an option. This should be available on a computer floppy disk. Otherwise, direct serial transfer of the sync relationships from the telecine system to the editing system may be possible.

A comprehensive telecine system provides both work products. The minimum one should accept is a computer-generated hard copy of the sync points. A hand-written copy must be carefully scrutinized, as vagaries in handwriting could severely affect the integrity of the data. More than one negative cut list has been affected by a hand-written telecine log that confused a B frame with a D frame.

The capability of providing a floppy disk that contains all the information regarding the telecine transfer is extremely important. In the case of a print-out only, the data must be entered manually into the editing system being used to correlate the material to be edited to the negative to be cut. If a computer disk is available that has all the sync points as a database file, this digital data can be entered automatically into the nonlinear editing system as long as the file structures are compatible. This avoids the manual entry of data when we enter the stage of moving information from telecine to nonlinear editing system.

EDITING AND DELIVERY ON VIDEOTAPE AND FILM

The next stage in creating the commercial is to load the nonlinear editing system with the required footage. The footage is then edited into the appropriate form to create the finished commercial. At this stage, one of the work products we require is complete: The Video EDL will be used to conform the original source tapes to create the final product.

We output the video EDL for the edited commercial, which consists of cuts, dissolves, and B rolls. The video EDL and the rough-cut output from the nonlinear system are taken to the conform room, and the commercial is finished on tape, yielding a videotape master that is now ready for duplication.

Film Cut Lists

Once the commercial has been edited, the video EDL and the sync relationships that have been established between the videotape timecode and the film edge numbers are used to create a series of lists that ensure the correct cutting and assembly of the film negative. Each list provides specific information to the negative cutter. While nonlinear systems may differ in the type of lists offered, the basic set of film lists includes the dailies report, pull list, assemble list, dupe list, and optical list. Additional film lists include the scene pull list and the change list.

Dailies Report
The dailies report is the list of sync points that is created during telecine. A dailies report is usually more sophisticated than the telecine report in that it contains ending edge numbers and the duration of film and video segments.

Pull List
The pull list shows the film segments that must be cut from each camera roll. The various pieces of film that must will be used in the conforming process exist on different camera rolls. The pull list provides the negative cutter with the information required to pull all the scenes required from each original camera roll. If a section of film has been used twice or is involved in an optical

effect, the pull list usually indicates that the one piece of original negative must be duplicated.

Assemble List

The assemble list indicates how the individual film segments should be ordered. The assemble list is, in essence, the final EDL. It is used in conjunction with the dupe list and the optical list to conform the negative.

Dupe List

The dupe list indicates all film segments that must be duplicated because the film frames have been used more than once. The dupe list is only used for those frames that are related to straight cuts.

Optical List

The optical list shows the specific film edge numbers used to create fades and dissolves optically. The optical list shows the scene start and end and the effect start, center, and end points. It is referenced by the assemble list.

Scene Pull List

The scene pull list is a scene-by-scene list that is pulled from the negative. The scene pull list differs from the pull list in that it lists the scenes that must be pulled but not the exact cuts within each scene. While the pull list indicates all segments that must be pulled from the individual camera roll, the scene pull list indicates all scenes that must be pulled from all camera rolls. The scene pull list is usually ordered by edge number.

Change Lists

Prior to negative cutting, when a film work print is used and screened, changes may be requested. The change list compares the original assemble list with the new assemble list and shows only the changes made to the original. In this way, the person conforming the work print can go only to the sections requiring changes rather than interpreting an entire assemble list.

Conforming the Negative

Having edited the commercial on the nonlinear editing system and conformed the source videotapes to create a finished videotape master, the next goal is to create a cut film negative to be used for theatrical distribution. Using the various film lists, the negative cutter pulls the different film segments from each camera roll. If film needs to be duplicated because it was used more than once in the commercial, the dupe list will indicate this, and dupes will be ordered from the lab. Using the optical list, fades and dissolves also will be ordered from the lab.

When all the required elements are available, the negative cutter assembles each segment in the order determined by the assemble list. The completed negative is then duped and color-graded, and additional film prints are struck. These are the prints that will be used for theatrical distribution.

Sound

If the video conform is being done before the negative cut, the audio will be on the source videotapes because it was transferred during the telecine. If the audio on the source tapes cannot be used because it is down a generation from the original, the sync sound is reassembled from the original sound recordings, usually on 1/4" audio tape or DAT.

The original recordings are referenced by either the timecode numbers or the slate IDs. An audio-only assemble list can be created because the telecine process included entries for sound timecode sync points. By using the timecode information relating to the audio, entire audio tracks are reassembled to create a first-generation soundtrack.

Once the negative has been assembled, the mixed soundtrack from the video conform stage is transferred to magnetic track and married optically to the finished film. Since the audio work has been completed, there is no reason to redo this effort.

This is the basic process of the film to tape to film method. Original film footage is shot, transferred to videotape, and edited on the nonlinear system. This yields a video EDL that is used to create the videotape master and a negative cut list that is used to conform the original negative for printing and distribution. The negative cut list has little tolerance for error; it must be accurate within one video field.

Again, it should be noted that the automatic key number reading systems may not be available on every telecine system because telecines must first be outfitted with the required bar code readers. Until telecine systems that offer automatic file output for sync relationships are widely available, a combination of manual entry of data and automatic creation of data will continue. It remains necessary to be aware of all the potential pitfalls that exist in the conforming of film negative to the nonlinear edit system's negative cut list.

FILM TO TAPE TO FILM TO TAPE

The film to tape to film to tape (FTFT) process is used when the end product is not a conformed negative but a color-corrected videotape master. The most common use of FTFT is for short format work such as television commercials.

Why is FTFT used? Consider a television commercial shot on film. A commercial shoot can yield quite a range of footage, sometimes from a few thousand feet of film to tens of thousands of feet of film; it is not unusual for 60,000 feet of 35mm film (approximately 11 hours) to be shot for a commercial. Instead of spending undue time and expense color-correcting all the footage, a one-lite transfer is made. Then, after the commercial is edited on the nonlinear system, only the film used in the final commercial is retransferred. This negative is pulled, color-corrected, and transferred to tape. This second videotape transfer is used during the online conform process.

The benefit of the FTFT method is that time and money are saved. In addition, using a computer program to calculate the correlations between the timecode numbers of the original film-to-tape transfer and the retransfer eliminates the possibility of human error in calculating these sync relationships.

The four-step process proceeds as follows:

1. A one-lite film-to-tape transfer is performed, establishing sync relationships.

2. The commercial is edited on a nonlinear system, creating a video EDL and a scene pull list.

3. The scene pull list is used to pull only the negative corresponding to the material used in the final video EDL.

4. Once the negative has been pulled, the negative is retransferred to videotape, and more time and greater care are given to color correction. This second videotape transfer has far less material than the original one-lite because only the selected film segments used in the final program are retransferred.

When the retransfer is complete, a new videotape source reel has been created. This second videotape usually has different time code numbers than the original source tapes, the ones used to create the original video EDL. A constant for both source tapes are the edge number sync points. By substituting the new source timecodes for similar edge numbers, the FTFT process creates a new video EDL that is then used to conform the finished video master from the new source tape.

Without FTFT capability, the process of determining what new timecode numbers relate to the original timecode numbers must be done manually. On occasion, methods even include "eye-matching" the retransferred material to the original material used in the program. Eye-matching means comparing by eye the material originally used to the newly retransferred material; while unsophisticated, eye-matching is a common practice. FTFT methods, utilizing computer programs and automatic edge number readers, will eliminate the need for eye-matching techniques.

EDITING AT 24 FPS

Much of the difficulty encountered when the media of film is transferred to the media of videotape is due to the difference in playback speeds. Regardless of whether the playback speed, and therefore the editing process, is different from the shooting speed, there will always be a need for some method of film-match back because, even if it were possible to edit at 24 fps, neither PAL nor NTSC television systems would be able to display material running at 24 fps.

Several benefits are gained from editing at 24 fps, and the nonlinear editing system should be evaluated as to whether it offers this capability. The first benefit is that the material shown on the nonlinear system's screen is running at its native speed. The movement originally seen in the film camera is preserved.

The second benefit of editing at 24 fps is a savings of overall storage time. If 24 frames instead of 30 are stored for each second of material, 20% more material can be stored for each second of footage. This can be significant. If we are storing 12 kB per frame at 30 fps on a 1 GB disk, the calculations are as follows:

12 kB × 30 fps = 360 kB/sec × 60 seconds/minute = 21.6 MB/minute

Approximately 46 minutes can be stored on the disk.

If, instead, we store the same material at 24 fps, the calculations change as follows:

12 kB × 24 fps = 288 kB/sec × 60 seconds/minute = 17.3 MB/minute.

Approximately 58 minutes can be stored on the disk, 12 minutes more than if we were editing at 30 fps. While this may not appear to be a significant increase in capacity, recall that nonlinear systems, especially digital systems, may offer larger capacity disks, and more disks may be accessible to the user at one time. Fourteen 2 GB disks in use at one time would provide 336 minutes, or an increased capacity of 5.6 hours over the same system running at 30 fps.

A third benefit of editing at 24 fps is realized solely in the digital nonlinear system. If fewer frames are stored each second, from 30 to 24, the space allocated for these six frames can be applied to an increase in the kilobytes per frame of the compressed image. In this way, there can be a choice: continue with the same picture quality but increase capacity, or increase the picture quality while leaving storage constant. If, for example, the system is allocating 12 kB per frame and there are six fewer frames, this represents 72 kB that can now be applied to the 24 frames. As a result, the 24 frames will be of higher kB per frame and will display an increase in picture quality.

An additional benefit can be gained from letter-boxing the film as it is transferred to the digital nonlinear system. Letter-boxing, which preserves the aspect ratio in which the film was originally shot, consists of two black bands across the top and bottom of the screen. This is necessary because film footage shot at a 1.85 ratio (or higher) must now temporarily exist in the 1.33 ratio of video. These two black bands consist of zero information to a JPEG-based digital video compression scheme. Therefore, the kB per frame that would normally have been concerned with the areas of picture above and below the frame can now be applied to the material within the letter box. Higher-quality pictures at the same kB per frame are the result.

The need to be able to transfer pictures back and forth easily among the film medium, the video medium, and the computer medium will continue simply because each medium has specific strengths that are not shared by the other media. The ability to move as invisibly, that is, as "losslessly," as possible among the various forms will become yet another feature set that the nonlinear edit system must offer. As more image manipulation techniques are founded in the digital domain, the use of nonlinear editing systems to transfer among film, video, and computer data under a common timebase will increase.

17

Evaluating Electronic Nonlinear Editing Systems

Evaluating the electronic nonlinear editing system, whether it belongs to the videotape-, videodisc-, or digital-based wave, will be an investigation on many levels. Before the capabilities of a specific system can be fairly judged, it is necessary to understand what the individual's requirements are based on the type of projects that will be undertaken.

Being able to determine what features and capabilities are needed will naturally lead to some of the questions that are discussed below. For example, how much footage must be accessible at any one time? How much disk storage should be included? What audio quality will be required? What type of work products will be needed at the completion of editing?

An important element is the amount of time that it will take to learn and master a certain system. This is a common concern raised by editors and their employers when an electronic nonlinear editing system is considered for purchase.

TECHNICAL CONCERNS

Before considering any nonlinear system for use, it is necessary to determine what the system will be used for and how it will be used within the editor's environment. Will the system be used primarily for commercials, episodic television, or feature films? Each answer may have significant ramifications on what type of system to consider.

Following are some common questions that often arise when a nonlinear system is being considered for use.

Anticipated Use

What type of work will primarily be done on the system?

Determine whether the system will be used for commercials, episodics, features, corporate industrials, documentaries, animations, and so on. It may well be that there is no primary type of work, although this is unusual. Commonly, the system is used for a very specific type of job. Alternatively, the reason that a nonlinear system is being considered in the first place may be to begin working in different editing genres.

If the system is being considered for a specific type of programming, such as commercials, you should immediately seek out acquaintances or facilities that are also known for expertise in the post-production editorial of commercials. Are nonlinear systems being used in these companies, and if so, what systems are being used? Why are they being used? Unless the facility is completely unapproachable and unwilling to share any information, asking questions of individuals who have already purchased nonlinear systems will be an invaluable aid in making a decision.

Storage

How much storage will be required for the type of jobs that will be done? Is late-arriving material anticipated? Will editing require simultaneous access to all material? Can transferred footage be easily removed? Will the system be expected to handle several projects with short turnaround times between projects?

Start with the overall amount of circled take footage a project will average. Here are two scenarios, each somewhat extreme: a commercial and a feature film. First, a particularly complex commercial is being shot with high shooting ratios, say 60,000 feet of 35mm film (11 hours) for a 30-second commercial. Of the 11 hours of film, three hours of selected footage will need to be transferred to the nonlinear system. Second, a feature film is being shot with a high shooting ratio. This yields 350,000 feet of 35mm film (64.75 hours) of selected footage, which the editor, at some point, must see.

For the commercial, if the average project is going to consist of three hours of dailies, a system configured with three to four hours of storage time will be required. If late-arriving material is a possibility, storage media that does not require processing time should be considered and the system should be evaluated with regard to how convenient it is to integrate such material.

If simultaneous access to all material is required, there is no choice. A system configured with at least three hours of storage is necessary. This may become very difficult to achieve when the total amount of footage becomes very large. Even the various digital systems, some of that offer 30 hours of footage that can be accessed at one time, will not be able to satisfy more intense storage demands.

More likely, you will have to work in stages. Consider the situation if only one hour of storage space is available. The process is to load one hour's worth of footage, edit a section of the project, and keep only the material that is appropriate. The unused material is discarded, thereby freeing up space for additional footage to be transferred into the system. The convenience of working in this mode is a valid question to consider.

If footage is transferred into the nonlinear system and, after working with the material, the editor decides that the material will not be used, can the material be easily removed or deleted and replaced by additional material? The ability to free up storage easily in this manner can be extremely useful, especially when the limits of the overall simultaneous storage are being reached.

Another consideration is system time. Especially in the fast-paced scheduling of editing commercials, systems are used constantly to edit projects for different clients. If a project has not yet been completed when the editing room must be relinquished to another project, what will happen to the footage stored on the system for the first project? In these cases, the media must be removable and should not be fixed in place. Until removable disks became available for the digital systems, this was a major concern. If the fast removal of disks is required, this capability should be investigated.

For the feature film described above, almost 65 hours of footage will be used to create a two-hour finished film. For all the waves of nonlinear systems, whenever there is such a large amount of footage, there will be the need to work in sections.

A major concern that editors express about nonlinear systems is that they will not have immediate access to all of the material that has been shot. Immediate access and simultaneous access are two very different things, and they should be treated as such. *Immediate access* means that all of the film that was printed exists in the appropriate form for the nonlinear system (either on videocassette, laserdisc, or digital disk). *Simultaneous access* means that all of the film that was printed is in the nonlinear system at the same time.

When working with a large amount of footage, the choices are to transfer all material to the different media of the nonlinear system. Many loads of videotape are created, many laserdiscs are made, or many computer disks are required.

Simultaneous access to all of the footage is not yet possible. Since videotape-based systems average 4.5 hours per load, laserdisc-based systems average 3 hours per load, and digital-based nonlinear systems at comparable picture quality average 6.5 hours per load, the film must be edited in stages, accessing as much footage as the system allows at one time. Over the next few years, improvements in digital video compression and disk arrays will yield greater storage capacities.

One way to approach the film is to look at the overall amount of printed footage, in this case, almost 65 hours. For a two-hour movie, the film is broken down into 12 ten-minute reels. Next, consider the number of scenes in each of the 12 reels. For example, scenes 1 to 13 could comprise reel 1. Next, look at the amount of printed footage for these scenes. Do the scenes fit on the system at one time? If so, all the footage for a reel can be put into the system, and these scenes can be cut. If the footage shot for scenes 1 to 13 does not fit on the system at one time, scale the range back, say, from scenes 1 to 6. Working in this manner should not be a problem; after all, there is a limit to the amount of film that can be run through the basic film flatbed.

Simultaneous access is really not as much of an issue as the editors' concerns would seem to suggest. Instead, immediate access is of primary concern. When an editor cuts a project and completes the editor's cut, it will usually be time for the director to join the process. If the director wants to replace a take with another that was printed, the editor must be able to have immediate access to the alternative take.

If the alternative take does not exist in the form that the nonlinear system requires, time has to be taken to get the take in that form. This means either transferring to tape, laserdisc, or computer disk. Because the editor never knows what footage will be used, it is important that all the printed footage be ready and in the form that the nonlinear system requires. As a result, a great deal of material is stored on a shelf for possible use: either loads of videocassettes, laserdiscs, or computer disks. While there is a cost associated with having all this storage available for immediate access, it must be remembered that, with the budgets associated for feature films, the overall amount of money spent for the film itself and for the nonlinear editing storage media tends to be the most inexpensive part of the budget.

In terms of storage costs, the only viable solution for the digital-based systems is to use optical disc cartridges, which are much less expensive than magnetic disks. To have immediate access to 65 hours of footage stored on magnetic disks involves a significant cost compared to tape and laserdisc. When enhanced methods of digital video compression and better data transfer rates of optical discs arrived in 1991, using opticals for long form work became a much easier economic issue. Optical discs, per minute of storage, are quite comparable to the cost of laserdisc pressings.

Audio Quality and Number of Channels and Tracks

With the digital wave of systems, there can be a variety of choices regarding audio. What audio resolution is required? How many audio channels must be simultaneously monitored? How many audio tracks should the system offer? The importance of audio to the editor can influence which systems are considered. The digital wave offers multiple channels, that is, multiple voices that can be simultaneously monitored, and multiple audio tracks, that is, several sound tracks, usually up to 24, that can be utilized during the editing process.

For many projects, being able to have four to eight tracks of audio available is a desired feature. Similarly, for many projects, two tracks of sync dialogue are all that the editor needs.

The editor must judge how important these items are for the project being done. If there is a budget for further sound work, then having these features may not be important. If, however, a portion of the budget is saved because more sound work was completed during the editing stage, then a benefit has been realized. Perhaps this money can be used to achieve a desired effect which could not have been done due to budget constraints.

Editorial Work Products

What does the editor expect the system to provide at the completion of the editing stage? Does the work product consist solely of a video EDL? Or are a number of work products absolutely essential, including a film cut list, digital audio at 48 kHz transferred to DAT and then taken to the audio sweetening session, and fully capable auto-assembly features with switcher control? These are all issues that must be considered. If it is important to leave the nonlinear system with more work products to reduce the time spent in the finishing suite, the work products of the system should be carefully evaluated.

Unfinished versus Finished Results

Investigating the nonlinear system starts with the type of job that the system will be used for, but it must include the type of product that the system will eventually provide. This category is different from the editing work products that the system offers. Will the output from the system be the final version of the program? This question has become an issue as digital-based systems have appeared. When digital video compression methods are used at compression ratios that cause little degradation from the original image, can the direct output of the system be used for final program delivery? For the tape-based systems, picture quality is limited to 3/4". For the laserdisc-based systems, picture quality ranges between 3/4" and 1" videotape. Digital systems, depending upon how the Q factors and kB per frame rates have been manipulated, are capable of providing pictures that do not show degradation from the original input material.

If the project being edited can make use of this direct system output, a careful judging is required. One way of doing this is to take footage of a project recently completed, transfer it to the nonlinear system, and edit it. Then, output the project to videotape and ask the client, whether the quality would be acceptable.

OPERATIONAL CONCERNS

Evaluation of the nonlinear editing system is not based solely on technical merits. Because the proper system must be chosen for the job, most editing systems are first judged by what they can and cannot do, not by how intuitive they are to learn and to use. If it is difficult to learn, even the most capable system will not attract as many users as a system that is less capable but extremely intuitive to learn and to operate.

Following are some common issues that arise when considering nonlinear editing systems.

Computer Capability

All electronic nonlinear editing systems use a computer as the processing unit. To what extent does the nonlinear system require the operator to be aware of the computer's role in the system's operation? Intuitive systems place the role of the computer largely in the background of editing operations.

When editors approach electronic nonlinear systems, regardless of the wave, with trepidation, a fear of technology and of computers usually ranks high on the list of reasons to resist such a move. There is no computer in a film flatbed, just a series of motors and film and magnetic track transports. When something goes wrong, it is usually a failed braking mechanism or a burned-out bulb.

Computers are a reality for the electronic nonlinear editing system. Some systems make the computer obvious; some do not. When the nonlinear system is being judged, it is important to be conscious of how the manufacturer has designed the system to take advantage of the power of computers and to avoid the reluctance that some users have to approaching technology. This is especially important for the digital-based wave. Instead of many different machines being present, as is the case with tape- and laserdisc-based systems, there is instead a computer display showing footage while the user interacts, not with machinery, but with computer software.

Editing Methods

Are the editing methods intuitive, or do they change the editor's normal work methods? When considering the nonlinear system, be conscious of how long it takes to understand the basic working and editing principles of the system. This, of course, will vary greatly from individual to individual, and that is precisely the point. Choose the system that is closest to the two requirements that must be met: the technical aspects of the desired system and the working methods of those who will operate the system. All too often, a facility purchases the newest technology only to lament that the editorial staff does not like the system.

Operating the System

There are very few instances in which film or online videotape editors insist on "test driving" their flatbeds or online edit controllers. Since the evolution of these systems has been slower than their nonlinear counterparts, less change occurs from year to year. Nonlinear systems, however, can vary dramatically.

There should be no resistance on the part of the manufacturer to allowing the editor to try the system before deciding whether to purchase it. One recommended test is to try to cut familiar footage into an edited sequence. There is no better method of investigating the technical and operational capabilities of a system.

Training Programs

A number of interesting developments are occurring with regard to learning the nonlinear systems. For example, videocassette tutorials are used in conjunction with workbooks. These can be very helpful in self-teaching situations. In addition, many nonlinear systems include modems that can be used to communicate with the manufacturer and to download software and software release notes. Additionally, computer bulletin boards that connect user to user also contribute to a sharing of information. Tips and techniques can be shared from all over the world.

Another benefit of modem-capable systems occurs with manufacturers who have incorporated methods of remote diagnostics. Remote diagnostics allow the manufacturer's customer support department to judge the state of the editing system through a normal phone line connection. This can be extremely helpful if the user is experiencing a problem for which a solution is not apparent. Being able to emulate the user's environment via the modem connection can be extremely helpful. Several manufacturers have this capability.

Benefits

Presumably, there is a reason why the nonlinear system is being considered in the first place. The benefits most often cited when electronic nonlinear editing systems are used are increased creativity and flexibility in trying different ways of cutting a scene. The economic benefits are more difficult to judge. At times, the amount of time saved by using a nonlinear system is quite significant because the project is completed in less time than would normally be required. Sometimes, however, the overall time spent editing the project is not less than that of conventional methods, but the perception is that a better product was achieved because there was time to try alternative ideas.

THE HUMAN SIDE OF THE NONLINEAR EDITING EXPERIENCE

When a new system is being considered, the evaluation is largely based on the technical aspects of the system, but there is also a human element that must be considered as the editor moves from either film or videotape to the electronic nonlinear editing system. The human element that is usually considered has to do with the ergonomics of the system. Are the buttons in the right place? Are they large enough? Are they too close together? Are the computer screens at the proper height for the editor?

These are all important points that must be investigated but the human side of nonlinear editing has nothing to do with the comfortableness of the system. Rather, what we are concerned with is the mental preparation required by film and videotape editors who make the transition from the familiar world of editing to the world of electronic nonlinear editing, which can be quite different.

For those editors who are coming from the world of linear videotape editing, there are techniques and rules that must be forgotten. When a professional editor has spent years learning the rules and years refining techniques based on those rules, simply forgetting them can be quite difficult. Nevertheless, there are things that the linear editor must be prepared to forget, and there are capabilities to which the editor must be open. This is also true for the film editor. Even though the film editor has been working in a nonlinear environment, it has not been an environment that has offered the ability to experiment easily with multiple versions.

If there is one thing that the editor must understand as she moves from linear to nonlinear, it is that she has become comfortable with nonlinear editing when she accepts that the beginning could be the end. This may sound cryptic, but it is essential to understanding why nonlinear editing should be used in the first place.

The major benefit of nonlinear editing is that there is no structure during the first and most important stage: trying to make sense of a great deal of material. Further, this is not just limited to the first stage; there is no structure until the editor decides, in her mind, that there will be a structure. The beauty is that the structure can be broken until the very last moment. This is precisely what makes it possible to turn in the best possible cut. It is unnecessary to commit until everyone involved agrees to the commitment.

Thus, the first thing that you must forget is that it is necessary to start at the beginning and work your way to the end. It is possible to start anywhere; start in the middle, start at the end. Start with the easiest material to cut, or start with the hardest. The nonlinear editing system is simply another tool that is available. The editor must be open to the notion that what he thought might be the beginning of a sequence may turn out to be better positioned somewhere else. The ability to throw out one's first impression about where material should be in the finished program and to investigate putting the material in a new location is a critically important mode of thinking that the editor must develop.

The second thing that you must realize is that everything is changeable. This may sound like an obvious fact, given the concept of virtual recording and the undo capabilities of the nonlinear editing system, but for the editor who has for years been accustomed to planning the next few edits while on the current edit, accepting the idea that everything can be changed can be a big leap. The editor must be open to these concepts.

The nonlinear editing process starts out with a winnowing of what you don't want to use. This is the footage that is not worth the videotape it is recorded on. Everything else, the material that you are absolutely sure you will use, and the material that might have a place in the program, should be loaded into the system, regardless of how remote that possibility of its use.

The editor must understand that, on the nonlinear system, sequences are made by honing and not by construction. The editor must be prepared to "go for tonnage," which means that shots should be ordered regardless of their length (the space that they occupy or the time that they represent). Instead, the editor should be prepared to pursue the flow of the images and to get that flow correct. Later, the amount of space or time allotted for the individual shots can be trimmed until they are the appropriate length. This is an especially difficult thing to ask the videotape editor to do because the transition editing concepts are quite influential in trying to pick the exact points where two shots should cut. Nevertheless, at this stage, it is a habit that must be broken.

If you are using a nonlinear editing system properly, you are trying ideas, experimenting with clips in different locations, changing the space that clips occupy, and attempting to hone the sequence as it moves toward completion. You can encourage serendipity by starting in a direction in which you do not expect a specific result. There can be many pleasant discoveries but only if you are ready for the expedition.

You can never judge how crazy an idea may sound until you've tried it on the nonlinear system. If you don't try the idea, you'll never know. Again, there is no penalty for trying these ideas. After all, it is assumed that the reason that you are working on the nonlinear system in the first place is to give creativity a better chance to emerge. But creativity is not going to emerge if you do not open your mind to trying those ideas that, at first, may sound very implausible. You must not think, "It's probably not going to work. Let's move on." Instead, you must say, "Let's try it, and if it doesn't work, we haven't lost what we have right now. We may come up with something."

When editing linearly, you do not usually have the opportunity of trying these "crazy ideas." Instead, you must pick the best idea since you probably will have only one opportunity to edit the program. In nonlinear, you have the opportunity to exercise these spontaneous notions.

It is essential that the editor try things and not fear reworking a scene.

Recutting is a natural process in creating the finished piece. However, in the case of the videotape world, the mere prospect of recutting carries with it such negative connotations that it becomes a dreaded task. The videotape editor must develop the attitude that, through the recutting process, the piece will become sharper and more defined.

It is often a difficult task for the editor to face cutting the hardest material in a program. Of course, what represents the hardest material depends upon the editor's perspective. It may be that a performance was particularly bad and the editor is faced with trying to edit a scene by using bits and pieces from many different takes. Or it may be that the coverage on a scene is very extensive. Facing hours and hours of footage for, say, a boxing scene, can be quite a demanding and difficult task.

Some editors will dive into the challenge of cutting the hardest material in a show, and they do it first. On the nonlinear system, it makes sense to start with the hardest material. Starting with the hardest material affords great benefits to the editor. These sequences can get old faster, and they can be fixed sooner. Knowing that everything is changeable, we can let the experience and results of cutting this material ripen in precisely the same fashion that Frank Lloyd Wright suggested that creativity takes time.

In the world of film and videotape editing, how much creativity can be afforded to cutting a scene often depends upon how much time you have. How much time you have to cut often translates into how much time you can afford—how much time and how much money. Given the undeniable fact of the economics of film and video post-production, it is sensible to allow the hardest parts of a project to have more of a life span from the time they are initially edited to the time that they are completed and inserted into the finished program.

When the hardest parts are given more life, it means that the editor has more time to live with these sequences. The ultimate goal is that it becomes possible to invest yourself to a greater degree in the work, certainly more deeply and more meaningfully. With more time to contemplate what is right and what is wrong about these difficult sequences, it is then possible to make the difficult sections better and to turn in a better cut.

The ability to take the time and to go back and rework such sections has always separated film and videotape editing. The film editor is able to go back and make changes, but the videotape editor cannot enjoy this reworking in any easy fashion. The videotape editor faces the task of having to deal with the hard sections as they come up. The videotape editor must edit these sequences as best as she can because the opportunity to rework them will not present itself. But now, armed with the ability to go back at any time and revisit cut sequences, the editor must be prepared, mentally, to look forward to, and not dread, recutting a sequence.

At the same time, the discipline that the videotape editor develops in choosing the best method of cutting a scene because it is difficult to make changes should be useful, especially when multiple versions are being made. The great benefit of nonlinear editing is that the virtual recording concept means that multiple versions are easy to achieve. It is up to the editor to know when the scene has been worked and reworked and is finished. Just because it is easy to make another version does not mean that it is necessary to do so. The editor's taste must take over despite the convenience of trying something slightly different. The editor must know when to say "enough is enough." As it gets easier and easier to try something different, it may become harder and harder to know when to stop.

If you begin working on a nonlinear editing system, you must be prepared to consider changes to the normal work flow to which you have become accustomed. By virtue of the different work products that are possible from the system, you may have more time to spend editing. For example, if the system can provide real-time review copies of the sequence, there will be no need to perform a tape-to-tape auto-assembly. The time saved can be put toward honing a sequence that still does not please you.

As another example, if the presentation is going to be a film work print showing and if the primary soundtracks can be taken directly from the system, there won't be a need to conform the audio tracks each time the work print is viewed and changes are requested. It is standard practice to have film assistants maintain the status of the work print and the sound for that work print in conjunction with what the editor is doing on the electronic nonlinear system. Here is an entire step that can be slashed away from the process. The result is more time and money available for the primary goal: to achieve the best possible cut.

The editor must be prepared to try new methods that will affect the usual work flow, whether the program is a film finish or a videotape finish. It is absolutely inevitable that the known and accepted flow of work in the film and videotape post-production process will change as the digital nonlinear wave matures.

You must also be ready to realize and accept the concept that there is a nonlinear exchange between you and the product. Although the system offers nonlinearity, you must be ready to take advantage of what nonlinear means and affords to the editing task. There are individuals who work on a nonlinear editing system but who bring to the system a linear philosophy. You must be ready to edit nonlinearly. Too often, an editor continues to work in a linear fashion, tied to the very structure of starting at the beginning and working to the end of the sequence. Granted that these are hard habits to break; not all editors will succeed at realizing the benefits of the nonlinear system until the constraint of always starting at the beginning is curbed.

Nonlinear editing means that there may not be as pressing a need for requiring a structure for the material. When footage is shot for a television show and the editor is armed with the footage and the script, the script provides the editor with the essential information that dictates how the material will be ordered. The editor chooses how much space the shots should occupy, but the overall flow of images is largely influenced by the structure imposed by the script. When such a blueprint is available, it can be both an asset and a liability. It is important that the editor uses the script as an outline and feels free to build the scene according to her own tastes.

When there is no set structure to the material that the editor is given, editing on the nonlinear system can be invaluable. In the case of documentaries, where structure usually evolves after working with the material for some time, the ability to work in small sections and then reorder these sections to create the actual structure of the documentary represents a task that the nonlinear system provides quite nicely. For the editor used to working in a structured format, nonlinear editing allows him to wade through a mass of footage and cut a path without feeling anxiety about not having a blueprint from which to work.

Nonlinear editing, raised to its highest level, is an experience that is unlike film or videotape editing. There are no penalties for trying ideas. There are many paths to creative success, and you can explore as many paths as you want. The only thing that is mandatory is a mind set that is ready to accept the challenge of exploration.

BENEFITS FOR THE EDITOR AND CLIENT AND PREDICTIONS FOR THE FUTURE

What are some of the benefits that digital nonlinear editing brings to the editor and client? What impact will digital technology have on the creative aspects of film editing, videotape editing, and program creation? Two experts in the field of digital editing give us their opinions.

William J. Warner, founder of Avid Technology, Inc., notes,

There are a few key benefits for a client editing on a nonlinear system. First is speed. The time it takes to try something is so much less, which means it's easier to try ideas and compare them. This means greater flexibility and creative freedom. Second is more results. It's easy to have multiple versions to suit specific situations.

From the editor's point of view, nonlinear editing lets him or her interact more smoothly with the client. The editor can take input from the client and turn it into reality more quickly. And they can also present their own ideas more quickly, so the client gets more of the creative talent that they're looking for in the editor.

The financial people will be happy that nonlinear often leads to cost savings. A show or a commercial can often be done in less time *and* will benefit from significantly tighter and better editing.

There's one benefit that few people admit to but all agree on. Nonlinear editing is *more fun*. It's fast, it's interactive, and people can work together in a way they never could before. Your time in editing is more enjoyable, more productive, and the results are visibly better.

Curt A. Rawley, president and CEO of Avid Technology, Inc., predicts,

In general, the next few years will see a revolution in the way producers, directors, and editors accomplish the desired results. It's much more than just editing. Right now we are experiencing a highly fragmented, relatively serial approach to video and film production. From both a media type and process step perspective we have different specialists involved with different elements (media types) at different steps in the creation of a product. There is no unifying force of commonality—and digital changes all that —or at least it holds the promise of potentially changing much of it.

The new tools that will evolve from the digital technology will afford the opportunity to break down the barriers that have existed between the highly skilled, but just as highly separated, specialists involved. The audio editor pretty much started over from scratch once the picture editing was complete. The script writer had no way of databasing elements of the story to the eventual media clips being edited except by manual paper techniques, or if a word processor had been used it certainly couldn't interface to an analog edit controller, directly accessing the disk-based media—there was no disk-based media! Very shortly all that will change.

The islands of expertise will no longer be separate and isolated. They will be able to link to one another through the power of the computer.

Warner notes,

"Nonlinear can, and will, handle all types of programs eventually. The only barriers are economic. For shows that have huge online storage requirements, the costs of storage will need to continue to come down to make it cost effective. And for programs that want to go directly from the nonlinear editing system out to tape or broadcast, the technology needs to go the next step in quality. Given that storage costs *fall* by half about every 18–24 months, the technology curve is very favorable. Furthermore, the processing power, and hence the picture quality, tends to *double* about every 18–24 months, so again the trend is very much in favor of nonlinear."

Continues Rawley,

"So where is the digital technology going in this broadcast industry and what changes can we expect? The fundamental way we work will be changed, but only where there are real needs and solutions that rationally fit in the user's environment. There is no question that with the advent of video and audio becoming native computer data types, the threshold of digital solutions powered by the computer has been crossed.

We can expect that many of the old ways of doing repetitive or iterative processes (like working different versions of a cut, or trying different audio dubs, or reworking the 6 A.M. news piece for the noon report) will be made easier by the computer becoming an integral part of the way we work. We are going to a faster, less expensive, more flexible way of working, and we are going to see the technology of video and audio become more accessible to broad markets by orders of magnitude.

We are not going to a place where the computer replaces the craftsman; the creative skill and talent of the editor will not be programmed into a box. Creative skills will be enhanced and accentuated and not made obsolete by powerful new tools, and there will be less propensity to bury oneself in the black art of electromechanical device management.

TALKING WITH EDITORS

How Do Digital Nonlinear Systems Contribute to the Production Process and Benefit the Client and Editor?

Tony Black, ACE, Washington, DC:

The best contribution is the instant feedback the editor gets when he changes a shot or trims an edit or just tries something completely different from what he just looked at. The flow of ideas and creativity is not constricted or interrupted when changes are made. There are no more time penalties for trying something new. The client can suggest a cut or idea to the editor and very quickly see the results. Even when a director or producer screens results in several pages of changes to be made in the show, the next cut can be completed and screened within a few hours if necessary.

The bottom line for me is that my editing is better on a nonlinear system than any other system I have tried. Digital nonlinear systems put nonlinear editing within my reach.

Peter Cohen, editor, Los Angeles

The nonlinear system's greatest contribution is the freedom of choice that it allows. My clients feel that they wind up with a better product for the money by working in a nonlinear world.

Alan Miller, editor, New York:

The two major benefits are creativity and speed. An editor can be more creative, have more options, and have the ability to try more ideas while also accomplishing the goal of a finished program in less time. Nonlinear editing doesn't make you a better editor, but it takes the burden of stress and frustration that always accompanied traditional editing and reduces it enough so that the real focus can be on making the best show possible.

Basil Pappas, editor, New York:

The client can save substantial time if they are well prepared and are not predisposed to exploring unnecessary creative options. Jobs that have alternate versions can save huge amounts of time in the hands of an experienced editor. I have many clients that come in under budget all the time and do so because they have developed a thorough understanding of how to take advantage of digital nonlinear editing's potential. For me, I compare time when discussing film editing to digital nonlinear editing. I can spend less time and get a more complete and creative look, provided all parties have a relatively similar approach. The nature of the process is collaborative. . . . If everyone is in sync, I've seen incredible amounts of creative editing produced in record times, often in much less time than would be required on film. The only decision I have had to make is whether I'll ever need to cut film again. The answer is: Why should I? In effect, I still cut film, and I only keep the film room handy to prep material and to accommodate jobs that require film matchback.

I've traded my synchronizer and splicer for a timeline and a mouse. I couldn't be happier. It has changed the course of my career, and probably extended it!

18

The Future of Nonlinear Editing

It has been a relatively short period since audio tape was edited with a razor blade and videotape was edited by using tracing powder to locate track pulses. It has been a remarkably short cycle that has brought three distinct approaches to electronic nonlinear editing. With technology progressing so quickly and so relentlessly, questions abound: How long will the third wave last before there is a fourth wave? How long will the first, second, and third waves continue to coexist? What will characterize the fourth wave?

The basic quest of nonlinear editing systems in the past has been singular: How fast can the footage be displayed to the editor? All three approaches—videotape, laserdisc, and digital—have sought to reduce the time required to get to the different pieces of footage.

Two additional questions were asked: How easy will it be for the editor to try options, and what tools are required for the editor to manipulate the footage in the way that it needs to be arranged to convey the wanted meaning? Here, the solutions are more variable. They are a function of the environment that the system designers create for the editor to work within.

An additional and important question is whether there is a way for the work products that result from the nonlinear session eventually to replace some of the tasks of the finishing session? This is the question that invites the most conjecture and that bears the most examination.

EVOLUTION OF THE THIRD WAVE

One must look at the movement of technology. In the late 1980's, we began to witness the adoption of smaller and more powerful editing systems based on digital technology. Technology itself is moving away from the analog world and moving toward the digital world. Manipulating digital signals and enhancing the digital signal processing paths from signal creation to signal distribution bring forth a series of tasks and technologies that have worldwide impact on a variety of areas. The movement of technology toward digital solutions is unceasing.

As the number of digital options increases—including digitally linking electronic paint systems, using digital nonlinear editing systems for file transfers, and originating images, not on charged-coupled device cameras, but on digital video cameras that create files that are then directly used by the digital nonlinear editing system—analog hardware and analog processing methods will be less frequently used.

The third wave of nonlinear editing systems, the digital wave, is at the steep curve of its development cycle. There are a substantial number of improvements to be made, and the third wave will remain for some time to come. As these improvements are achieved, the coexistence of the three waves will diminish.

The two most basic improvements that are necessary for digital nonlinear editing systems are enhanced picture quality with less storage requirements and increased economies of operation. Picture quality akin to the original is, of course, the goal, but better pictures must be accompanied by a reduced requirement for the storage of those pictures. The term *improved economies of operation* refers primarily to cost. The cost of the media required for acceptable quality pictures for whatever final work products will be taken from the digital system must not preclude using the system.

When these two factors equalize, the third wave of nonlinear systems will dominate, but the third wave will never replace the analog methods that exist today. Just as the word processor could not eliminate the use of the typewriter, the third wave will not be able to obviate the traditional methods of editing film and videotape. Instead, the third wave will take its place alongside these traditional methods.

Digital video compression represents a technology that is also in its infancy. There are vast areas of improvement to be made on a variety of levels. Discrete cosine transforms have heretofore represented the method by which compression is usually achieved. Future compression schemes will replace this extremely lossy stage. The word *future* is something of a misnomer. The technology required to further the capabilities of digital nonlinear editing systems is progressing rapidly. Better compression methods, faster disk drives of increased capacity, and lower costs for computer platforms translate into a furthering of the third wave of nonlinear systems.

Boundaries

Electronic nonlinear editing systems, regardless of the wave, have always been characterized as having certain unpublished boundaries. That is, each wave has its niche, but no one system is capable of easily handling all types of projects. Some systems are acknowledged to be very good for the episodic television market, while others are thought to be excellent for television commercials. These boundaries are usually a direct result of economics. If a system requires a tremendous number of laserdisc or videotape players or a great number of computer disks to accomplish a job, people begin to wonder if it is cost effective to do the project on a nonlinear system as opposed to the normal method, whether it be film or offline and online videotape editing.

The digital wave of nonlinear editing has the best chance of removing these traditional boundaries. Improvements in compression technology will shatter these boundaries because the quality of pictures and the cost of storing these images are moving in a positive fashion.

Equipment manufacturers are taking note of the increased interest in digital nonlinear editing systems. In 1992, for example, over 2,000 digital nonlinear editing systems were in use. When we examine the variety of digital linear and nonlinear systems that are appearing in the marketplace and the certainty with which manufacturers of linear editing systems are positioning themselves to offer nonlinear solutions, it is clear that there will be an increased development of digital nonlinear editing systems.

Manufacturers will have to wrestle with the subject of building proprietary hardware or utilizing industry standard equipment. The latter is a major commitment since most traditional manufacturers of equipment in the film and television industries are accustomed to designing both software and hardware. Placing an emphasis on using readily available hardware and offering software-intensive solutions is a very large issue. The advantage of proprietary hardware is clear: The system can be tailored to the desires of the manufacturer. The advantage of industry standard hardware is equally clear: The hardware can usually be changed more easily than redesigning and remanufacturing proprietary systems.

Improvements to video compression techniques will benefit more than digital nonlinear editing systems. The range of products that will become possible is fascinating. There will be completely digital productions in which material is shot with digital video cameras that pass along digital files to the digital nonlinear editing system. The output of this system will offer excellent picture and sound quality and will be suitable for finished results directly from the system. In this scenario, videotape will be used for one purpose: distribution.

Incorporation of Digital Video Compression into Existing Product Lines

One tangible result of the use of digital video compression in providing a different emphasis on a product's capabilities is in the area of the digital audio workstation. Digital audio workstations, used to accomplish a variety of tasks, such as laying track and mixing, must reference to some picture. This picture has always been provided by a copy of the locked picture playing back and slaving to the digital audio workstation. While the audio workstation offers fast access times, the workstation must function within the confines of the slowest piece of machinery in the modern digital audio suite: the shuttling time of the linear tape machine.

However, using digital video compression, the digital audio

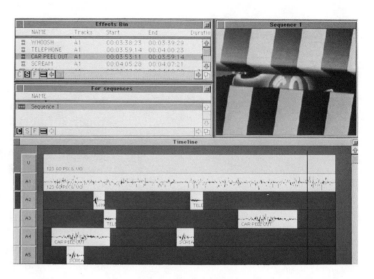

Figure 18–1 Incorporating compressed digital video playback of picture along with multi-track digital audio editing capabilities creates a unique way of accessing picture for the audio editor. Courtesy of Avid Technology, Inc.

workstation need not be dependent upon the linear videotape player (Figure 18–1). Using a digitized picture, the digital audio workstation suddenly gains a shuttle time of picture that is approximately equal to a search time of audio. Were it not for the advancement in technology, such applications would not be possible at an affordable price due to the cost of storing data-intensive video files.

Metamorphosis

Digital nonlinear editing systems have the capability of metamorphosing. Because of this, they have the unique ability of offering increased capabilities through software-intensive methods. Integrating features, such as electronic paint capability, character generator, animation files, digital video effects, and so on, results in the establishment of an editing system that mutates into something much more than just an editing system.

Digital nonlinear editing has the capability of affecting the manner in which files are created from the writing stage, to the storyboarding stage, to the design of costumes, to the planning of set construction, to the automatic capture of picture and sound on set, and on through the entire post-production process. Information, always lost from one stage to the next, is preserved, except that now the information does not exist as a multitude of log sheets, notes, and comments. Instead, information is data, actual bits of digital data that originate, are compiled and transformed during the pre-production and post-production stages, and are finally released and distributed.

TRANSFORMATION OF OFFLINE AND ONLINE EDITING

The transformation of offline and online videotape editing will not be a lengthy process. It will occur rapidly. It is clear that, with the increased capabilities being built into the digital nonlinear edit system, the distinction between offline and online has been considerably blurred. Offline videotape editing will first become absorbed completely by the digital nonlinear system. With the correct combination of image quality, storage capability, and operational tools, the digital nonlinear system will offer more benefits and equal or better economies of operation over offline linear videotape editing systems.

Online videotape editing will not be a place where creative ideas are attempted. Rather, online will be transformed and emerge in much the same fashion as the film laboratory. More and more projects will be brought to the online room with increasingly defined work products: edit lists and direct output tapes that will be followed as a blueprint. The list will not be interpreted or changed. Online will become much more of a quality-assurance center. The dedicated hardware of the online room will be gradually absorbed by the digital nonlinear system. Some capabilities will occur more rapidly than others, and some may be impossible to achieve. However, over time, what can be achieved in the online room and what can be achieved in the digital nonlinear system will begin to blur.

In 1992, a new phrase was coined: *nonlinear online*. Nonlinear online systems seek to offer the benefits of online with nonlinear capability at full-resolution video quality. The amount of video that can be stored by these emerging systems is low, approximately seven to 15 minutes, but the important item is that the phrase has been coined. The implications for the future of online are clear: Become nonlinear, or face a lessened demand for online services.

ALTERNATIVE APPLICATIONS

The technology that enables pictures and sounds to be stored in digital form will allow a series of applications and products to emerge. A wide variety of applications are already being discussed, analyzed, and designed.

Among these applications are digital databases that contain both static and moving examples of footage contained in the database. Being able to search a database for a shot that is needed by a documentary filmmaker in another country and being able to see a representation of the footage instantly without first having to find a videotape, shuttle to the correct timecode location, and then judge the shot can be an enormous time saver.

The filmmaker can send a request via modem to the footage supplier. The stock footage house can search the database and print out a series of color still frames that can be faxed back. The filmmaker can judge what shots will be useful. The desired goal would be not to have to look at a fax, but rather for the filmmaker to search the database and download the footage that is needed and then pay a fee for the material that is down-loaded. Large media servers are necessary. As improvements are made in the methods of transferring data from network servers to computer, these products will appear. There are certainly a wide range of implications for educational and industrial applications.

Broadcasting

Digitizing satellite feeds, editing a news insert in the remote news truck, and sending the digital material to the news station are processes that are being developed. Digital pictures and sounds, available to the editor, the news writer, the news producer, and the news director, will mean that scheduling the news will become a visual exercise, whereas today, it is largely an exercise in moving text around until a clear idea of the final story is achieved. Broadcasting, too, will change as the promos, commercials, and programs that are aired migrate away from videotape and toward digital video.

This is a source of controversy, to be sure, as there is some concern that digital compressed video will not be acceptable for direct broadcast. However, initial reactions to early tests have shown that excellent quality can be achieved through the use of JPEG compression, while preserving either the 50 fields of PAL or the 60 fields of NTSC for each second of video at a compression scheme of 70 kB or more per frame. At these rates, it becomes difficult to distinguish the compressed file from the original material. Television programming will eventually be broadcast from digital disk.

Being in Two Places at Once

Digital video compression techniques are already being used to solve the age-old problem of trying to be in two places at one time. Some owners of digital nonlinear editing systems have had fiber optic lines installed between the editing facility and the client's facility. The client is able to observe, in real time, the editing process from a location of three miles. The main disadvantage is the cost since fiber optic lines represent a considerable expense. However, if the lines are already in place, they can be used to transmit this information.

The ultimate goal is not only to be able to observe the editing process, but also to be able to edit from remote locations over standard audio lines. It is not inconceivable that two editing systems, linked by the standard dial-up phone line, will be able to communicate while moving pictures and sounds are transferred, in real time, between computers.

AFFECT OF DIGITAL VIDEO COMPRESSION AND DIGITAL NONLINEAR EDITING ON THE SHOOTING PROCESS

The shooting process has two ways of recording events. What passes in front of the camera is recorded on film or on videotape. In the case of a film shoot with a video tap, the images from the film camera are split to a small CCD camera and are then recorded to tape. In this way, the film director can review the take at any time.

A shooting set has a variety of equipment to support these recording processes. However, editing systems are traditionally not found on the shooting set. While films are often edited on location, they are done so in a location different from the shooting set. This has already changed with the emergence of the digital nonlinear editing system. The process of shooting does not change, but the process of reviewing what has been shot is augmented by the ability to edit what has been shot. By taking a video feed from the film camera or from the videotape recorder, it becomes possible to digitize directly into the digital nonlinear edit system.

This can be very advantageous to the writer, director, and producer. If there is a question as to whether a series of shots will edit together successfully, the digitized pictures can be quickly edited on the set. The parties can then decide if they must shoot more material or if they can move on. Being able to see what takes do and do not work as little sequences are edited is easily achieved by the digital nonlinear editing system. This is due to the immediacy of being able to capture and edit material as well as to the usual portability of the equipment. What impact will this capability have on the creative process if one can see the results of what is being shot quickly after it has been shot?

Changing the Work Flow

It ain't over till it's over.

—Yogi Berra

It isn't done until the files are erased.

—Digital nonlinear picture editor, sound editor, director, writer, producer, etc.

Heretofore, there has been an established series of steps that characterize where a project is at any point in time. The project may be in offline or it may be in online; it may be in dubbing or it may be in graphics design. In any case, while the process of making films and videos is nonlinear, the process of moving a project through to completion is linear. A picture is "locked" such that intensive sound work can be done.

As digital nonlinear systems proliferate, there will be a drastic change in the way that creative and business professionals in the film and video industries plan the path for a project's completion.

Treating pictures and sounds as data means opening up the entire process of how a project is completed. That process itself can become nonlinear in nature. If the music composer finds that a scene can be improved by having just a bit more time to ring out a particular passage, the picture editor can move the picture transition accordingly. The process can be simultaneously collaborative: The picture editor can work with picture files while the composer is working with the audio files for those picture files. Being able to work concurrently and not having to wait until the picture is locked will be a very significant development in furthering the creative aspects of the post-production process.

WIDESPREAD SYSTEM AVAILABILITY AND USE

Film and videotape editing have required special technical and aesthetic skills. Lacking an understanding of how the technical side of editing is accomplished has prevented people from trying these art forms.

Digital nonlinear systems used for editing film and television programming are one small piece of the puzzle. What these machines represent is simply the first generation of systems that provide a unique set of tools in a computer platform and that, depending upon the elegance of the user interface, are easy to learn and operate. Also, they are usually quite inviting; after all, to the novice, the online video editing suite can look quite intimidating.

Traditionally, achievements in the technology of computers have had an affect on the products that later become available to the general consumer. While professional editing products will have greater capabilities and will cost more, digital nonlinear editing systems for the consumer will, indeed, appear. These systems will give those who have not been classically trained as film and video editors the experience of putting together pictures and sounds.

DEFINING THE THIRD, FOURTH, AND FIFTH WAVES

The Third Wave: Digital Offline

The third wave of digital nonlinear editing must be defined as *digital offline*. Although the third wave of nonlinear editing has achieved a great deal of growth, the wave is in its infancy with regard to the sophistication of the dedicated peripheral equipment that one finds in the top-of-the-line online editing suites. Although image improvements are continually being made and additional software features are being added to this wave, digital offline is the best description because the wave exists as a bridge between analog data and digital data.

The Fourth Wave: Digital Online

The fourth wave of digital nonlinear editing will accomplish the tasks that have heretofore been the exclusive domain of the online videotape editing suite. Finished results from the digital online system will mean that the entire program can be completed within one system. This is starting to develop with the third wave, but the premium on storage needed to achieve acceptable-looking pictures for finished results can be a concern if the program being edited requires the loading of a large amount of footage. However, the fourth wave, which will most likely be targeted at a normal compression ratio of 10:1, will achieve results that will be indistinguishable from the original image.

Digital online will bring forth a variety of software packages within the one main system. Digital video effects, electronic paint, integrated character generation, digital audio editing and mixing, and a variety of the features available in the online suite will be found in the digital online system. The fourth wave will transcend the online suite in that there will be tasks that will be easier to achieve in the digital online system than in the online room. Faster computers, better digital video compression schemes, and larger-capacity computer disks will contribute to make digital online an inevitable step in digital nonlinear editing's evolution.

The fourth wave will set the stage for the integration of the digital media manager, which will come to fruition in the fifth wave.

The Fifth Wave: Digital Uncompressed Media Management

The fifth wave of digital nonlinear editing will not involve compressed video signals. Rather, the information to be used in the presentation will originate as digital data. There will be no need to transfer from the analog world to the digital world. These signals will originate in the form that the digital media manager expects for file importation.

The fifth wave will not be concerned solely with editing. This wave will be compatible with information regardless of its origin. Files will transfer back and forth and among systems that will originate the pictures and sounds, catalog the signals, edit the signals, and transmit the signals. The digital media manager affects how individuals work with information and how disparate forms of media are combined into meaningful wholes. Clearly, leaps in the current technologies that are employed by the third wave must occur for the fifth wave to be realized. The most compelling facts here indicate that these technologies are progressing rapidly. The limitations of the third and fourth waves can be surmounted.

The DMM begins with the concept of images and sounds captured on film, on videotape, or on computer disks as raw material to be combined. The usual process has routes in film and video editing but has been changed and is inherently better than these individual disciplines because the project can be worked on interactively by a group of specialists up until the last possible moment of transmission.

Even though the DMM will process information regardless of its origin, the difficulty of putting together a presentation from disparate sources will be lessened. Rather, presentations will be created from sources that have a common frame of reference: they will have been treated by the DMM and exist as data that can be shared from system to system.

Film and videotape editing methods will change as a result. The creative decisions will continue to be the most important aspect of the craft, and the technical challenges will become easier to achieve.

19

Electronic Nonlinear Systems

The area of electronic nonlinear editing is growing rapidly. Although the introduction of new systems in the videotape- and videodisc-based waves has diminished since 1990, the digital-based category has seen continual growth and the introduction of many new systems. This chapter is intended to serve as a guide through the various waves of systems while providing a brief history and description of each. Inquiries for additional information should be directed to the manufacturer.

This chapter is divided among the following categories: videotape-, videodisc-, and digital-based electronic nonlinear systems. Several manufacturers are represented in more than one category, and entries are in alphabetical order. Also included is a fourth category concerned with developing technologies that may have an impact on the production and editorial facets of the film and video industries. This fourth category, an explanation of the systems and approaches on the periphery of editing, is solely concerned with computer-based applications.

THE VIDEOTAPE-BASED WAVE

BHP TouchVision

Developed in 1986, the BHP TouchVision system utilizes VHS videocassettes. The system is standard with six source machines and monitors, although more playback machines (up to 12) can be added, depending upon the editing requirements, making the system quite accessible for multiple camera requirements.

Original film or videotape footage is transferred to videocassette, providing approximately 4.5 hours of material accessible from one complete videotape load. Originally introduced on an IBM PC–compatible platform, a Macintosh computer platform is now used; a higher emphasis is placed on graphics, as the interface is largely graphical, providing the editor with icons of film tools. System playback is at 30 fps. The system provides a video EDL, a film cut list, and a screening version of the program.

Popular in Hollywood, TouchVision has been used on over 22 feature films, including *Blind Fear*, *Rocky V*, and *Truth or Dare*;

television movies-of-the-week such as *Laura Lansing Slept Here* and *Liberace*; television series such as *Young Riders*, and television commercials. TouchVision is preferred for use on long format projects. Over 25 TouchVision systems are installed in the United States and Canada, and rental programs are also available.

For more information, contact TouchVision Systems, 1800 Winnepac Avenue, Chicago, IL 60640.

Ediflex I and II

Developed in 1984 and introduced in 1985, the Ediflex uses multiple VHS machines to offer random access to material. Up to 12 machines can be used on the system, each with its own separate monitor.

Original film or videotape footage is transferred to videocassette, providing approximately 4.5 hours of material accessible from one complete videotape load. Originally introduced on an IBM PC–compatible platform, the original interface has been enhanced to include more graphical menus and icons. The system relies heavily on using the script to relate footage on the system to lines in the script. A light pen is used to correlate footage to script and to gain access to material. Two channels of audio can be mixed. The system playback is both NTSC and PAL and is switchable to 24 fps. The system creates a video EDL, a film cut list, and a screening version of the program.

Largely used in Hollywood, Ediflex has been primarily used on television episodics such as *Matlock*, *Thirtysomething*, and *Father Dowling Mysteries*; musical concerts such as *Sting in Tokyo*, and television commercials. A new emphasis is being placed on feature films but Ediflex is currently preferred for television episodic work. There is a thriving rental business and the system is also available for purchase.

The Ediflex system is also evolving. Ediflex II, introduced in 1992, offers the following enhancements: the storage of material on rewriteable laserdiscs for improved access times, higher image quality, and improved sound. A timeline with user-definable colors has also been incorporated. Input control includes light pen, mouse, and keyboard.

For more information contact: Ediflex Systems, 1225 Grand Central Avenue, Glendale, CA 91201.

Montage I & II

Developed in 1984, the Montage was the first videotape-based nonlinear system. Montage II appeared in 1987. It utilizes up to seventeen Super Beta Hi-Fi VCRs to offer random access to material. Fourteen black and white monitors show digitized black and white representations of the head and tail frames of footage, allowing the editor to see seven clips at once. Approximately 4.5 hours of material are accessible from one complete videotape load.

The editing console includes an electronic Chinagraph marker that can be used to write directly on the digitized frames. While footage is being loaded into the system, 17 identical copies are made. These dailies are "tagged" with scene and take numbers. During this process, the editor can review the footage being loaded. Editing can begin immediately afterward.

Montage makes use of a dedicated controller knob, which the editor can manipulate to "scroll" through footage. The system playback is both NTSC and PAL. The system creates a video EDL, a film cut list, and a screening version of the program. In addition, print-outs of the digitized images used in the sequence are also available.

Most popular in Hollywood, Montage has been used on hundreds of projects, including feature films such as *Full Metal Jacket*, *Family Business*, and *The Godfather Part III* and episodics such as *Star Trek: The Next Generation* and *Mission: Impossible*. Montage systems have also been used for television commercials. The system is available for rental or purchase.

The Montage II system is evolving. The Montage IIH is a hybrid system that combines the videotape base of Betamax cassettes with the digital base of computer disks. Depending upon the type of work being done, the editor can choose to work on tape or on full motion digitized video. Introduced in 1992, the Montage IIH can help to address the void between access times of the videotape-based and digital-based systems.

For more information, contact Montage Group Ltd., One West 85th Street, New York, NY 10024.

THE LASERDISC-BASED WAVE

CMX 6000

The CMX 6000, introduced in 1986, utilizes write-once laserdiscs and multiple playback machines to offer random access to material. Each CLV laserdiscs holds approximately 27 to 30 minutes of footage; configurations of two to 12 machines can hold from 30 minutes to three hours of sync dailies. Approximately four hours of silent material can be stored per laserdisc load.

The control panel is a dedicated keyboard with a touch pad for accessing footage from the log. Touching the appropriate clip displays the material from disc to source monitor. The monitoring system includes two screens. One displays source material as pictures and serves a secondary role as the text display for the footage contained on the system. The other displays the sequence being created. Material transferred to disc requires a premastering stage in which selected material is first transferred to videotape and then used to create the laserdiscs.

Multicamera features were added to the system in 1987, and a recordable laserdisc module to accommodate late-arriving material was introduced in 1988. Dissolves are shown through the use of an on-board video switcher.

Since it is based on laserdiscs, the system enjoys the particular benefit of being able to display only certain frames. While both

NTSC and PAL playback is available, as long as the 2-3 pulldown sequence has been identified, the CMX 6000 is capable of not showing the extraneous video frame when working with film material transferred to 30 fps video. In this way, 24 fps playback is simulated from the 30 fps laserdiscs. The system creates a wide range of edit lists: a video EDL, a film cut list, video- and audio-only lists, and a screening version of the program.

The CMX 6000 has been used on feature films, such as *Harley Davidson & The Marlboro Man*, *Get Back*, *Graffiti Bridge*, and *The Sheltering Sky*; television miniseries such as *Lonesome Dove* and *The Burden of Proof*; television series such as *Eerie, Indiana*, *Law and Order*, and *Major Dad*; and many television commercials. The majority of work is one-hour episodic and feature films. The system is available for rental through various post-production facilities.

The CMX 6000 evolved to incorporate rewriteable laserdiscs in 1991. At the National Association of Broadcasters (NAB) show in April, 1992, CMX showed a prototype version of the next generation of the CMX 6000, a product tentatively called, *Cinema*, which is a digital-based system patterned after the third wave of nonlinear systems. This system uses digital video compression to provide random access to material.

For more information, contact CMX Corporation, 2230 Martin Avenue, Santa Clara, CA 95050.

EditDroid I & II

The EditDroid, introduced in 1984, was the first laserdisc-based nonlinear system. Introduced at the same time as the tape-based Montage system, EditDroid I initially labored under the initial costs of creating laserdiscs and the cost of the system. In 1989, after several years of redesigning the system, EditDroid II was released.

EditDroid I originally used only CLV laserdiscs. Playback machine configurations, like those for the CMX 6000, are variable, and approximately three to four hours of footage are available with a complete load depending upon whether the material is sync or without sound.

When the EditDroid II was introduced, a greater emphasis was placed on utilizing a combination of storage methods. Rather than limit the playback material to read-only laserdiscs, a combination of videotape machines can be used. EditDroid II is somewhat of a hybrid in that sense. A basic EditDroid is configured at six laserdisc machines and is expandable to 12 machines. A hybrid configuration combines laserdiscs with, usually, three 3/4" videotape players. Two machines serve as sources for late-arriving material, and the third machine serves as the record machine for screening copies.

The system is based on a Sun 3/80 computer. The editor uses a touch pad with a dedicated knob for controlling and moving through footage. The computer display graphically represents picture and track and their relationship to one another. The monitoring system can expand to include many screens, especially if the editor is working on a multicamera project.

Both NTSC and PAL playback is available, and as long as the

2-3 pulldown sequence has been identified, the EditDroid II can also work in 24 fps mode. The system creates a video EDL, a film cut list, and a screening version of the program.

The EditDroid has been used on feature films such as *The Doors*; television series such as *Knots Landing*, *LA Law*; and many television commercials. Work is divided among one-hour episodics, feature films, and commercials.

For more information, contact LucasArts Editing Systems, 3000 Olympic Boulevard, Suite 1550, Santa Monica, CA 90404.

Epix

The Epix system, introduced in 1989, originally used write-once laserdiscs but has since evolved into a system that uses both videodisc players and recorders. Integrating late-arriving material is a simple process of recording the material to disc. For this laserdisc-based system, configurations and storage capabilities are similar to those of the CMX 6000.

The control panel is a dedicated keyboard with a jog/shuttle knob. The interface is a combination of text and graphics; the structure of the program is depicted visually as a color-coded timeline. The monitoring display is standard with two monitors: one for source material, the preview monitor, and the other for the sequence being edited, the program monitor. A third monitor displays the graphical displays for the editor.

The official name of the system, Epix Hybrid Editing System, has a great deal to do with the system's ability to act as a nonlinear system or as a conventional linear editing system. This is useful when multiple machines need to be played back to simulate multicamera situations. Up to 16 serial transports are available. As a result, the Epix has enjoyed success on multicamera television shows. The on-board internal switcher can perform cuts and dissolves. An external switcher option provides wipes and keys. The system playback is available in NTSC and PAL. The system creates a video EDL, a film cut list, and a screening version of the program.

The Epix system has been used on feature films such as *All I Want For Christmas* and *Critters 3*; television series such as *Harry and the Hendersons*, *Top Cops* and *A Different World*; and a variety of television commercials. The system is available for purchase or rental through various post-production facilities.

The Epix system evolved to include rewriteable laserdiscs in 1991. At the same time, a companion product, ETrax, an eight-channel digital audio workstation, was developed. ETrax offers eight hours of mono CD-quality sound on a 760 MB disc, using lossless compression schemes. Additions in 1992 included incorporating Pioneer, Sony, and Panasonic optical disc recorders.

For more information, contact Amtel Systems, Inc., 310 Judson Street, Unit 6, Toronto, Ontario M8Z 5T6, Canada.

Laser Edit

The Laser Edit system, introduced in 1985, is often omitted in lists of electronic nonlinear editing systems due to confusion as to whether the system is nonlinear or linear. In practice, the system has both nonlinear and linear aspects. Based on a dual-headed laserdisc player, the system was the first to offer noticeably improved access due to its unique design. The Laser Edit system utilizes write-once laserdiscs and one 3/4" videotape recorder. Configurations and overall storage times are similar to those of the other laserdisc-based systems.

There is often confusion as to how this system operates. The initial stages of working are very similar to those of other laserdisc-based systems. Footage is transferred to write-once laserdiscs. Multiple laserdisc machines are used to simulate random access. However, the process of accepting edits is where this system deviates from the nonlinear philosophy.

Everything leading up to actually choosing the edit to be made is similar to the different nonlinear systems, as the editor searches through footage while judging the possibilities of editing from material on the different discs. However, when the editor finally decides which shot is going to be used, the material is recorded to 3/4" videotape. The program is being recorded to tape as the editor decides to accept the order of the edits being made. Herein lies the confusion: Is the Laser Edit a nonlinear system or a linear system?

The answer is that the system is a bit of both. During the first stage of editing, the editor has random access to material. When changes are required, the 3/4" videotape becomes a source, the material that is not involved in the changes is recorded directly from the first master tape to a new master tape, and the material involved in the changes originates from the laserdiscs.

While this mode of operation may appear to be limiting, the Laser Edit system has been popular with multicamera programs, especially the half-hour television show. The system is available in NTSC playback mode. The system's control panel combines a video and an audio switcher to create dissolves. The system creates the following work products: a video EDL, a film cut list, and the master tape created during the editing process, which is, in itself, the screening version of the program.

The Laser Edit system has been used on numerous television series such as *Cheers* and *Wings* and has also been used for television commercials. The system is available for rental through various post-production facilities.

For more information, contact Spectra Image Systems, 820 North Hollywood Way, Burbank, CA 91505.

THE DIGITAL-BASED WAVE

Avid Media Composer

The Avid Media Composer, introduced at the NAB show in 1989, utilizes JPEG digital video compression techniques to store material on computer disks (Figure 19–1). Storage rates vary, based on the image resolution selected and the size and number of storage devices. Using SCSI 2 interfaces, up to 21 storage

Figure 19–1 The Avid Media Composer. Courtesy of Avid Technology, Inc.

devices can be connected to the system. Storage devices include fixed magnetic hard disks, removable hard disks, and Panasonic phase-change optical discs. The company estimates that over 29 hours of picture and sound will be available to the user.

Based on the Macintosh computer, the system offers the following input devices: keyboard, mouse, trackball, and dedicated controller knob. The image resolution using JPEG video compression is at 640 × 480 pixels for NTSC and 640 × 576 for PAL. Audio resolution is at 22, 24, 44.1, and 48 kHz.

The interface includes two monitors: One displays the footage contained in bins, and the second displays the editing interface, which comprises a source monitor and a record monitor. A graphical timeline display is capable of showing the edited sequence in graphical or pictorial form. Footage is drawn from the bin display to the source monitor and edited into the record monitor, creating multiple versions. Features include 24 tracks of audio editing and mixing, slow and fast motion, software-displayed waveform and vectorscope, graphics import and positioning, cuts, precomputed dissolves, wipes, and keys.

The NTSC- and PAL-capable system plays back footage at 30 fps, 25 fps, and 24 fps. With provisions for automatically inputting film edge numbers, the system is capable of a variety of work products: video EDL, a film cut list, video- and audio-only lists, "digital cut" screening copies, and auto-assembly features. The optional switcher interface commands up to eight source decks and one record deck while controlling a video switcher.

The Avid Media Composer has been used on feature films such as *Let's Kill All the Lawyers*, *High Heels*, *Lost In Yonkers*, *Monreal Vu Par*, *Fool's Fire*, and *Life On the High Wire*; television series such as, *LA Law*, *Civil Wars*, *Secret Service*, *Silk Stalkings*, *Kids in The Hall*, *They Came From Outer Space,* and *Forever Night*; documentaries such as *Making Sense of the Sixties*, and *Coming Out of the Dark*; and many television commercials.

System use is divided among all categories of programming, from commercials to documentaries to feature films. The system is available for purchase and rental through various post-production facilities.

The Avid Media Composer series has evolved rapidly, and the number of systems sold has grown consistently. At the NAB show in 1992, the company announced worldwide sales in excess of 1,0000 systems and introduced AudioVision™, a digital audio workstation, and AirPlay™, a broadcast version of Media Composer for on-air uses.

For more information, contact Avid Technology, Inc., Metropolitan Technology Park, One Park West, Second Floor, Tewksbury, MA 01876.

DVision

DVision, created by TouchVision Systems, is a digital-based entry from the designers of the tape-based TouchVision system (Figure 19–2). DVision, introduced at the NAB show in 1991, utilizes Intel's RTV digital video compression techniques. Footage is stored on both magnetic disks and optical discs. As with all digital-based systems, storage rates vary. DVision can connect up to 21 SCSI devices, which the company estimates will offer over 27 hours of storage.

Based on the IBM PS/2® computer, the system offers the following input devices: keyboard, mouse, and trackball. The image resolution using DVI video compression is not quoted in pixels but is said to be "up to 3/4" picture quality."

The editing interface consists of two computer monitors. One monitor displays footage and a record time line. The second monitor, the picture monitor, displays the sequence being edited. The system allows the user to see both outgoing tail and incoming head at transition points. A key word search function permits catalog searches of footage.

Features include cuts, precomputed dissolves, wipes, audio editing and mixing, graphics import and positioning, and pitch audio during variable speed jogs. The NTSC system plays back footage at 30 fps. The system provides a video EDL, screening

Figure 19–2 DVision. Courtesy of TouchVision Systems, Inc.

copies, and A/B roll auto-assembly features. The system is available for purchase and rental through various post-production facilities.

DVision continues to develop new software. Version 2, introduced at the NAB show in 1992, includes enhancements to RTV picture quality and the improved creation of special effects. Company representatives envision the system evolving into a full editing suite in an easily portable package.

For more information, contact TouchVision Systems, 1800 Winnepac Avenue, Chicago, IL 60640.

EMC2

The EMC system was introduced at the 1988 SMPTE show in New York City as the first digital nonlinear offline editing system. A subsequent release, the EMC2, was introduced in May 1990 (Figure 19–3), and the integration of JPEG-based digital video compression occurred in October 1990. As with all digital-based systems, storage rates will vary. EMC2 supports Wren and Panther hard disks, Sony and Maxtor magneto-optical discs, and Panasonic phase-change discs. The company estimates that 24 hours of footage are available to the user.

Based on the IBM PC, the system offers the following input devices: keyboard, mouse, trackball, and dedicated editing pad. The image resolution using JPEG video compression is at 720 × 484 pixels for NTSC. Audio resolution is at 16 and 48 kHz.

The interface is based on one monitor that combines the functions of the source and record monitors. This monitor is also capable of displaying the logging program used to organize and storyboard footage. Below the main editing display is a timeline where each track of video and audio is shown graphically.

Features include an unlimited number of audio channels for building and mixing sound tracks, digital video effects, overlays, transparencies, slow motion, cuts, precomputed dissolves, wipes, and keys. The NTSC- and PAL-capable system plays back footage

Figure 19–3 EMC2. Courtesy of Editing Machines Corporation.

at 30 and 25 fps. The system creates a variety of work products: a video EDL, a film cut list, a screening copy, and an auto-assembly function.

The EMC2 has been used on a variety of projects, including feature films such as *High Strung*, television programs, documentaries, and many television commercials. The system is available for purchase and rental through various post-production facilities.

At the NAB show in 1992, the company announced several new products, including EMC-Log, a logging package for PC and Macintosh computers, EMC-Tracks, a digital audio workstation, and EMC-Cuts, a linear version of the EMC2.

For more information, contact Editing Machines Corporation, 1825 Q Street NW, Washington, DC 20009.

Lightworks

Lightworks, introduced at the SMPTE show in 1991, utilizes JPEG digital video compression techniques to store material on computer disks (Figure 19–4). The company estimates that a minimum of two hours of sync material, expandable to 40 hours, depending upon image and sound quality chosen, are available to the user. Storage devices include fixed and removable magnetic hard disks. Backup storage via optical discs is available.

Based on an enhanced 486 computer, the system offers the following input devices: keyboard, mouse, and dedicated controller knob. The image quality is described by the manufacturer as comparable to offline video or suitable for critical presentations at the best quality. Audio resolution is at 24, 32, 44.1, and 48 kHz.

The standard interface consists of one monitor on which footage is viewed and edited. An optional monitor displays a

Figure 19–4 Lightworks. Courtesy of OLE, Ltd.

full-screen output of any selected frame. Footage is displayed in a gallery and can be organized in various ways. A file card serves as a description for each shot. Other views include a "Strip-view" that displays the edited sequence in timeline form.

Features include eight tracks of audio editing with four independent output channels. Script input is available from most word processor formats. Cuts, dissolves, and wipes are displayed in real time and do not require precompute time. The NTSC- and PAL-capable system plays back footage at 30 fps, 25 fps, and 24 fps. The system is capable of a variety of work products: a video EDL, a film cut list, and screening copies. The image resolution using JPEG video compression is not quoted in pixel matrix.

Lightworks is targeted at episodic television programming and feature films. In addition, the system is in use for editing television commercials. The system is available for purchase and rental through various post-production facilities.

At the NAB show in 1992, the company announced the DIGIstation, intended for digitizing footage to a variety of media.

For more information, contact Lightworks, OLE Limited, 69 Wells Street, London W1P 3RB, England.

Montage III

Montage III, is a digital-based entry from the designers of the tape-based Montage system. Montage III, introduced at the NAB show in 1991, utilizes Intel's RTV digital video compression techniques. As with all digital-based systems, storage rates will vary. The company estimates that up to eight hours of online storage on 1.2 GB hard disks or 1 GB removable optical discs are available to the user.

Based on an Intel 486 computer, the system utilizes keyboard, mouse, trackball, or dedicated controller knobs similar to those of the tape-based Montage, along with dedicated keys and the familiar electronic Chinagraph marker. The image resolution using DVI video compression is not quoted in pixel matrix.

The editing interface consists of one computer monitor that serves as the main editing display. Shown on this main view are the various shots that comprise the edited sequence. Below this view is a graphical representation of the video and audio tracks in the sequence. The system also offers a shot database that can be used to organize material. An interface for linking script to pictures is also available. A second NTSC or PAL preview monitor displays the digitized picture for additional reference.

Features include cuts, precomputed dissolves, fades, and wipes, eight virtual audio tracks, expandable to 24, with two-channel audio monitoring at CD quality, and pitch audio during variable speed jogs. The NTSC- and PAL-capable system plays back footage at 30 fps, 25 fps, and 24 fps. The system provides a video EDL, a film cut list, screening copies, and auto-assembly features. The system is available for purchase and rental through various post-production facilities.

For more information, contact Montage Group Ltd., One West 85th Street, New York, NY 10024.

DEVELOPMENTS IN THE DIGITAL-BASED WAVE

Further development within the tape- and laserdisc-based waves has remained somewhat stagnant since 1989. The digital-based wave, however, has experienced rapid growth since 1991. A series of related digital products have appeared on the scene. These range from frame grabber units that operate as linear editing systems with digitized reference frames to software applications that only play back digital files.

Included below are some of the many new peripheral applications that have appeared. Some of these products are based on DVI and JPEG digital video compression. While it is not possible to include the tens of products that are rapidly appearing, some of the more notable entries are included herein.

QuickTime

QuickTime is a set of operating extensions to the Macintosh computer platform. QuickTime allows Macintosh computers to display time-dependent media such as video, audio, and animation and to combine these media with time-independent media such as text and graphics (Figure 19–5).

Essentially, QuickTime allows presentations to be created that combine pictures, sounds, and text. These QuickTime "movies" have the following attributes:

1. Frame rates vary, but the movies play back at considerably less than at normal play rates. Typical is 2 to 5 fps.

2. Image quality is currently based on software compression, and as a result, the pictures are noticeably inferior to either DVI- or JPEG-based compression. This attribute will change as DVI and JPEG boards become available for QuickTime applications.

3. Usual QuickTime images are displayed in a matrix of 100 × 160 pixels. A movie can be expanded to 200 × 320 pixels but will play at lower frame rates than the smaller display.

4. Using the internal Macintosh audio system, quality is 10 kHz, 8 bits per sample which is comparable to dial-up phone service.

QuickTime applications are unique because they are both hardware smart and hardware independent. If a QuickTime "movie" is created, it can be played on any Macintosh, and the movie's playback speed will be upgraded or downgraded according to the hardware platform being used. A faster operating CPU allows the movie to be played back at its normal speed; a less capable CPU causes the movie to be played back at less than its true speed.

QuickTime was created as a presentation tool to combine media that formerly could not be combined into a single presentation on a standard computer platform. Various applications

Figure 19–5 This QuickTime movie player interface shows a sequence ready to be played.

are appearing that offer QuickTime-based editing. These programs should be carefully judged based on what the user requires. In addition to the four attributes listed above, QuickTime applications using software compression have the following issues to address: As the frame demands grow and the QuickTime application cannot reliably deliver frames from the computer disk, there will be some slippage of audio to video frames as the application struggles to keep up with playback. Lip sync should be described as being "loose."

Vendors are beginning to bring to market various applications using QuickTime. These products usually consist of software and a hardware computer board. Applications arriving on the market should be tested as to whether they process SMPTE timecode or generate EDLs. Of course, if these work products are not desired, then these areas can be overlooked. Not all applications will offer these more professional output options. If the goal is to assemble a presentation from several different forms of media at less than full-frame rates, these applications can serve quite nicely.

One interesting application of using QuickTime as a presentation tool is to use it in conjunction with the digital nonlinear editing system. For example, if a commercial is being edited on a digital system and the system is capable of exporting an edited sequence into a QuickTime movie format, the sequence can be played elsewhere. It works in this way: At any time during the editing process, the commercial can be exported to a QuickTime movie. The image quality is downgraded (this will vary, but typical is approximately 5 kB per frame), and audio quality is limited to 10 kHz, with only one audio channel being available.

The QuickTime movie version of the commercial would easily fit on a 1.2 MB computer floppy disk. The calculations are as follows:

5 fps (the rate of playback of the movie) × 5 kB/frame (the size of the file) = 25 kB/sec × 30 seconds (entire commercial) = 750 kB for the picture portion + 10 kHz, 8 bits per sample (for audio)/sec × 30 seconds (entire commercial) = 300 kB for the audio portion.

750 kB + 300 kB = 1.050 MB

By storing the 1.050 MB of data on a floppy disk allows this QuickTime version of the commercial to be played on any Macintosh computer. For individuals who cannot take part in the editing process but who wish to get an idea of what is being accomplished in editing, being able to see a version of the commercial on their personal computer can be extremely convenient. Recall that the QuickTime movie does not require special hardware purchases.

When it is necessary to show someone who is not at the edit what is being done, transmitting even a downgraded version of the work in progress can be quite helpful. As a "proof of concept" tool, QuickTime can be quite useful as an auxiliary tool to the digital nonlinear editing system.

DVI and JPEG Frame-Based Edit Systems

Another emerging category of products that is being offered for sale is concerned with a limited form of nonlinear editing. These packages comprise a hardware computer board and a set of software applications. All other hardware necessary for a more complete program must be provided by the user.

In general, these systems allow the user to digitize representative frames, usually head and tail frames. The frames can be labeled with text descriptions and can be storyboarded and rearranged. Since only head and tail frames are usually available, not all frames are displayed during the editing process. However, the individual shots can be arranged as if all frames were present.

Excellent character generation capabilities may be integrated into the system, allowing the user to place titles over the representative frames in the sequence. The goal of these systems is two-fold. The first goal is to provide a video EDL. If a more complete program is desired, systems in this category can perform an auto-assembly process in which the original source tapes are used in conjunction with additional hardware that the user has supplied and configured. The result of this process is to edit with less than all frames, arrange them accordingly, and build either an EDL or an auto-assembled tape.

Products in this category are beginning to enjoy some popularity, especially for those applications where seeing every frame is not as important as getting the flow of the presentation in proper order. It would be somewhat unfair to list a series of products in this category since they are advancing rapidly as they strive to offer more frames per second and be truly nonlinear in the sense that access to all frames each second is provided. Many systems are well on their way to moving from representative frames to normal play rates.

When evaluating systems in this category, inquiries should be made as to the number of frames displayed per second, the type of equipment necessary for the finished results that the user desires, and the upgrade path of the system being considered for purchase. Products in this category fill a void, especially for the corporate communicator who has heretofore had to work in the linear videotape editing environment but who may not need to see every digitized frame for each second.

Glossary

A/B Roll An editing system comprised of three videotape machines: two source machines and one record machine. The A/B roll system allows the editor to make transitions, such as dissolves and wipes, and involves the use of a video switcher.

Access Time The amount of time from when an inquiry for information is made until the information becomes available.

AES/EBU American Engineering Society/European Broadcasting Union.

Analog Electrical signals which vary constantly. In analog recordings, the changes to the recording medium are continuous and analogous to the changes in the waveform of the originating sound or are analogous to the reflectance of the original sound.

Answer Print The first version of the entire film with optical effects and complete mixed optical sound track.

ASCII American Standard Code for Information Interchange. Represents the manner in which binary definitions are assigned to numbers and letters. It provides a standard for exchanging different types of files; ASCII code.

Asymmetric Compression Asymmetric compression techniques require a greater amount of processing power to compress a signal than is required to decompress the signal.

Audio Layback When the final audio mix is completed, the process of recording it back onto the original video master. The completed tracks are laid back onto the videotape.

Audio Layup The process of transferring audio from the master tape in progress to a multitrack audio editing system to add additional sounds and complete a sound mix.

Auto-Assembly The process of using an edit controller to implement the edit decision list to create a videotape master. A cuts-only auto-assembly process involves one source machine and one record machine.

Bandwidth The number of bits per second of material. The computer processing unit is tasked with processing a number of bits per second when digitizing; that number becomes a limiting factor. The computer can process only a certain number of frames and a certain amount of information for each frame every second.

Betacam A recording videotape process that utilizes a variation of component techniques.

Betacam SP The "superior performance" version of Betacam videotape. It uses magnetic particle videotape.

Bins In film editing, a canvas container used to hang film strips. In nonlinear editing systems, the location where footage is stored is called the *bin*.

Bit The smallest piece of information in the digital world. The term is short for "binary digit."

CAV Laserdisc Constant angular velocity. A laserdisc that is capable of slow motion, step frame, and freeze frame. Used for laserdisc-based nonlinear systems, CAV discs offer 30 minutes per side (54,000 frames).

Checkerboard Editing A nonlinear method of assembling a master videotape to maximize efficiency. The most popular forms of checkerboarding are B and C mode auto-assembly.

Chroma Subsampling A technique used to reduce the file size of an image by reducing the amount of color information it contains.

Circled Takes In film, the takes that will be printed from the exposed camera negative.

CLV Laserdisc Constant linear velocity. A laserdisc that is not usually capable of slow motion, step frame, and freeze frame. Used for playback purposes, CLV discs offer 60 minutes per side (108,000 frames).

Coding The way in which information is represented in file size reduction programs.

Compression To reduce in volume and to force into less space.

CTDM Color time division multiplexing. A basic technique used in the record and playback processes of Betacam videotape.

Dailies The circled takes that have been printed. Dailies are reviewed every day that a film is in production. Also called *rushes*.

Data Transfer Rate The amount of information that a computer storage drive can write and read in a certain amount of time. Also called *read/write speed* and *transfer speed*.

DAW Digital Audio Workstation. A workstation that uses optical discs or magnetic disks to offer random-access nonlinear audio editing.

Decimation The removal of a great proportion of elements that make up a whole. This form of subsampling occurs at the pixel level.

Destructive An operational process that causes original audio or video files to be altered. Nondestructive processes do not alter the original files.

Dial-Up Phone Service An ordinary telephone in a home or office. A modem is used to transmit and to receive digital data. The channel that the data can travel through is small: 56 kbits/sec.

Digital The conversion of an analog signal into a binary form. In digital recordings, digits are used to represent quantities, and digits in a rapid sequence represent varying quantities.

Digitize To convert continuous analog information to digital form for computer processing.

Direct Output The component or composite video signals and analog or digital audio signals that are available directly from the nonlinear system.

Disk Default The average amount of information that can be stored based on the digitizing parameters chosen and the capacity of the computer disk.

D1 A SMPTE standard for recording digital videotape recordings. D1 is a component recording process.

Double System The film recording method in which picture and sound are recorded as two separate elements.

Down Convert To convert convert video files from normal play rates to less than normal rates. This type of conversion is primarily used to increase storage capacity.

D2 A standard for recording digital videotape recordings. D2 is a composite recording process.

D3 A recording process for digital composite 1/2" videotape.

Dual Finish Cognizant The desire to leave the editing process with lists that can be used to create a finished videotape master and a cut film negative for both videotape and film releases.

Dual-Sided Laserdiscs The laserdisc that results when two single-sided discs are glued together. If a player with a single laser is used, one must flip the disc over to play the other side. With a dual-laser machine, the disc is simply left to rotate, while the two laser beams seek frames on each side of the disc.

Dubbing In film, when the entire film becomes available for final audio mixing. As a reel is projected, all the various sound elements for that reel are mixed together. In video, the process of making copies of the master tape for distribution.

DVA Code Discovision Associates code. These five-digit codes are encoded onto the glass laserdisc master by the laser. They represent the manner by which the disc can be searched. DVA code is related to the timecode of the premaster tape.

DVE Digital video effects device used to alter and manipulate images.

DVI Digital video interactive. A compression method consisting of a programmable chip set and software. DVI supports both still images and motion video and can be both asymmetric and symmetric in nature. A current implementation can decode and encode in DVI as well as JPEG.

Edge Numbers Identifying numbers for 35mm and 16mm film. Film negative has edge numbers that identify the film frames. In 35mm four-perforation film, an edge number appears every 16 frames, or once every foot. In 16 mm film, there are 40 frames for each foot of film; edge numbers appear every 20 frames, or every half foot.

EDL Edit decision list. In videotape editing, a list that indicates how a program was put together. The EDL is based on SMPTE timecode, and it forms the basis for the interchange of information between the offline and online stages. A minimal form of EDL shows the timecode numbers of the source tapes used and the transitions between images.

A Mode EDL An A mode edit decision list is ordered based on the record in times. The list is sequential, and for each edit that is done, the master tape is recorded from start to end in a linear fashion. An A mode list takes the most time to execute.

B Mode EDL A B mode edit decision list introduces the concept of "checkerboard" auto-assembly. In a B mode list, the order is determined by the source tape and the record in time. A B mode list does not minimize source shuttle time, but it does minimize record shuttle time. Therefore, a B mode list is used when the source reel is short, and the record master is long.

C Mode EDL A C mode edit decision list is also a checkerboard list. A C mode list orders the edits based on source reel and source in times. A C mode list minimizes source shuttle time, but not record shuttle time. Therefore, a C mode list is used when the source reel is long, and the record master is short.

D Mode EDL A D mode edit decision list is similar to an A mode list, but all transitions other than cuts are placed at the end of the list.

E Mode EDL An E mode edit decision list is similar to a C mode list, but all transitions other than cuts are placed at the end of the list.

Ethernet A form of local area network. It consists of a coaxial cable that can extend approximately 1.25 miles and can offer up to 1,000 nodes (computers and peripheral devices). Ethernet is rated at a bandwidth capability of 10 Mbit/sec, but practical throughput is about 1 Mbit/sec.

FDDI Fiber digital data interconnect. A transmission method that uses fiber optics to transmit and receive signals. FDDI is often offered in two configurations: bandwidths of 100 Mbits/sec and 200 Mbits/sec. A 100 Mbit/sec FDDI network will most likely deliver about 2 MB/sec.

Film Cut List The film counterpart of an edit decision list. Instead of providing numbers that relate to the timecodes of the videotapes, the film cut list indicates which film edge numbers should be cut from the original film negative.

Final Cut The point at which the picture portion of the editing is complete, or "the picture is locked."

Fixed Frame Size In compression methods, an implementation in which there is a fixed amount of data that the compression algorithm will allow for each frame. It will not expand this amount of information if the frame contains more data than the algorithm is set to process.

Flash Convertor A device used to convert analog signals to digital signals. With the flash convertor, it is possible to convert frames of video into data that can then be interpreted by computers.

Floppy Disks A thin, flexible magnetic medium encased in a protective plastic shell. Floppy disks are erasable and can be used many times. They come in three sizes: 8", 5.25", and 3.5".

FPS Frames per second. The normal playback of 35mm and 16mm film is 24 fps.

Fractals A compression technology that utilizes fractal patterns to represent every possible pattern that can exist. By dividing a picture into small pieces, these smaller sections can be searched and analyzed fairly quickly. On a smaller level, instead of trying to find the pattern for the entire picture, the search is for smaller patterns of pixels.

Framestores A digital device that stores from one to several frames of video.

4 FSC 4x subcarrier frequency. The sampling frequency that is most often used for digitizing composite video signals (14.3 MHz NTSC and 17.7 MHz PAL).

GUI Graphical user interface. The operating environment defined by computer software programs.

Hybrids A nonlinear system that combines a variety of storage methods, including videotape, laserdisc, optical disc, and magnetic disk.

Ink Numbers The process of coding numbers on the mag track and film print. Also known as *inking*.

IPS Inches per second. The term is used to describe audio tape consumption. The normal recording and playback speed of 1/4" reel-to-reel audio tape is 15 ips.

Interframe Coding A method for compression in which certain film or video frames are dependent upon previous or successive frames. Interframe coding offers a major benefit of MPEG compression: a significant savings of storage compared to JPEG methods.

Intraframe Coding A method for compression in which each film or video frame being independent of previous or successive frames. In intraframe coding, each JPEG-compressed frame contains all the information that it needs to be displayed.

ISDN Integrated services digital network. A transmission method that is designed to combine various sorts of information, from telephone transmissions to fax to images and sounds, into one digital network. At 64 kbits/sec, the bandwidth is not much greater than that of dial-up service.

JPEG Joint Photographic Experts Group. A form of hardware-assisted compression. JPEG is based on still images and employs discrete cosine transforms, which are lossy algorithms. When a file is compressed using a JPEG-based processor, information about the original signal is discarded and lost.

KeyKode™ A numbering system introduced by the Kodak Corporation that consists of the usual film edge number information but also includes a bar code alternative. This bar code, which is deciphered by the KeyKode reader, is physically housed on the telecine.

L Cut A transition in which audio and video are not cut together as in a straight cut. In an L cut, either audio or video precedes the straight cut. Also called *overlap* or *split edit*.

Linear Editing A type of editing in which the program is assembled from beginning to end. If changes are required, everything downstream of the change must be rerecorded. The physical nature of the medium dictates the method by which the material placed on that medium must be ordered.

List Cleaning The process of removing unnecessary information in an edit list to maximize the efficiency of the final conforming process.

Lossless Compression The process of compressing information without irretrievably losing any of the data that represent that information. To be lossless, a great deal of analyzing must be done.

Lossy Compression The process of compressing information that results in a loss of some portion of the data in the original message.

Magnetic Disks A computer storage method. Magnetic disks are erasable and highly reliable. They typically have a fast access time and offer a data transfer rate that is faster than that of any other computer disk technology.

Magneto-Optical Discs A computer storage method that combines two technologies: magnetic and optical recording methods.

Mag Track Magnetic track. Sprocketed magnetic audio tape used in the double-system method of film editing. The picture elements are contained on the film, and the sound for those pictures are contained on the mag track.

MIDI Musical instrument digital interface. The interface responsible for the translation of musical information into digital terms.

Mix-To-Pix The process of mixing audio to the final video master. Mixing to the picture is accomplished as sections of the program are played on the video monitor.

MPEG Moving Picture Experts Group. A form of hardware-assisted compression. Whereas JPEG is based on still images, MPEG is based on motion. It is a lossy compression method.

MII A videotape format that utilizes a component recording process.

Multiple Versions The nonlinear editing system's ability to provide multiple versions of a sequence without requiring additional copies of footage or degrading the signal by losing generations.

Nagra Professional-grade reel-to-reel audio tape recorder that is most commonly used for field recording.

Nonlinear Editing A type of editing in which the program need not be assembled from beginning to end. The physical nature of the medium and the technical process of manipulating that medium do not enforce or dictate a model by which the material must be physically ordered. Changes can be made regardless of whether they are at the beginning, middle, or end of the sequence being edited.

NTSC National Television Standards Committee, which standardized the color television transmission system in the United States. Motion video is normally played back at 30 frames per second. (Actually, it is 30 fps for non–drop frame and 29.97 fps for dropframe.) The scan rate is 525 lines at 60 Hz.

Nyquist Limit One-half of the highest frequency at which the input material can be sampled. Also called *Nyquist rate*.

Offline An editing process that does not result in a finished product. The resulting program is in a form of preview.

Online An editing process that results in a finished product that is ready for final viewing and distribution.

Opticals Transitions in film, other than cuts, that require two or more pieces of film to be printed together. Optical effects, such as dissolves and wipes, are created with an optical printer.

PAL Phase alternate line. A color television standard in which motion video is normally played back at 25 fps. The scan rate is 625 lines at 50 Hz.

Phase-Change Optical Discs A computer storage method based on the ability of a material to exhibit two properties: amorphous and crystalline.

Pixel A single point of an image's makeup.

Pixel Matrice The number of pixels that are contained vertically and horizontally over the span of the viewing screen.

Playlist A list of items to be played back in a certain order. The playlist is the principle behind virtual recording.

Predictive Coding A method of creating routines that attempt to complete a model by analyzing the existing sequence.

Pre-master Tape A tape that is used to make laserdiscs. Original footage is transferred to the premaster tape, and then this tape is played directly into the laserdisc mastering system.

Px64 A proposed standard for motion video compression in any transmission applications designed around systems that operate at 64 kbit/sec. P refers to the number of channels that could be used in tandem and that would transmit the data at 64 kbits each.

Quantization The loss that results from the process of sampling.

Random Access The ability of an editing system to find a section of material without having to proceed sequentially through other material to reach that location.

Raster A pattern of scanning lines covering the area upon which an image is displayed.

Release Print The final version of a film project from which additional film copies are made to distribute to theatres.

Run Length Encoding A process used to determine messages by defining the message as a string of zeroes and ones that would run for a certain length of time before changing characteristics.

Sampling Measuring an analog signal at regular intervals. A sample is a smaller part of a whole that represents the nature or quality of that whole.

Sampling Theorem A theorem that states that a signal must be sampled at least twice as fast as it can change in order to process that signal adequately. Sampling the signal less frequently can lead to aliasing artifacts.

Scaling The process of reducing the size of the matrix for an image by removing pixels.

Scrubbing Representing the analog waveform of audio to the human ear with an accompanying change of pitch as the audio playback is decreased or increased. Also called *audio scrub with pitch change*.

SCSI Small computer systems interface. A chain consisting of a 50 pin cable and a protocol and format for sending and receiving commands. It is used to connect computers and peripheral devices. Pronounced "scuzzy."

SECAM Séquential couleur à Mémoire. A television standard which, like PAL, has a normal playback of 25 fps with a similar scan rate. It is primarily used in Eastern Europe and France.

Single System The videotape recording method in which picture and sound are simultaneously recorded onto the single element of the videotape.

SMPTE Society of Motion Picture and Television Engineers.

Spatial Aliasing The distortion of the perception of where items are positioned in two- and three-dimensional space.

Storyboard A visual arrangement of shots that can be easily rearranged to experiment with the flow of a sequence.

Straight Cut   A transition in which audio and video are cut together. Also called *both cut*.

Subsampling   A technique in which the overall amount of data that will represent the digitized signal has been reduced.

Symmetrical Compression   A compression technique that requires an equal amount of processing power to compress and decompress an image. In applications designed for editing, the compression of a frame must occur in real time. Decompressing that same frame must also occur in real time.

T1 and T3   Common carriers of signal that require dedicated systems installed at either end of two sites that will be in communication with each other. T1 links have a bandwidth of approximately 1.5 Mbit/sec; T3 links have a bandwidth of 45 Mbit/sec.

Temporal Aliasing   The distortion of the perception of movement over time.

TBC   Timebase corrector. An electronic device used to correct and stabilize the playback of a video signal.

Timecode   A signal that is recorded onto videotape and that identifies each video frame. Timecode takes the form of hours, minutes, seconds, and frames. Longitudinal timecode is recorded onto an audio track. Vertical interval timecode (VITC) is recorded onto a section of the video track. Address track timecode is recorded simultaneously with picture recording.

Trafficking Problems   The conflicts that can arise when there are many items to be cued and played back and too few playback devices.

Trim   Material not used for a take. Everything before the start frame being used is the head trim. Everything after the end frame being used is the tail trim.

Turing Machine   A general-purpose machine that is capable of any operation if given the appropriate software instructions.

Variable Frame Size   In compression methods, an implementation in which there is no fixed amount of data for each frame. If action occurs from frame to frame, thereby increasing the complexity of those frames, the compression algorithm expands accordingly and allows more data to pass.

Virtual Recording   A "recording" process for nonlinear editing systems that is actually a "record keeping" process that lists the edits that will eventually be made instead of actually transferring video and audio signals from several source tapes to a record tape.

Wavelets   A compression technology in which a picture is scaled and a set of wavelet functions (transforms) are run that seek to encode error. The information in the original picture is compared to the differences in the scaled version. The goal is to store just one of four portions of the picture and to quantize the other three portions of the picture.

WMRM   Write many, read many. In laserdisc, optical disc, and magnetic disk technology, an erasable medium.

WORM   Write once, read many. A medium that can only be written once but can be read many times. It is nonerasable.

Window Dub   Timecode information placed as an overlay window over copies of source footage as a step in the offline process. Also called *burn-in dub*.

Bibliography

1. Le Gall, Didier J. *The MPEG Video Compression Algorithm: A Review*. San Jose: C-Cube Microsystems, 1991.

2. Purcell, S. *The C-Cube CL550 JPEG Image Compression Processor*. San Jose: C-Cube Microsystems, 1990.

3. Zettler, William, John Huffman, and David C.P. Linden. Application of Compactly Supported Wavelets to Image Compression. Cambridge: Aware, 1991.

Index

Abekas A42, 29
Abekas A53D, 29
Abekas A62, 31, 184, 185
A/B roll systems of videotape editing, 17, 18, 50
Access
  carrier sense multiple (CSMA), 236
  immediate, 295, 296
  random. *See* Random access
  simultaneous, 294, 295–96
Access time, disk, 220
  floppy disk, 222
  on laserdisc players, 86
  magnetic hard disk, 223, 226
  Magneto-Optical disc, 229–30
  phase-change optical disc, 231
AES/EBU (American Engineering Society/European Broadcasting Union) inputs and outputs, 119
AirPlay, 324
Algorithm, 167
Aliasing, 170–71, 273
Alpha channel, 264
A mode list, 51
Ampex Corporation, 15, 28–29
Ampex 100, 30
Amplitude definition, 159–60
Analog, defined, 155
Analog compression, 185–86
Analog-to-digital (A/D) convertor, 118, 158
Analog video, migration toward digital processing, 27
Answer print, 14
Anticipated use of nonlinear editing systems, 293–94
Apple Computers, 161–62
  Macintosh, 225, 328–29
AppleTalk, 235
Artifacts, 170, 173

ASCII (American Standard Code for Information Interchange) code, 165–66
Assemble list, 287
  audio-only, 288
Asymmetrical vs. symmetrical compression, 200, 205, 206–7
Audio, transfer to laserdisc, 84–86, 88
Audio buffering techniques, 97, 98
Audio capabilities, high-quality, 118
Audio channels, number of, 296
  digitization and, 119
Audio dissolves (crossfades), 98
Audio editing, 14
  advanced, in digital-based systems, 264, *265*
  completing audio for video as separate stage, 18–19
  in digital-based system, 150–51
  graphical user interface and, 95–99
  in online room, 47
  second wave, 99
  traditional film sound editing, 95–96
Audio frequency rate, *176*
Audio interchange file format (AIFF), 270
Audio layback, 19
Audio layup (audio layover), 18
Audio mixing capabilities, 98–99
Audio-only assemble list, 288
Audio playback in laserdisc systems, 96–98
Audio quality, 296
  digitization and, 117–19
Audio sampling rates, 177
Audio signals, 8
  digitization of, 176–77
  migration toward digital processing, 27
Audio sweetening, 100

Audio switching mechanisms, 98
Audio timecode (TC), 251, 282, 283, 285
Audio tracks
  number of, 296
  virtual, 137
AudioVision, 324
Audio workstation, digital, 309–10
  dedicated, 176
Auto-assembly, 60–61
  DVI and JPEG frame-based systems to provide, 330
  as output in digital-based system, 107, 139
  videotape copy of auto-assembled show, 140
Automatic film logging software, 132
Automatic key number readers, 283–86
Avid Media Composer, 322–24

Bandwidth, 177, 178
  computer network, 235, 236
  editing full-resolution full-bandwidth digital video, 184–85
  FDDI, 236–37
  ISDN, 235
  T1 and T3, 236
Bandwidth requirements, calculating, 228–29, 237
Bar code reader, 283, 284
Baud rate, 234
Bays (video editing room), 17
B (bidirectional) frames, 204
Bell & Howell's Professional Division, 77. *See also* BHP TouchVision
Bell Laboratories, 155–57, 186–87
Benefits of electronic nonlinear editing systems, 80–81, 299, 300, 304, 305–6

Betacam CTDM storage methods, 195–97

Betamax tape, 68

BHP TouchVision, 67, 68, 77–79, 81, 317–18

Bias against videotape editors, 56–57, 63–64

Bidirectionally predicted B frames, 204

Binary, defined, 165

Bit, the, 165–66
  defined, 177

Black, Tony, 32, 62, 63, 141, 258, 305–6

Blow-ups, 29

B mode list, 51–52

Bowling of laserdisc, 87

Broadcast industry, transmitting video data for, 238–40

Broadcasting, digital video compression applied in, 312

Buffering techniques, audio, 97, 98

"Bug" notes, 153

Burn in copies, 39

Bus, computer, 225
    computer back plane, 189–90, 191
    memory, 190

Butt splicer, 12

Buzzing, 97

Byte, 177

Camera and sound reports, 251

Camera roll (CR), 250

Capacity, disk, 219
    floppy disks, 222
    laserdiscs, 83
    magnetic hard disk, 223, 226

Carrier sense multiple access (CSMA), 236

CAV disc, 83, 85, 91

CBS, 65

CCIR 601, 184

CCITT, 193, 203, 207

CDI (compact disc interactive) system, 187

CD-ROM, 200, 227

Change list, 139, 287

Channels, audio, 119, 296

Character-generation on laserdisc-based systems, 91

Checkerboarding, 52

Chinagraph marker, 90

Chroma subsampling, 191–92, 195–96

Chrominance, 195–96, 197

Cinema, 320

Circled takes, 9

Clapsticks, 9, 10, 285

Client, benefits of digital nonlinear editing to, 304, 305–6

Clip, editing on digital-based system, 108–9, 112, 144, 145–50
    adding and deleting material without affecting length of sequence, 149–50
    rearranging order of, 146–48
    splicing shot together, 145
    trimming, 145–46

CLV laser videodiscs, 83, 84, 85

C mode list, 52, 53

CMX Corporation, 36, 65

CMX 50 editing system, 36

CMX 600, 65–67, 76

CMX 6000, 87, 88, 89–90, 319–20

CMX 300 editing controller, 36

CMX 3400 edit list, 269

Coding
    Huffman, 167–69, 199–200
    interframe, 203–4
    intraframe, 202, 203, 204
    LZW (Lempel-Ziv-Welch) encoding, 169
    predictive, 204
    run length encoding, 199

Coding (inking), 10

Cohen, Peter, 33, 62–64, 141, 306

Color test pattern, 45

Color time division multiplexing (CTDM) storage methods, 195–97

Color under, 186

Commercial footage, image complexity and storage requirements for, 122–23

Compact disc interactive (CDI) system, 187

Compact disc-read only memory (CD-ROM), 200, 227

Compatibility, file
    achieving, 269–70
    incompatibility vs., 260–61, 269

Compositing in digital-based systems, 180, 265–66

Compression, 103, 113, 166–67, 178–80, 183–217
    analog, 185–86
    coding techniques and, 168–69
    defined, 155
    developing and emerging technologies, 207–9
    digital, of audio signal, 177
    digital video, 117, 186–88, 233, 239–41, 308
      characterizing results of, 210–13
      digital video interaction (DVI), 117, 187, 206–7

effect on shooting process, 312–13
      evolution of, 210
      hardware-aided, 193, 214–17
      hardware and software methods, 188
      incorporation into existing product lines, 309–10
      JPEG. *See* JPEG compression
      MPEG, 193–94, 203–6, 207, 239–40
      RTV, 324–25, 327
      software-only, 188–93, 213–15
      state-of the art (1989–1992), 213–17
    fixed vs. variable frame size, 200–202
    for high-definition television, 210
    lossless, 175, 178, 188
    lossy, 173, 175, 179–80, 195
    pixel matrix used in, 183–84
    PX64, 207
    quantization and, 197–200
    symmetrical vs. asymmetrical, 200, 205, 206–7
    transmission of video data and, 233, 239–41

Computer(s)
    limitations of, 162–63
    networks, 235–36
    storage devices, 219–22
    video and, 161–65

Computer back plane bus, 189–90, 191

Computer capability of editors, 298

Computer-controlled video editing systems, 26–27. *See also* Electronic nonlinear random-access editing system

Computer database of footage, 252–57
    pictorial representations, 253
    searching through, 254–55
    statistical information, 253

Conforming the negative, 14, 287–88
    on laserdisc-based system, 100
    online editing stage as conform stage, 45, 48

Constant angular velocity (CAV) laser videodiscs, 83, 85, 91

Constant linear velocity (CLV) laser videodiscs, 83, 84, 85

Continuity sheet, 248, *250*

Continuous-tone images, 193

Control track editing, 36

Convertor
    analog-to-digital (A/D), 118, 158
    digital-to-analog (D/A), 118, 159
    flash, 157–60, 163–64

Convolving function in wavelet compression, 209
Correlation of film and videotape, 276–77
Cost(s)
  of laserdiscs, 87
  of online videotape editing, 44
  storage, 226–27, 296, 305
Cranking, 276
Creativity, 301
  in film vs. videotape editing, 56–57
  limitations of offline editing for, 49–50
Crossfades, audio, 98
CTDM storage methods, Betacam, 195–97
Cue tones, 16
Curie point, 228
Cuts-only systems of videotape editing, 17
Cutting film, 12–14, 301–2
  negative, 14, 139–40
  of videotape, 15–16

Dailies, 9–10
Dailies report, 286
Database
  automatic creation of, 284–85
  computer, of footage, 252–57
  digital, 311
Data error masking techniques, 174–75
Data rate, 220
  formula for determining, 221–22
Data transfer rate, 219–20
  for floppy disks, 222
  for magnetic hard disks, 223, 224
  for Magneto-Optical discs, 228–29
  for phase-change optical discs, 231
DCTs (discrete cosine transforms), 194, 196–97, 308
Decimation, 173, 174, 181, 195–96
Dedicated digital audio workstations (DAWs), 176
Dedicated systems vs. software modules, 261–62
Dedicated word processors, 2–3
Diagnostics, remote, 299
Digi slates (clapsticks), 9, 10, 285
Digital, defined, 155
Digital audio workstation, 309–10
  dedicated, 176
Digital-based systems, 103–53, 322–30
  Avid Media Composer, 322–24
  boundaries shattered by, 308–9
  components of, 106
  defined, 314

development of, 263–68, 328–30
digitizing material, 112–19
  footage, 113–14, 133–35
  parameters, 116–19, 133–34
  playback speeds, 114–15
DVision, 81, 130, 324–25
editing on, 143–53
  cutting the negative, 139–40
  editing stage, 108–9, 112, 135–38, 144–51
  film to videotape transfer, 112, 132–33
  implication of MPEG for, 204–6
  input stage, 143–44
  interface, 144
  output stage, 138–39, 151
  producing finished products, 140
  sample procedure for, 131–40
  systemic issues in, 152–53
  time savings in, 151–52
  transferring to film, 139
editors' reactions to, 141–42
effect on shooting process, 312–13
EMC2, 325–26
evolution of, 307–10
future of, 307–10, 314
general design of, 126–31
  editing interface, 126, 127, 129–31
  representing footage, 126, 127–29
general objectives, 104–5
Lightworks, 326–27
logging for. See Logging process
metamorphosis capabilities, 310
Montage III, 81, 327
as offline or online, issue of, 270–71
paradigms of, 108–12
storage devices for, 219–31
  computer devices, 219–22
  determining data rate and storage capacity, 221–22
  disks, types of, 222–31
storage of material in, 119–24
  in early systems, 121–22
  image complexity and, 122–24
  storing to disk, 115–16
user interface, 124–26, 135
work flow in, 105–8
Digital compositing devices, 180, 265–66
Digital databases, 311
Digital disk recorders, 30–32
Digital framestores, 28
Digital imaging, 274

Digital manipulation
  product and capabilities based on, 180–81
  of video, 27, 28–29
Digital media manager (DMM), 3, 259–71, 315
  dedicated systems vs. software modules, 261–62
  digital-based wave, 263–68
  file compatibility, achieving, 269–70
  file incompatibility and, 260–61, 269
  horizontal and vertical editing concepts, 268–69
  introduction to, 259–62
  tape- and laserdisc-based waves, development of, 262–63
Digital media processors, evolution of, 3–4
Digital nonlinear editing, benefits of, 2
Digital offline, 314
Digital online, 314–15
Digital Optics System (ADO), 28–29
Digital paint systems, 180
Digital still stores, 29–30, 180
Digital support of video, 30–32
Digital-to-analog (D/A) convertor, 118, 159
Digital uncompressed media management, 315. See also Digital media manager (DMM)
Digital video, editing full-resolution full-bandwidth, 184–85
Digital video compression. See under Compression
Digital video effects (DVEs), 19, 20, 28–29, 43, 180, 264–65
Digital video interaction (DVI) compression, 117, 126, 187, 206–7
Digitization, 103
  of audio signal, 176–77
  bandwidth and storage in context of, 177–78
  defined, 155
  flash convertors and, 157–60, 163–64
  of footage, 113–14, 133–35
  in Montage, 75
  parameters, 116–19, 133–34
  real-time, 164–65
  sampling and, 155–57, 169–75
  terms relevant to, 155
Digitize, defined, 155
Digitizing rate, 116–17
Direct-access systems, 79. See also Videotape-based systems

Direct finished output from electronic
    nonlinear system, 61–62
    from digital-based system, 107–8
Dirty lists, 42
Disc copies (laserdiscs), 86
Discovision Associates (DVA) Code,
    85–86
Discrete cosine transforms (DCTs),
    194, 196–97, 308
Disc skipping, 87
Disk(s)
    characteristics of, 219–20
    portability of, 120–21
    storage capacity, 119–20
    storing to, 30, 115–16
    types of, 222–31
        choosing, 221, 222
        disk arrays, 225
        floppy disks, 222–23
        magnetic hard disks, 223–27
        optical discs, 134, 227–31, 296
Disk arrays, 225
Disk default, 122, 123
Display, interlaced vs. noninterlaced,
    161
Dissolves
    audio (crossfades), 98
    in combined text and graphic form,
        94
    in digital-based system, 137–38
    in laserdisc editing systems, 90
    in videotape editing, 18
DMM. See Digital media manager
    (DMM)
D mode list, 52–53
Documentation, 152–53
Dougherty, Paul, 32, 62, 63, 81, 141,
    257
Down captured, 221
Dual-finish cognizant system, 67
Dual-headed laserdisc machines,
    88–89
Dual-sided laserdiscs, 86–87
Dual-system approach to shooting
    film, 9
Dubbing, 14, 19
Dubs, window, 39
Dupe list, 287
Dupe work print, 26
Duplication procedure in nonlinear
    editing system, 147
DVEator, 29
DVEs, 19, 20, 28–29, 43, 180, 264–65
DVI compression, 117, *126*, 187,
    206–7
DVI frame-based edit system, 330
DVision, 81, 324–25
    editing interface display of, *130*

Eastman Kodak, 278, 283
Edge numbers, 10, 251, 278–79, 283
    automatic key number readers,
        283–86
    optical list of, 287
    as sync points, 282
Ediflex I and II, 59, 67, 68, 76–77, 79,
    81, 318
Edit decision list (EDL), 26, 37, 38
    different forms of, 51–54
    dirty lists, 42
    as journaling method of online edit
        session, 266–67
    list cleaning, 42
    from offline process, 38, 41, 42
        online editing from, 45–46
    video, 60, 67, 151, 286
        DVI and JPEG frame-based
            systems to provide, 330
        as output in digital-based
            system, 138–39
        variety of formats, 261
EditDroid I and II, 87, 93, 320–21
Editing
    on digital-based systems. *See under*
        Digital-based systems
    with digital disk recorders, 31
    horizontal and vertical, 268–69
    needs of program and choice of
        editing system, 24
    offline vs. online, 24
    at 24 fps, 289–91
    See also Film editing; Videotape
        editing
Editing interface. *See* User interface
Editing methods, evaluation of, 298
Editing modes, 51–54
Editing process, 7–24
    coming together of film and video
        editing, 6
    development of different tech-
        niques, 7
    film editing procedures, 9–15
    formats and standards, 7–8
    increasing complexity of, 4–6
    linear vs. nonlinear, 1, 21–24
    videotape editing procedures,
        17–20
Editorial work products. *See* Work
    products
Editors
    benefits of digital nonlinear editing
        to, 299, 304, 305–6
    computer capability of, 298
    on digital-based systems, 141–42
    on film and video editing cultures,
        62–63
    on logging software, 257–58

mental preparation required of,
    nonlinear editing and,
    299–303
videotape
    bias against, 56–57, 63–64
    training and learning process of,
        55–56
    views of videotape-based systems,
        81–82
EDL. *See* Edit decision list (EDL)
EECO Company, 16
E-Flex, 28
EIAJ (Electronics Industries Associa-
    tion of Japan) format, 35
"Electronic bin," 127–28
Electronic linear videotape editing, 67
Electronic nonlinear editing systems,
    58–62
    benefits of, 80–81, 299, 300, 304,
        305–6
    common stages of, 58–59
    evaluating, 293–306
    work products of, 60–62
    See also Digital-based systems;
        Laserdisc-based systems;
        Videotape-based systems
Electronic nonlinear random-access
    editing system, 25–33
    definition of, 25–26
    digital manipulation, 27, 28–29
    digital storage, 28, 29–30
    digital support, 30–32
Electronic painting system, 5–6
Electronic transcriptions, 15
EMC2, 325–26
E mode list, 53, *54*
Entropy of message, 178
Episodic footage, image complexity
    and storage requirements for,
    122–23
Epix, 87, 321
Erasability, 86, 105, 121
Ergonomics of system, 299
Error masking techniques, 174–75
Ethernet, 235–36
ETrax, 320, 321
Ettlinger, Adrian, 76
European Broadcasting Union (EBU),
    16
Evaluation of electronic nonlinear
    editing systems, 293–306
    benefits of systems and predictions
        for future, 299, 304, 305–6
    editors on contributions of systems,
        304–6
    human side of editing experience
        and, 299–303
    operational concerns, 297–99

technical concerns, 293–97
External disk storage, 191
Extracting in digital-based system, 135
Eye-matching, 289

Facsimile (FAX) machines, 4, 159, 233, 240
Fast motion (undercranking), 276
Fear of technology and computers, 298
Fiber digital data interconnect (FDDI), 236–37
Fields, 115
Fifth wave (digital uncompressed media management), 315
File compatibility, achieving, 269–70
File incompatibility, 260–61, 269
File organization, 152
Film
    correlation of videotape and, 276–77
    formats and standards, 7–8
Film chain, 280
Film clips, 12
    editing on digital-based system, 108–9, 112, 144, 145–50
Film cut lists (negative cut list), 60, 67, 76, 139, 140, 286–87
Film editing
    areas for improvement of, 57
    coming together of video editing and, 6
    as nonlinear editing, 21–22
    operational aspects of, 55–64
        creative process, 56–57
        electronic nonlinear editing systems, 58–62
        procedures, 9–15
        traditional audio editing, 95–96
Film script, logging process for, 246–52
    additional notes, 251–52
    camera and sound reports, 251
    continuity sheets, 248, 250
    film transfer log sheet, 250–51
    lab (telecine) report, 249
    marking up script, 248, 249
Film speeds, 275–76
Film splicer (butt splicer), 12
Film synchronizer, 92, 93–94, 111
Film transfer log sheet, 250–51
Film transfer process, 273–91
    editing at 24 fps, 289–91
    film to tape (FT), 275–77
        correlation of film and videotape, 276–77
        on digital-based system, 112, 132–33

film and videotape speeds, 275–76
film to tape to film, 277–88
    edge numbers, 278–79
    editing and delivery on videotape and film, 286–88
    pulldown, 279–81
    sync point relationships, 281–86
    telecine process and, 278
film to tape to film to tape (FTFT), 288–89
to laserdisc, 84–86, 88
videotape masters from assembled film negative, 274
Final cut, 14
Finished vs. unfinished results, 296
First cut, 14
First wave. See Videotape-based systems
Fixed magnetic disks, 120–21
Fixed vs. variable frame size, 200–202
Flash convertor, 157–60
    evolution of, 163–64
    frequency definition and amplitude definition, 159–60
    method of, 158–59
Floppy disks, 222–23
Flops, 29
"For broadcast" programs, 259
Formats, editing, 7–8
4015 timecode reader/generator, 284
Fourth wave (digital online), 314–15
Fractals and fractal compression, 207–9
Frame offset number, 279
Framestores, digital, 28
Freeze frames, 29, 85
Frequency
    Nyquist limit, 171–72
    quantization frequency array, 198–99
Frequency definition, 159–60
Frequency modulation (FM) carrier, 84, 85
Full-resolution full-bandwidth digital video, editing, 184–85
Future of nonlinear editing, 307–15
    alternative applications, 311–12
    fifth wave, defined, 315
    fourth wave, defined, 314–15
    shooting process, effect of digital systems on, 312–13
    third wave, 307–10, 314
    transformation of offline and online editing, 310–11
    widespread system availability and use, 314

General-purpose interface (GPI), 19, 20
Generation loss or no loss, concept of, 175
Genlocking, 28
Gigabyte (GB), 177
Glossary, 331–36
Graphical user interface (GUI), 78, 93–99, 131
Graphic treatments for television commercials, 5
Grass Valley Group, 29

Hard disks, 223–27
    evolution of capacity and cost of, 226–27
    multiple (disk arrays), 225
Hardware-aided compression, 193, 214–17. See also JPEG compression; MPEG compression
Hardware and software compression methods, 188
Head trim, 12
High-definition television (HDTV), 184, 210
High-quality audio capabilities, 118
Horizontal editing, 268–69
Horizontal play in Ediflex, 77
Huffman coding, 167–69, 199–200
Human side of nonlinear editing, evaluating, 299–303
Human visual system (HVS) studies, 197
Hybrid laserdisc system, 89

IBM, 2
I (intraframe) frame, 204
Image complexity, storage requirements and, 122–24
Image quality, 121
    digitization and, 117
Imaging, digital, 274
Immediate access, 295, 296
Improved economies of operation, 308
Incompatibility, file, 260–61, 269
Inking (coding), 10
Ink number, 282
In point, 17
Input signal, importance of proper, 217
Integrated services digital network (ISDN), 234–35
Integration of media, 5
Intel Corporation, 206
Interface, user. See User interface
Interframe coding, 203–4
Interlaced display, 161

International Organization for Standardization (ISO), 193, 203
International Telephone and Telegraph Consultative Committee (CCITT), 193, 203, 207
Intraframe coding, 202, 203, 204
ISDN, 234–35
ISO, 193, 203
ISO synchronous FDDI, 237

Jaggies, 170
Joint Photographic Experts Group, 193. *See also* JPEG compression
Journaling in digital-based systems, 266–67
JPEG compression, 117, 193–202, 209, 240, 312
   digital-based systems using, 322–27
   distinctions between MPEG and, 203
   hardware-aided
      from 1991–1992, 214–16
      second-generation (1992), 216–17
   intraframe coding of frames under, 202
   quantization and, 197–200
   as symmetrical in nature, 200
JPEG frame-based edit systems, 330

Kaleidoscope, 29
kB/frame, 220–21
Kerr Rotation, 228, 231
Keying in digital-based systems, 264
KeyKode, 278, *279*, 283, *284*
Key numbers. *See* Edge numbers
Kilobyte (kB), 177
Kinescopes, 15

Lab (telecine) report, 249
Laserdisc-based systems, 83–102, 319–22
   appraising, 100
   CMX 6000, 87, 88, 89–90, 319–20
   development of, 262–63
   EditDroid I and II, 87, 93, 320–21
   Epix, 87, 321
   evolution of, 100–102
   graphical user interface, 93–99
      audio editing and, 95–99
   LaserEdit, 88, 321–22
   laser videodiscs, 83–87
      other characteristics of, 86–87
      transferring video and audio to, 84–86, 88
      types of, 83–84, 105, 227

pre-visualization tools, 91–92
   theory of operation of, 87–89
   trafficking, 89, 91
   typical design of, 89–90
   work products, 99–100
LaserEdit, 88, 321–22
Late-arriving material, 105, 294
Learning nonlinear systems, 299
Left page notes (continuity sheet), 248, *250*
Le Gall, Didier J., 205
Letter-boxing, 290
Lightworks, 326–27
Linear online, digital system using, 106–7
Linear videotape editing, electronic, 67
Linear vs. nonlinear editing, 1, 21–24
List cleaning, 42
Local area network (LAN), 235–36
Logging process
   automatic film logging software, 132
   in digital-based systems, 243–58
      avoiding duplicative work, 256–57
      computer database for footage, 252–57
      editors on logging software, 257–58
      evolution of process, 257
      film script, 246–52
      during or after shooting, 255–56
      preparing for editing and, 256
      purpose of, 243–44
      video log sheet, 244–46
   in Ediflex, 76–77
   in electronic nonlinear editing systems, 59
   in offline videotape editing, 39, *40*
Lossless compression, 175, 178, 188
Lossy compression, 173, 175, 179–80, 195
Luminance, 195–96, 197
LZW (Lempel-Ziv-Welch) encoding, 169

Magnetic hard disks, 223–27
   evolution of capacity and cost of, 226–27
Magneto-Optical (MO) discs, 227–30
Mag track, 9
   coding, 10
Mastering formats, 45
Master tape, 17
   from assembled film negative, 274
   building, 23

Matrix encoding, 186
Mean time between failure (MTBF), 223
Mechanical copying of laserdisc, 86
Media portability, 120–21
Megabyte (MB), 177
Memorex, 65
Memory bus, 190
Miller, Alan, 33, 63, 64, 142, 258, 306
Mind set, nonlinear, 10–11, 153, 299–303
Mixing capabilities, audio, 98–99
Mix to pix, 19
Modem (modulator/demodulation), 165–66, 233, 234, 299
Monitor positions, 90
Montage (Montage Picture Processor), 59, 67, 68, 75–76
   I, 318–19
   II, 79, 81, 318, 319
   III, 81, 327
Moving Picture Experts Group, 203. *See also* MPEG compression
Moviola, *12*
MPEG compression, 193–94, 203–6, 207, 239–40
   as asymmetric compression, 205
   delivering methods, 205–6
   intraframe and interframe coding in, 203
MPEG Extended (MPEG II), 207
Multiple laserdisc machines, random access and, 88
Multiple versions of sequence, 26, 73–74
Multiple videotape machines, nonlinear editing by using, 68
Musical instrument digital interface (MIDI), 267–68

Narrow-window analog TBCs, 27
NEC, 28
Negative
   conforming the, 14, 139, 287–88
   cutting, 14, 139–40
Negative cut list (film cut list), 60, 67, 76, 139, 140, 286–87
Networks, computer, 235–36
News gathering footage, 123–24
Noise, video, 217
Noninterlaced display, 161
Nonlinear editing, 1–3
   benefits of, 80–81, 300
   defined, 1
   future of, 307–15
   linear editing vs., 1, 21–24
   See also Electronic nonlinear

editing systems; Electronic nonlinear random-access editing system
Nonlinear mind set, 10–11, 153, 299–303
Nonlinear online systems, 311
"Not for broadcast" programs, 108, 259
NTSC (National Television Standards Committee), 15, 183
  component digital, 184
  master, creating, 274
  standards conversion to PAL from, 273
  standards for playback speed, 114, 115
  television system, 92
  videotape, correlation between film and, 276–77
NuVista+ card, 164
Nyquist limit, 171–72

Offline editing, 24, 57
  digital, 314
  transformation of, 310–11
  videotape, 35, 37–42, 49–51, 310–11
  See also Digital-based systems; Laserdisc-based systems; Videotape-based systems
offline-online link, 36–37
One-lites (dailies), 9–10
One-step read/write process of phase-change optical disc, 230–31
One-strikes, 87
Online editing, 24
  digital, 314–15
  direct finished output from nonlinear system and avoidance of, 61–62
  linear, digital system using, 106–7
  transformation of, 310–11
  videotape, 35, 43–48, 55, 310–11
Online edit suite, 43–44
  audio editing in, 47
  equipment in, 43–44, 260
Operating system
  documentation on, 152–53
  software program vs., 152
Optical copying of laserdisc, 86
Optical discs, 134, 227–31, 296
  Magneto-Optical (MO) discs, 227–30
  phase-change optical (PCO) discs, 230–31
Optical effects
  in digital-based system, 137–38

with digital video effects, 29
  in film editing, 13–14
  in laserdisc editing systems, 90
  in Montage, 75–76
Optical list, 287
Overcranking, 276
Overlap, 13, 94
Oversampling, 175

PAL (phase alternate line)
  master, creating, 274
  playback speed, 114, 115
  RGB picture, 183
  standards conversion from NTSC to, 273
  transfers, 279
  videotape, correlation between film and, 276–77
Paper edit list, 39–40, 41
Pappas, Basil, 33, 62, 63, 82, 141, 258, 306
Persistence of vision, 202
Phase alternate line (PAL). See PAL (phase alternate line)
Phase-change optical (PCO) discs, 230–31
Picket fence effect, 173
Picture overlap (split edit), 13, 94
Picture quality, 308
  laserdiscs and, 99–100
Pinnacle, 29
Pioneer 8000, 86
Pixel, 157, 161
Pixel confetti, 201
Pixel matrices, 157, 162
  used in compression, 183–84
  of video vs. computer systems, 161
Pixel subsampling techniques, 173, 174
Pixel-to-byte conversion, 158–59
Playback
  on laserdisc-based systems, 91
    audio playback, 96–98
    real-time, 77
Playback and recording capabilities, simultaneous digital, 30–32
Playback speeds, 114–15
Playlist, 68–72, 73, 90
Pointers, using (virtual recording), 72–73, 302
Portability, media, 120–21
P (predicted) frames, 204
Prebuild, 72, 73, 89
Predictive coding, 204
Preroll point, 17
Presentation copies of program, 60
Presentation-level video (PLV), 206

Preview of videotape, 17
Pre-visualization tools in laserdisc system, 91–92
Programming, range of, 259–60
Pull, 251
Pulldown, 279–81
  2–3 and 3–2, 280, 281
Pulldown mode, 280–81, 284
  as sync points, 282
Pulldown sequence, 284
Pull list, 286–87
  scene, 287, 289
Punch frame, 282
Punching the printed film, 282–83
PX64, 207

Quantel Harry, 184
Quantization, 169–70, 173
  compression and, 197–200
Quantization factor (Q factor), 198–99
Quantization frequency array, 198–99
Quantized table elements (frequencies), 198
QuickTime, 328–29

Random access, 22
  digital systems and, 105
  electronic nonlinear random-access editing system, 25–33
  multiple laserdisc machines and, 88
  transferring original footage to same or different material and, 58–59
Raster, 161
Rawley, Curt A., 304
RCA, 187
Read/write speed. See Data transfer rate
Real-time digitization, 164–65
Real-time playback, 77
Real time video (RTV), 206–7
  digital video compression, 324–25, 327
Recording, virtual, 72–73, 302
Record keeping in linear offline editing, importance of, 49
Recutting, 301
Reediting of videotape, 23
Release notes, 153
Release print, 14
Remote diagnostics, 299
Removable magnetics (RMAGs), 223, 224
Repositions, 29
Representations of footage, digitized, 126, 127–29

Resolution
audio, in laserdisc systems, 98
digital video compression and, 211, 212–13
editing full-resolution full-band-width digital video, 184–85
Reverse motion (reverse printing), 91
Rough cut, 14
RTV, 206–7
RTV digital video compression, 324–25, 327
Run length encoding (RLE), 199

Sample points, defined, 155
Sampling, 169–75
audio sampling rate, 176, *177*
decimation, 173, *174*
early experiment with, 155–57
error masking techniques, 174–75
Nyquist limit, 171–72
selective removal of samples, effect on message, 172–73
Sampling theorem, 170, 176
Sarnoff Labs, 187, 206
Scaling, 191
Scaling function in wavelet compression, 209
Scan across in Ediflex, 77
Scan conversion, 273
Scene pull list, 287, 289
Scene (SC) numbers, 251
Screening copy, 67, 139
Script
Ediflex system based on, 77
film, logging process for, 246–52
as outline in nonlinear editing, 303
video log sheet, 244–46
Script integration, 255
Scrubbing the track, 96–98
SCSI (small computer systems interface), 223
SCSI 1, 225
SECAM (séquential couleur à mémoire) playback speed, 114, 115
Second wave. See Laserdisc-based systems
Sequence editing, 111
on digital-based system, 144–50
multiple versions of sequence, 26, 73–74
Sequencer, MIDI, 268
Servers on network, 235
Shooting of film, 9
effect of digital systems on, 312–13
logging process during and after, 255–56

Silent movie, 5
Simultaneous access, 294, 295–96
Single-system recording, 15
Slate (clapsticks), 9, 10, 285
Slop disc, 89
Slow motion (overcranking), 276
Small computer systems interface (SCSI), 223
SCSI 1, 225
Smart slates (clapsticks), 9, 10, 285
SMPTE-A transfer method, 280
SMPTE-B transfer method, 280
SMPTE timecode, 16, 35, 36, 86, 113, 134, 261
Society of Motion Picture and Television Engineers (SMPTE), 16
Software modules, dedicated systems vs., 261–62
Software-only compression methods, 188–93, 213–15. *See also* MPEG compression
Software program
documentation on, 152–53
operating system vs., 152
Software synchronizers. *See* Timeline (software synchronizers)
Software user interface. *See* User interface
Sound and camera reports, 251
SoundDesigner 2 (SD2) format, 270
Sound effects (sfx), 95–96. *See also* Audio editing
Sound overlap, 95
Sound rolls, 9
Soundtrack, 288
35mm magnetic, 9, 10
Sound transfer report, 9
Spatial aliasing, 170, *171*
Splicing, 12–13
in digital-based system, 135, 145
of new material into existing sequence, 148–49
videotape, 15, *16*
Split edit (picture overlap), 13, 94
SqueeZoom, 28
Standardization of digital video compression methods, 240
Standards, editing, 7–8
Steenbeck, 12
Sticks (clapsticks), 9, 10, 285
Still stores, digital, 29–30, 180
Storage
ability to free up, 295
color time division multiplexing (CTDM), 195–97
costs, 226–27, 296, 305
in digital-based systems, 119–24

early systems, 121–22
image complexity and requirements for, 122–24
storing to disk, 115–16
digitization and, 177–78
external disk, 191
formula for determining capacity, 221–22
on laserdisc, 89
technical concern over, 294–96
terms, 177
Storage devices for digital editing systems, 219–31
computer storage devices, 219–22
determining data rate and storage capacity, 221–22
disk types, 222–31
Storyboard as Montage work product, 76
Storyboarding, 128–29, 144
Straight cut, 13
Submaster, copying footage to, 23–24
Subsampling, 170, 172–73
chroma, 191–92, 195–96
decimation, 173, *174*, 181, 195–96
picket fence effect, 173
results of, 192–93
Suites (video editing room), 17
Sweetening, audio, 100
Switcher, 18
Symmetrical vs. asymmetrical compression, 200, 205, 206–7
Sync block (synchronizer), 12
Synchronization
film synchronizers, *92*, 93–94, 111
genlocking, 28
Synchronizing (syncing) dailies, 9–10
Sync point relationships, 281–86

T1 and T3, 236, 238
Tail trim, 12
Takes, circled, 9
Take (TK) numbers, 251
Take-up reel, 13
Tape generation, losing, 23
TBC, 27, 195, 217
Telecine machine
automatic key number readers in, 283–86
pulldown using, 280
Telecine process (film to video transfer), 275–77
on digital-based system, 112, 132–33
film to tape to film transfer and, 278
logging process and, 256

work products of, 285–86
Telecine report, 249
Television, high-definition, 184, 210
Television commercials
    early digital nonlinear systems for
        editing, 121–22
    film to tape process for, 275
    graphic treatments for, 5
    as horizontal-based storytelling,
        268–69
    image complexity and storage
        requirements for, 122–23
Temporal aliasing, 170–71, 273
Temporary mixdown (temp mix), 99
Text-based user interface, 130–31
35Mm film
    edge numbers, 10
    playback speed, 114–15
35mm four-perf (perforations per
    frame) film, 9
35mm magnetic sound track (mag
    track), 9
    coding, 10
35mm motion picture, 5
Timebase correctors (TBC), 27, 195,
    217
Timecode, 16, 18, 26, 27, 67
    audio, 251, 282, 283, 285
    integrating and orchestrating
        equipment via, 19–20
    SMPTE, 16, 35, 36, 86, 113, 134,
        261
    as sync points, 282
    video, 251, 283
Timeline (software synchronizers), 92,
    93–94, 111–12, 144
    audio waveforms on, 150–51
    customizing, 146, 147
Timeline views, 136–37
Time savings, digital-based editing
    systems and, 151–52
Titling abilities on laserdisc-based
    systems, 91
TouchVision, 67, 68, 77–79, 81,
    317–18
TouchVision Systems, 324
Tracks, audio, 137, 296
Trafficking, 72, 73, 89, 91
Training programs, 299
Transfer, film. See Film transfer
    process
Transfer speed. See Data transfer rate
Transition-based editing, 109–10, 112,
    145–46, 301
Transmission of video data, 233–41
    for broadcast industry, 238–40
    compression and, 233, 239–41

methods of, 233, 234–37
    of pictures and sound among
        editing systems, 237–38
Trim, 12
Trim bin, 12
Trimming a shot
    in digital-based system, 136,
        145–46
    in film editing, 13
    in videotape editing, 18
Truvision, Inc., 163, 164
Turing Machine, 4, 262, 271

Undercranking, 276
Unfinished vs. finished results, 296
User interface
    in digital-based systems, 124–26,
        135
        editing interface, 126, 127,
            129–31
        graphical, 78, 93–99, 131
        in Montage, 75
        text-based, 130–31
    in videotape-based system, 74–75

Variable vs. fixed frame size, 200–202
Vertical editing, 268–69
VHS, 39, 68, 99
Video
    presentation-level (PLV), 206
    real time (RTV), 206–7
Video edit decision list (EDL), 60, 67,
    151, 286
    DVI and JPEG frame-based systems
        to provide, 330
    as output in digital-based system,
        138–39
    as work product in Montage, 76
Video log sheet, 244–46
Video noise, 217
Video reference frame, 283, 284
Videotape
    correlation of film and, 276–77
    cutting, 15–16
    development of, 15–16
    formats and standards, 7–8, 35
    transfer to laserdisc, 84–86, 88
Videotape-based systems, 65–82,
    317–19
    appraising, 79–80
    BHP TouchVision, 67, 68, 77–79,
        81, 317–18
    CMX 600, 65–67, 76
    development of, 262–63
    economic benefits of, 80–81
    Ediflex I and II, 59, 67, 68, 76–77,
        79, 81, 318

editors on, 81–82
editors' views of, 81–82
evolution of, 81
as first wave of electronic nonlinear
    editing systems, 67–68
Montage, 59, 67, 68, 75–76
    I, 318–19
    II, 79, 81, 318, 319
    III, 81, 327
multiple versions, 73–74
playlist concept, 68–72, 73
trafficking, 72, 73
user interface, 74–75
virtual recording, 72–73
Videotape editing
    areas for improvement of, 57
    based on transition, 109–10
    coming together of film editing
        and, 6
    computer-controlled, 26–27. See
        also Electronic nonlinear
        random-access editing system
    electronic linear, 67
    linear process of, 21, 22–24
    offline, 35, 37–42, 49–51, 310–11
    offline-online link, 36–37
    online, 35, 43–48, 55
    operational aspects of, 55–64
        creative process, 56–57
        electronic nonlinear editing
            systems, 58–62
        procedures, 17–20
Videotape editors
    bias against, 56–57, 63–64
    training and learning process of,
        55–56
Videotape masters from assembled
    film negative, creating, 274
Videotape speeds, 275–76
Video timecode (TC), 251, 283
VID I/O, 163
Viewing copy in Montage, 76
Virtual audio tracks, 137
Virtual recording, 72–73, 302
Vision, persistence of, 202
Vital, 28

Warner, William J., 304
Wavelet compression, 209
Wavelet space, 209
Werner, Tom, 66
Wide-window digital TBCs, 27
Window dubs, 39
Wipes in videotape editing, 18
WMRM laserdiscs, 86, 89, 105
WMRM optical discs, 227
Word processing, 1, 2–3

Work products, 296
    in digital-based system, 138–39
    from Ediflex session, 77
    of electronic nonlinear editing
        systems, 60–62
    from laserdisc-based systems,
        99–100
    in Montage, 76

    of telecine session, 285–86
    of TouchVision system, 79
WORM (write once, read many)
    laserdiscs, 86, 89, 105,
        227
Wright, Frank Lloyd, 1, 302
Write many, read many (WMRM)
    laserdiscs, 86, 89, 105

Write many, ready many (WMRM)
    optical discs, 227

Xerox, 161

YUV (luminance and color difference
    components), 195–96, *197*

Zero packer, 199